Selected Solutions for

FUNDAMENTALS OF PHYSICS

—

Third Edition

Third Edition Extended

David Halliday
University of Pittsburgh

Robert Resnick
Rensselaer Polytechnic Institute

Prepared by

Edward Derringh
Wentworth Institute of Technology

JOHN WILEY & SONS New York • Chichester • Brisbane • Toronto • Singapore

ISBN 0 471 81996 4

Printed in the United States of America

10 9 8 7 6 5

PREFACE

This solutions supplement is intended for use by students. It contains solutions to about 28% of the exercises and problems appearing at the end of the chapters. It is designed to be used with the textbook at hand: the problem statements are not reproduced, and there are many references to numbered equations and Sample Problems in the textbook.

Selection of the exercises and problems to be included, and presentation of the solutions was left to me; I alone am responsible for the contents of this volume.

Edward Derringh

July 9, 1988

199 Depot St,
Duxbury, MA 02332

CONTENTS

5E

(a) Since 1 μm = 10^{-6} m, it follows that

$$10^6(1 \text{ μm}) = 10^6(10^{-6} \text{ m}),$$

$$10^6 \text{ μm} = 1 \text{ m}.$$

Combine this result with 1 km = 10^3 m to get

$$1 \text{ km} = 10^3 \text{ m} = 10^3(10^6 \text{ μm}) = 10^9 \text{ μm}.$$

(b) Calculate the number of microns in one centimeter:

$$1 \text{ cm} = 10^{-2} \text{ m} = 10^{-2}(10^6 \text{ μm}) = 10^4 \text{ μm}.$$

Therefore 1 μm = 10^{-4} cm, so the fraction sought is 10^{-4}.

(c) From Appendix F, 1 yard = 3 ft = 3(0.3048 m) = 0.9144 m; thus,

$$1 \text{ yard} = 0.9144(10^6 \text{ μm}) = 9.144 \times 10^5 \text{ μm}.$$

6E

(a) The circumference is given by C = 2πR (see Appendix G) where R is the radius of the earth. The radius R in km is

$$R = (6.37 \times 10^6 \text{ m})(10^{-3} \text{ km/m}) = 6.37 \times 10^3 \text{ km},$$

so that

$$C = 2\pi(6.37 \times 10^3 \text{ km}) = 4.00 \times 10^4 \text{ km}.$$

(b) Since the earth is a sphere, its surface area A is given from (see Appendix G)

$$A = 4\pi R^2 = 4\pi(6.37 \times 10^3 \text{ km})^2 = 5.10 \times 10^8 \text{ km}^2.$$

(c) Again, the needed formula, this time for the volume V of a sphere, can be found in Appendix G:

$$V = \frac{4}{3} \pi R^3 = \frac{4}{3} \pi(6.37 \times 10^6 \text{ m})^3 = 1.08 \times 10^{21} \text{ m}^3.$$

Note that the problem statement did not call for specific units for the answer to this part, so we chose to obtain the answer in m^3, the meter being the unit in which the radius of the earth was given; this minimizes the chance of errors.

12P

The volume V of antarctic ice equals the area A times the average thickness D of the ice: $V = AD$. But $A = \pi R^2/2$ (one-half the area of a circle; see Appendix G). To obtain V in cm^3 convert R and D both to cm:

$$D = 3000 \text{ m} = (3 \times 10^3 \text{ m})(10^2 \text{ cm/m}) = 3 \times 10^5 \text{ cm},$$

$$R = 2000 \text{ km} = (2 \times 10^3 \text{ km})(10^5 \text{ cm/km}) = 2 \times 10^8 \text{ cm}.$$

Hence,

$$V = \tfrac{1}{2}\pi R^2 D = \tfrac{1}{2}\pi(2 \times 10^8 \text{ cm})^2(3 \times 10^5 \text{ cm}) = 1.88 \times 10^{22} \text{ cm}^3.$$

14P

Use the conversions found in Appendix F:

$$1 \text{ acre} \cdot \text{ft} = (43{,}560 \text{ ft}^2) \cdot \text{ft} = 43{,}560 \text{ ft}^3,$$

$$1 \text{ acre} \cdot \text{ft} = 43{,}560(2.832 \times 10^{-2} \text{ m}^3) = 1234 \text{ m}^3.$$

Since 2 in = 1/6 ft, the volume V of water that fell on the town during the storm is

$$V = (26 \text{ km}^2)(10^6 \text{ m}^2/\text{km}^2)(\tfrac{1}{6} \text{ ft})(0.3048 \text{ m/ft}) = 1.3208 \times 10^6 \text{ m}^3.$$

In terms of acre·ft, this volume is

$$V = \frac{1.3208 \times 10^6 \text{ m}^3}{1234 \text{ m}^3/\text{acre} \cdot \text{ft}} = 1070 \text{ acre} \cdot \text{ft}.$$

Note that the duration of the rainstorm (30 min) does not enter into these calculations.

20E

(a) By Appendix F, 1 m = 3.281 ft; therefore

$$3 \times 10^8 \text{ m/s} = 3 \times 10^8 \left(\frac{3.281 \text{ ft}}{10^9 \text{ ns}}\right) = 9.843 \times 10^{-1} \text{ ft/ns},$$

$$3 \times 10^8 \text{ m/s} = 0.9843 \text{ ft/ns}.$$

(b) Converting m to mm and s to ps gives

$$3 \times 10^8 \text{ m/s} = 3 \times 10^8 \left(\frac{10^3 \text{ mm}}{10^{12} \text{ ps}}\right) = 0.3 \text{ mm/ps}.$$

22E

Use the chain-link conversion method of Section 1-3 to obtain

$$(365.25 \ \frac{days}{year})(24 \ \frac{hours}{day})(60 \ \frac{min}{hour})(60 \ \frac{seconds}{min}) = 3.156 \ X \ 10^7 \ s/y.$$

27P

Use chain-link conversions:

$$(3 \ X \ 10^8 \ \frac{m}{s})(1 \ \frac{AU}{1.5 \ X \ 10^8 \ km})(1 \ \frac{km}{10^3 \ m})(60 \ \frac{s}{min}) = 0.12 \ AU/min.$$

31P

The last day of the twenty centuries is longer than the first day by

$$(20 \ centuries)(10^{-3} \ s/century)$$

which is 0.020 s. Thus, the average day during the twenty centuries is $(0 + 0.020)/2 = 0.010$ s longer than the first day. Since the increase occurs uniformly, the total cumulative effect T is

$$T = (average \ difference)(number \ of \ days),$$

$$T = (0.010 \ s/average \ day)(365.25 \ days/y \ X \ 2000 \ y),$$

$$T = 7305 \ s = 2 \ h \ 1 \ min \ 45 \ s.$$

32P

The moon revolves around the earth and the earth revolves around the sun in the same direction (counterclockwise as seen from the north). The lunar month is the period of revolution relative to the sun, and can be measured as the time required for the moon to pass from new moon to new moon, since the phases of the moon are determined by the position of the moon relative to the earth-sun line. The sidereal month, the period of revolution relative to the stars, can be measured as the time required for the earth-moon line to sweep through 360° with respect to the stars. The earth-moon lines at the start and end of one sidereal month are parallel, for the stars are so far away compared to the size of the earth's orbit about the sun, that lines drawn from the earth to any one particular star are parallel regardless of where the earth is in its orbit. In the diagram on p.4, I,A and F,A show the positions of the earth and moon one sidereal month apart. B is where the moon would have to be if instead the diagram showed the passage of one lunar month. The months differ by the time needed for the moon to cover the angle θ shown in the lower part of the sketch. We can see that it will take the moon some extra time to cover this distance, i.e., that the lunar month is longer than the sidereal month, because the moon revolves in the same direction that the earth does. Now, this angle θ corresponds to 1/12th of a revolution of the earth about the sun (since the two positions of the earth are shown one month = 1/12 year apart), and therefore also to 1/12th of a revolution of the moon about the earth. Thus the time sought is (30 d)/12 = 2.5 days (since the moon takes about 30 days to revolve about the earth).

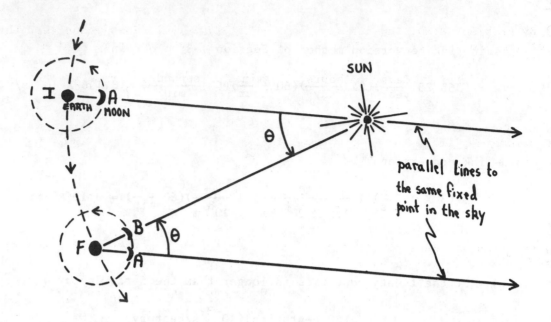

35E

By the definition of average mass, the number of atoms is found from the relation

$$\text{(number of atoms)(average mass per atom)} = \text{bulk mass} = \text{mass of earth},$$

$$N(40\ u) = M_{\text{earth}},$$

$$N(40)(1.66 \times 10^{-27}\ \text{kg}) = 5.98 \times 10^{24}\ \text{kg},$$

$$N = 9.01 \times 10^{49}.$$

38P

(a) Since $1\ g = 10^{-3}\ kg$ and $1\ m^3 = 10^6\ cm^3$ (see Appendix F), it follows that

$$1\ g/cm^3 = (10^{-3}\ kg)/(10^{-6}\ m^3) = 1000\ kg/m^3.$$

(b) The mass M of 5700 m^3 of water is

$$M = (5700\ m^3)(1000\ kg/m^3) = 5.7 \times 10^6\ kg.$$

Hence, the mass flow rate R is

$$R = \frac{\text{mass}}{\text{time}} = \frac{5.7 \times 10^6\ \text{kg}}{(10\ \text{h})(3600\ \text{s/h})} = 158\ \text{kg/s}.$$

41P

(a) By Problem 38(a), the density of iron is 7870 kg/m^3. This means that there are 7870 kg of iron in 1 m^3 of iron. Therefore, the number of iron atoms in 1 m^3 of iron is

$$\frac{7870 \text{ kg/m}^3}{9.27 \times 10^{-26} \text{ kg/atom}} = 8.490 \times 10^{28} \text{ atoms/m}^3.$$

The volume available for each atoms is, then,

$$\frac{1}{8.490 \times 10^{28} \text{ atoms/m}^3} = 1.178 \times 10^{-29} \text{ m}^3/\text{atom}.$$

Now set this volume equal to $4\pi R^3/3$, the formula for the volume of a sphere in terms of its radius R; solve for R:

$$1.178 \times 10^{-29} = \frac{4}{3}\pi R^3 \rightarrow R = 1.41 \times 10^{-10} \text{ m}.$$

Thus, the distance between neighboring atoms is $2R = 2.82 \times 10^{-10}$ m = 0.282 nm.

(b) Repeating the calculation for sodium:

$$\frac{1013 \text{ kg/m}^3}{3.82 \times 10^{-26} \text{ kg/atom}} = 2.652 \times 10^{28} \text{ atoms/m}^3;$$

$$\frac{1}{2.652 \times 10^{28} \text{ atoms/m}^3} = 3.771 \times 10^{-29} \text{ m}^3/\text{atom};$$

$$3.771 \times 10^{-29} = \frac{4}{3}\pi R^3 \rightarrow R = 2.08 \times 10^{-10} \text{ m};$$

$$2R = 4.16 \times 10^{-10} \text{ m} = 0.416 \text{ nm}.$$

3E

Use Eq.1. We are given that $\bar{v}$ = 160 km/h and Δx = 18.4 m and asked to find the time Δt. The distance units in the data, km and m, are different: we must use either km or m. Also, the time unit given is hours, but the ball takes only a short time to reach the plate (certainly not hours). If we convert km/h to m/s then we have consistent distance units and a reasonable time unit. From Appendix F (the table for Speed), we find that $\bar{v}$ = 160 km/h = 160(0.2778 m/s) = 44.45 m/s. With this, Eq.1 gives

$$\Delta t = \Delta x / \bar{v} = (18.4\ m)/(44.45\ m/s) = 0.414\ s = 414\ ms.$$

6E

At constant speed $\bar{v}$, the time Δt required to travel a distance Δx is given by Eq.1 and is $\Delta t = \Delta x / \bar{v}$. At 65 mi/h, the time needed is Δt = (435 mi)/(65 mi/h) = 6.692 h; at the slower speed of 55 mi/h the time required is Δt = (435 mi)/(55 mi/h) = 7.909 h. Thus, the time saved in traveling at the higher speed is 7.909 – 6.692 = 1.217 h = 1 h 13 min.

8E

The total distance traveled is 80 km, and requires a total time given by $t = x_1/v_1 + x_2/v_2$ = 40/30 + 40/60 = 2 hours. The average speed, by definition, is (total distance)/(total time), giving (80 km)/(2 h) = 40 km/h.

11P

(a) The position at t = 1 s is found by replacing t with 1 in the equation given for the position; that is,

$$x(1) = 3(1) - 4(1)^2 + (1)^3 = 0.$$

Similarly,

$$x(2) = -2; \quad x(3) = 0; \quad x(4) = 12\ m.$$

(b) When t = 0 the position of the object is x(0) = 0; at t = 4 s the position is +12 m. Hence, the displacement during this time interval is

$$displacement = x(4) - x(0) = +12\ m,$$

the plus sign indicating a displacement in the +x-direction.

(c) By definition, Eq.1, the required average velocity is

$$\bar{v} = \frac{x(4) - x(2)}{4 - 2} = \frac{12 - (-2)}{4 - 2} = +7.0\ m/s = \frac{\Delta x}{\Delta t}.$$

16E

Use Eq.8 to calculate the average acceleration. If we choose the positive direction as the direction of the first velocity given, then we have $v_1 = +18$ m/s, $v_2 = -30$ m/s; also $t_2 - t_1 = 2.4$ s (t_1 and t_2 are not each given, only the elapsed time Δt). Therefore

$$\bar{a} = \frac{v_2 - v_1}{t_2 - t_1} = \frac{(-30 \text{ m/s}) - (+18 \text{ m/s})}{2.4 \text{ s}} = -20.0 \text{ m/s}^2.$$

The negative sign indicates that $\bar{a}$ points in the direction of v_2.

23E

(a) The displacements at the beginning and end of the interval are $x(0) = 0$, $x(3) = 240$ meters. Therefore,

$$\bar{v} = \frac{\Delta x}{\Delta t} = \frac{x(3) - x(0)}{3 - 0} = \frac{240 - 0}{3} = 80 \text{ m/s}.$$

(b) The instantaneous velocity v is given by

$$v = \frac{dx}{dt} = 50 + 20t,$$

so that $v(3) = 50 + 20(3) = 110$ m/s.

(c) The acceleration is $a = dv/dt = 20$ m/s^2 for all values of t.

29P

(a) Since at^2 must have length dimensions, a must have the dimensions (length)/(time)^2; similarly, b must have the dimensions of (length)/(time)^3; in British units, these are ft/s^2 and ft/s^3.

(b) With a = 3, b = 1 the given equation for position becomes

$$x = 3t^2 - t^3,$$

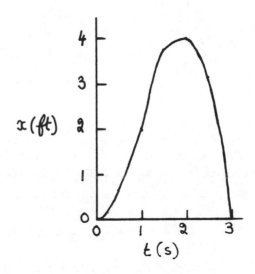

and this is shown on the sketch. At the maximum x-position the particle comes to a stop momentarily and turns around; therefore $v = dx/dt = 0$ there. Taking the derivative gives $0 = 6t - 3t^2$. This is satisfied for t = 0 and t = 2 s. The value t = 2 s gives the maximum x position.

(c) From t = 0 to t = 2 s the particle travels $x(2) - x(0) = 4$ ft. From t = 2 s to t = 3 s the particle travels $x(3) - x(2) = -4$ ft; we count this as +4 ft since we are adding distances regardless of direction of motion. Finally, from t = 3 s to t = 4 s it travels $x(4) - x(3) = -16$ ft, 16 ft in distance. Thus the total distance is 4 + 4 + 16 = 24 ft.

(d) Since $x(4) = -16$ ft, the displacement from $t = 0$ to $t = 4$ s is $x(4) - x(0) = -16 - 0 = -16$ ft.

(e) $v = dx/dt = 6t - 3t^2$, and (f) $a = dv/dt = 6 - 6t$; from these expressions the following table is derived by direct substitution of the successive values of the time:

$t(s)$	$v(ft/s)$	$a(ft/s^2)$
1	3	0
2	0	-6
3	-9	-12
4	-24	-24

30E

(a) The given data is: $a = 9.8$ m/s^2; $v_0 = 0$ (starts from rest); $x_0 = 0$ (starting point taken as the origin); $v = 3 \times 10^7$ m/s. Wanted: the time t. Use Eq.10:

$$v = v_0 + at,$$

$$3 \times 10^7 = 0 + (9.8)t,$$

$$t = 3.06 \times 10^6 \text{ s.}$$

(b) Wanted: distance x. Eq.15, containing only data given in the problem statement, is preferable to Eq.14 or Eq.16 which, for their use, require that the value of t found in (a) is correct. That is, using Eq.15 allows us to get (b) right even if (a) is wrong. By Eq.15, then,

$$v^2 = v_0^2 + 2a(x - x_0),$$

$$(3 \times 10^7)^2 = 0^2 + 2(9.8)(x - 0),$$

$$x = 4.59 \times 10^{13} \text{ m.}$$

33E

If the direction in which the muon is traveling is taken to be positive, then $v_0 = +5.00 \times 10^6$ m/s and, since the acceleration is in the opposite direction, $a = -1.25 \times 10^{14}$ m/s^2. At a distance x from the point ($x_0 = 0$) where the muon entered the field, the velocity $v = 0$ since the muon is brought to rest. By Eq.15,

$$v^2 = v_0^2 + 2ax,$$

$$0^2 = (+5 \times 10^6)^2 + 2(-1.25 \times 10^{14})x,$$

$$x = 0.100 \text{ m.}$$

34E

It is given that $v_0 = 1.5 \times 10^5$ m/s, $x - x_0 = 1.0$ cm $= 0.01$ m (we must use a single unit of displacement throughout), $v = 5.7 \times 10^6$ m/s. To find the acceleration a, use Eq.15:

$$v^2 = v_0^2 + 2a(x - x_0),$$

$$(5.7 \times 10^6)^2 = (1.5 \times 10^5)^2 + 2a(0.01),$$

$$a - 1.62 \times 10^{15} \text{ m/s}^2.$$

35E

(a) To avoid the use of negative time, label the instant "2.5 s earlier" as $t = 0$; the object's velocity at $t = 0$ is v_0. Then, $v = +9.6$ m/s at $t = 2.5$ s; also $a = 3.2$ m/s^2. By Eq.10,

$$v = v_0 + at,$$

$$9.6 = v_0 + (3.2)(2.5),$$

$$v_0 = 1.60 \text{ m/s}.$$

(b) With our choice of $t = 0$, the time called "2.5 s later" corresponds to $t = 5$ s. Using the result from (a), Eq.10 now gives

$$v = v_0 + at,$$

$$v = 1.60 + (3.2)(5) = 17.6 \text{ m/s}.$$

40E

The data must be converted into the same set of units. By Appendix F, the initial speed is $v_0 = 85$ mi/h $= 85(1.467$ ft/s$) = 124.7$ ft/s; similarly, the final speed is $v = 55$ mi/h $= 55(1.467$ ft/s$) = 80.69$ ft/s. The acceleration $a = -17$ ft/s^2 (slowing down). Eq.10 allows us to find the time t:

$$v = v_0 + at,$$

$$80.69 = 124.7 + (-17)t,$$

$$t = 2.59 \text{ s}.$$

44P

(a) We work the problem here using the British units. Let the vehicle be moving at a speed v_0 when, at $t = 0$, the driver slams on the brakes a distance $x - x_0$ from the barrier, which it strikes 4.0 s later at a speed v. That is, $v_0 = 35$ mi/h $= 51.33$ ft/s (see Appendix F), $t = 4$ s, $x - x_0 = 110$ ft. The acceleration a is found from Eq.14:

$$x - x_0 = v_0 t + \tfrac{1}{2}at^2,$$

$$110 = (51.33)4 + \tfrac{1}{2}a(4)^2,$$

$$a = -11.9 \text{ ft/s}^2.$$

(b) The speed at impact is given by Eq.10:

$$v = v_0 + at = 51.33 + (-11.9)(4) = 3.73 \text{ ft/s.}$$

45P

(a) Let the first point be at x_0 and the second at x. By Eq.16,

$$x - x_0 = \frac{1}{2}(v_0 + v)t,$$

$$60 = \frac{1}{2}(v_0 + 15)(6) \rightarrow v_0 = 5.0 \text{ m/s.}$$

(b) The acceleration can be found from Eq.10:

$$v = v_0 + at,$$

$$15 = 5 + a(6) \rightarrow a = 1.67 \text{ m/s}^2.$$

(c) Shift origins, so that now the point from which the car starts from rest is x_0, and the "first point" is now at x. By Eq.15,

$$v^2 = v_0^2 + 2a(x - x_0),$$

$$5^2 = 0^2 + 2(1.67)(x - x_0) \rightarrow x - x_0 = 7.49 \text{ m.}$$

47P

(a) Let t = driver reaction time, t' = breaking time at 50 mi/h and t" = breaking time at 30 mi/h. The cars must be brought to rest by the brakes over the distances (186 − 73.33t) and (80 − 44t) in the two situations; these distances are in ft, the initial speeds having been converted to ft/s via Appendix F. If a = magnitude of the acceleration then Eqs.14 and 10, applied to the two situations, require that

$$\tfrac{1}{2}at'^2 = 186 - 73.33t,$$

$$\tfrac{1}{2}at''^2 = 80 - 44t.$$

$$at' = 73.33,$$

$$at'' = 44.$$

These are 4 equations for the four unknowns a, t, t', t". Use the third equation to eliminate a from the first, and the fourth equation to eliminate a from the second to get

$$36.665t' = 186 - 73.33t,$$

$$22t'' = 80 - 44t.$$

Now divide these equations, and also divide the pair of equations directly above them; the results can be put into the form

$$\frac{36.665}{22}(\frac{t'}{t''}) = \frac{36.665}{22}(\frac{73.33}{44}) = \frac{186 - 73.33t}{80 - 44t},$$

$$2.7775 = \frac{186 - 73.33t}{80 - 44t} \rightarrow t = 0.741 \text{ s.}$$

(b) Next, solve for t'':

$$22t'' = 80 - 44(0.741) \rightarrow t'' = 2.154 \text{ s.}$$

Finally, the fourth equation will yield

$$a = \frac{44}{t''} = \frac{44}{2.154} = 20.4 \text{ ft/s}^2,$$

that is, the acceleration is -20.4 ft/s^2.

55E

(a) "Dropped" implies that the wrench was released with initial speed $v_0 = 0$. We let $y_0 = 0$ be the point from which the wrench was dropped. We are told that $v = -24$ m/s (recall that up is positive and the wrench is falling downwards). Eq.20 yields

$$v^2 = v_0^2 - 2g(y - y_0),$$

$$(-24)^2 = 0^2 - 2(9.8)(y - 0) \rightarrow y = -29.4 \text{ m,}$$

that is, the ground is 29.4 m below the release point.

(b) To find t using only the original data (not the answer to (a)), use Eq.18:

$$v = v_0 - gt,$$

$$-24 = 0 - (9.8)t \rightarrow t = 2.45 \text{ s.}$$

56E

(a) Assume that the ball was thrown from ground level $y_0 = 0$. It reaches height $y = +50$ m (up is positive); at its highest point the speed $v = 0$. To find the initial velocity v_0 use Eq.20:

$$v^2 = v_0^2 - 2g(y - y_0),$$

$$0^2 = v_0^2 - 2(9.8)(50 - 0) \rightarrow v_0 = +31.3 \text{ m/s;}$$

v_0 is positive since the ball is thrown upward.

(b) The ball is back on the ground after the time of flight has elapsed; i.e., $y = 0$. Use Eq.19 for y in terms of t:

$$y - y_0 = v_0 t - \tfrac{1}{2}gt^2,$$

$$0 - 0 = +31.3t - \tfrac{1}{2}(9.8)t^2,$$

$$0 = t(31.3 - 4.9t) \rightarrow t = 0, \; t = 6.39 \text{ s}.$$

The first answer corresponds to the time the ball was launched, the second to the time when it is back on the ground; i.e., the time of flight = 6.39 s.

60E

(a) Put $y_0 = 0$ at the point the ball was dropped; the initial speed is $v_0 = 0$. Up is positive, so after falling 50 m the position of the ball is $y = -50$ m. We are to find the time t; Eq.19 is appropriate:

$$y - y_0 = v_0 t - \tfrac{1}{2}gt^2,$$

$$-50 - 0 = 0 - \tfrac{1}{2}(9.8)t^2 \rightarrow t = 3.19 \text{ s}.$$

(b) After falling an additional 50 m, the position of the ball is $y = -100$ m. The time needed to reach this position is found as in (a); the result obtained is 4.52 s. This is the time to fall 100 m. Since 3.19 s was occupied in falling the first 50 m, the time needed to fall the second 50 m is 4.52 s - 3.19 s = 1.33 s.

61P

(a) The launching point of the ball is taken as the origin $y_0 = 0$. The data given is $y = +36.8$ m when $t = 2.25$ s. The unknown is v_0. Apply Eq.19:

$$y - y_0 = v_0 t - \tfrac{1}{2}gt^2,$$

$$36.8 - 0 = v_0(2.25) - \tfrac{1}{2}(9.8)(2.25)^2 \rightarrow v_0 = 27.4 \text{ m/s}.$$

(b) The desired velocity is v; with v_0 now known, use Eq.18:

$$v = v_0 - gt = 27.4 - (9.8)(2.25) = 5.35 \text{ m/s}.$$

(c) Find the greatest height y the ball will reach by setting $v = 0$; employ the value of v_0 found in (a) in Eq.20:

$$v^2 = v_0^2 - 2g(y - y_0),$$

$$0^2 = (27.4)^2 - 2(9.8)(y - 0) \rightarrow y = 38.3 \text{ m}.$$

Hence, the ball will travel an extra distance 38.3 - 36.8 = 1.5 m above the point that was mentioned in the problem statement.

72P

Call the speed of the ball upon striking the floor v and the speed of the ball as it leaves the floor u. Up is positive, so that the associated velocities are −v and +u. The ball is in contact with the floor for a time t. The average acceleration is given, by definition, by Eq.8:

$$\bar{a} = \frac{+u - (-v)}{t} = \frac{u + v}{t}.$$

As this is positive, the average acceleration is directed up. Since the speed with which the ball leaves the floor is the same as the speed with which it would strike the floor if dropped from a height of 3 ft, both u and v can be found from Eq.20 with $v_0 = 0$:

$$v^2 = 2gy,$$

using y = 4 ft for v, and y = 3 ft for u, g = 32 ft/s^2 for each. Solving gives v = 16 ft/s and u = 13.86 ft/s, so that

$$\bar{a} = \frac{13.86 + 16}{0.01} = 2986 \text{ ft/s}^2,$$

directed upward.

76P

(a) Let y be the height of the fall and t the time to fall this distance. Since the object falls from rest $v_0 = 0$ and Eq.19 gives

$$y = \tfrac{1}{2}gt^2,$$

$$\frac{y}{2} = \tfrac{1}{2}g(t - 1)^2.$$

Eliminating y between these equations yields

$$\frac{t}{t - 1} = \sqrt{2} = 1.4142 \rightarrow t = 3.41 \text{ s.}$$

(b) The height of fall is

$$y = \tfrac{1}{2}gt^2 = \tfrac{1}{2}(9.8)(3.41)^2 = 57.0 \text{ m.}$$

80P

Let y, Y be the distances of the first and second objects, from the common point of release, and let the second object be released at t = 0. Then,

$$y = \tfrac{1}{2}g(t + 1)^2; \quad Y = \tfrac{1}{2}gt^2.$$

For y − Y = 10 m,

$$\tfrac{1}{2}g(t + 1)^2 - \tfrac{1}{2}gt^2 = 10 \rightarrow g(t + \tfrac{1}{2}) = 10.$$

Using $g = 9.8$ m/s^2, this gives $t = 0.520$ s, so that the elapsed time since the first object was released is $0.52 + 1 = 1.52$ s.

81P

(a) The initial speed v_0 of the ball relative to the ground is $10 + 20 = 30$ m/s. The height y reached by the ball above the elevator $y_0 = 0$ is found from Eq.20:

$$v^2 = v_0^2 - 2g(y - y_0),$$

$$0^2 = (30)^2 - 2(9.8)(y - 0) \rightarrow y = 45.9 \text{ m}.$$

Since the ball was thrown from a point 30 m above the ground, the highest point reached by the ball is $30 + 45.9 = 75.9$ m above the ground.

(b) Let t = time needed by the ball to reach its maximum height. Then, by Eq.18,

$$t = (v_0 - v)/g = (30 - 0)/9.8 = 3.06 \text{ s}.$$

During this time the elevator has moved a distance $= v_{el}t = (10 \text{ m/s})(3.06 \text{ s}) = 30.6$ m up the shaft, so that ball and elevator are separated by $45.9 - 30.6 = 15.3$ m at the moment that the ball is at its highest point. Call T the time needed by the ball to fall back to the elevator. Relative to the elevator, the ball is projected downwards with a speed of 10 m/s from a height of 15.3 m. Choosing the origin as the point of maximum height reached by the ball and using Eq.19,

$$y - y_0 = v_0 T - \tfrac{1}{2}gT^2,$$

$$-15.3 - 0 = -10T - \tfrac{1}{2}(9.8)T^2 \rightarrow T = 1.02 \text{ s}.$$

(We chose the positive root of the quadratic.) Therefore the total elapsed time is $3.06 + 1.02 = 4.08$ s.

83P

(a) Choose the origin $y_0 = 0$ at the balloon at the instant of release of the package. Up is positive. The initial velocity of the package, presuming it is not tossed from the balloon, is the same as the velocity of the balloon at that instant; i.e., $v_0 = +12$ m/s. Note that the ground is at $y = -80$ m (80 m below the origin). By Eq.19,

$$y - y_0 = v_0 t - \tfrac{1}{2}gt^2,$$

$$-80 - 0 = 12t - \tfrac{1}{2}(9.8)t^2,$$

$$4.9t^2 - 12t - 80 = 0 \rightarrow t = -3.00 \text{ s}; 5.45 \text{ s}.$$

A negative time of flight makes no sense in this situation so $t = 5.45$ s.

(b) The velocity v at impact is, by Eq.18,

$$v = v_0 - gt = +12 - (9.8)(5.45) = -41.4 \text{ m/s}.$$

CHAPTER 3

3E

(a) Measure the length of the displacement vector $\vec{d}$; applying the scale, this length, or magnitude, is 370 m; measure the angle with a protractor to find that $\theta = 36°$ north of east.

(b) The distance she actually walks is 250 m + 175 m = 425 m, whereas the magnitude of her displacement is 370 m.

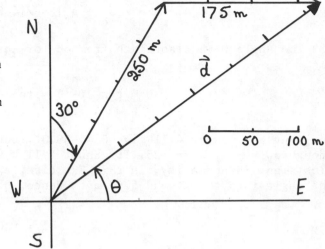

5E

The total displacement vector $\vec{d}$ is found, by the construction below, to have a magnitude (length) = 81 km, and this is directed at 40° north of east.

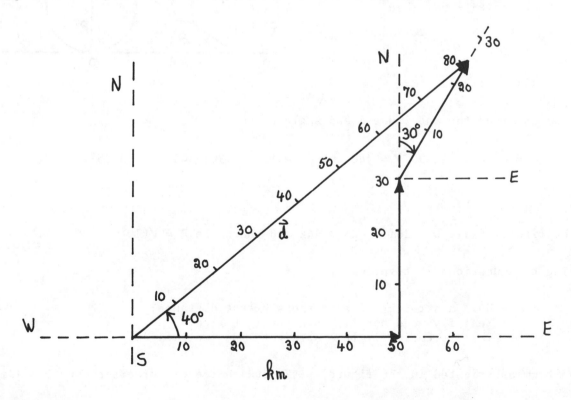

9E

By Eq.5,

$$a_x = a\cos\theta = 7.3\cos 250° = -2.50; \quad a_y = a\sin\theta = 7.3\sin 250° = -6.86.$$

10E

(a) Let's call the vector $\vec{a}$; we are told that $a_x = -25$, $a_y = +40$. By Eq.6, the magnitude of $\vec{a}$ is

$$a = \sqrt{\{a_x^2 + a_y^2\}} = \sqrt{\{(-25)^2 + (+40)^2\}} = 47.2.$$

(b) The angle θ is also given from one of Eqs.6:

$$\theta = \tan^{-1}(\frac{a_y}{a_x}) = \tan^{-1}(\frac{+40}{-25}) = \tan^{-1}(-1.60).$$

Now your calculator will give $\theta = -58.0°$. But, in so doing, the calculator has **acted as though** $a_y = -40$, $a_x = +25$. It cannot tell which component, x or y, carries the minus or plus sign. In actuality, in this Exercise, a_y is positive and a_x is negative. This puts the angle into the second quadrant. Hence, $\theta = 180° - 58° = 122°$. See Hint 3, p.42 of HR.

17P

If the wheel rolls without slipping, then the straight line distance PQ equals one-half the circumference of the wheel, since the wheel rolled through one-half of a revolution. Therefore,

$$\vec{d} = \tfrac{1}{2}(2\pi R)\vec{i} + 2R\vec{j} = \pi R\vec{i} + 2R\vec{j}.$$

With R = 45 cm,

$$\vec{d} = 141.4\vec{i} + 90\vec{j}, \quad cm.$$

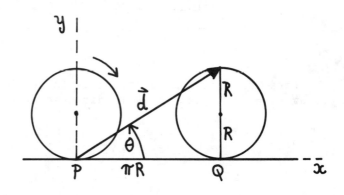

From these we can find the magnitude d and angle θ:

$$d = \sqrt{(141.4^2 + 90^2)} = 168 \text{ cm}; \quad \theta = \tan^{-1}(+90/+141.4) = 32.5°.$$

19P

(a) From the figure, p.17, $\vec{D} = 10\vec{i} + 12\vec{j} + 14\vec{k}$ and therefore $D = \sqrt{(10^2 + 12^2 + 14^2)} = 21.0$ ft.

(b) Answering the questions in turn:

> No: a straight line is the shortest distance;
> Yes: fly need not fly in a straight line;
> Yes: it could fly in a straight line (unlikely).

(c) For the room as oriented in the figure, p.17, and choice of corners shown, $\vec{D} = 10\vec{i} + 12\vec{j} + 14\vec{k}$.

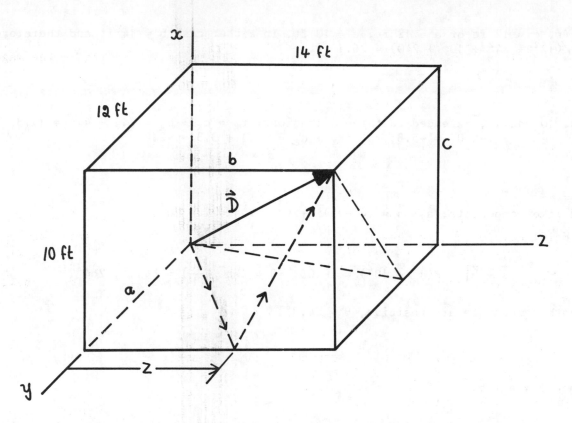

(d) For this calculation, call the lengths of the sides of the room a, b, c, as indicated above (do not assign numerical values yet), for the fly has a choice of walls on which to walk first. Two of the possible paths to the opposite corner are shown; focus on the one with arrows. The total length L of the path is

$$L = [a^2 + z^2]^{\frac{1}{2}} + [(b - z)^2 + c^2]^{\frac{1}{2}}.$$

To find the shortest path, set dL/dz = 0; this gives

$$(c^2 - a^2)z^2 + 2a^2bz - a^2b^2 = 0.$$

The two solutions follow from the quadratic formula (see Appendix G) and are

$$z_1 = \frac{ab}{a - c}; \quad z_2 = \frac{ab}{a + c}.$$

Now z represents a distance and must be positive. Likewise b − z also is a distance (see figure) and also must be positive. These latter distances are, using the solutions above,

$$b - z_1 = \frac{bc}{c - a}; \quad b - z_2 = \frac{bc}{c + a}.$$

Clearly, z_1 and $b - z_1$ both cannot be positive. Therefore, choose $z = z_2 = ab/(a + c)$. Substituting this into the equation for L above gives

$$L_{min} = L(z_2) = [(a + c)^2 + b^2]^{\frac{1}{2}} = [a^2 + b^2 + c^2 + 2ac]^{\frac{1}{2}}.$$

For the smallest possible L_{min}, select a,b,c so that 2ac is the smallest it can be; i.e., a = 10 ft, c = 12 ft or a = 12 ft, c = 10 ft. In either case b = 14 ft and therefore $L_{min} = \sqrt{(12^2 + 14^2 + 10^2 + 240)} = 26.1$ ft.

21E

Use Eqs. 10 and 11, extended to three dimensions: $r_x = c_x + d_x = 7.4 + 4.4 = 11.8$; $r_y = c_y + d_y = -3.8 + (-2.0) = -5.8$; $r_z = c_z + d_z = -6.1 + 3.3 = -2.8$.

22E

(a) Add like components: $\vec{a} + \vec{b} = (4 - 3)\vec{i} + (3 + 7)\vec{j} = \vec{i} + 10\vec{j}$.

(b) By Eq.6,

$$|\vec{a} + \vec{b}| = \sqrt{(1^2 + 10^2)} = 10.05; \quad \theta = \tan^{-1}(\frac{10}{1}) = 84.3°, 264°.$$

Since both components are positive, $\theta = 84.3°$.

27P

(a) Since $r = \vec{a} + \vec{b}$,

$$r_x = a_x + b_x, \quad r_y = a_y + b_y.$$

The components of $\vec{a}$ are

$$a_x = a\cos\theta_a = 10\cos30° = 8.660, \quad a_y = a\sin\theta_a = 10\sin30° = 5.000.$$

The angle θ_b counterclockwise from the +x axis to $\vec{b}$ is 30° + 105° = 135°. Hence,

$$b_x = b\cos\theta_b = 10\cos135° = -7.071, \quad b_y = b\sin\theta_b = 10\sin135° = +7.071.$$

Therefore $r_x = 8.660 - 7.071 = 1.589$; $r_y = 5 + 7.071 = 12.071$.

(b) The magnitude of $\vec{r}$ is

$$r = \sqrt{(r_x^2 + r_y^2)} = \sqrt{(1.589^2 + 12.071^2)} = 12.175.$$

(c) The angle is given from

$$\theta = \tan^{-1}(r_y/r_x) = \tan^{-1}(12.071/1.589) = 82.5°.$$

28P

See the sketch, p.19. We take the east direction as the x axis and north as the y axis. The components of the three successive putts are

$$p_{1x} = 0, \quad p_{1y} = 12 \text{ ft};$$

$$P_{2x} = 6\cos 315° = 4.243 \text{ ft};$$

$$P_{2y} = 6\sin 315° = -4.243 \text{ ft};$$

$$P_{3x} = 3\cos 225° = -2.121 \text{ ft};$$

$$P_{3y} = 3\sin 225° = -2.121 \text{ ft}.$$

The putt $\vec{P}$ that should have been made in the first place is the vector sum of these, so that

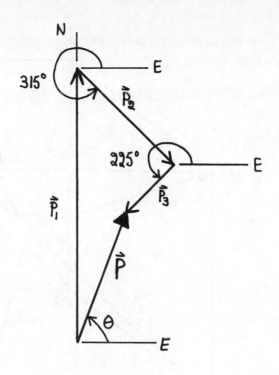

$$P_x = 0 + 4.243 - 2.121 = 2.122 \text{ ft};$$

$$P_y = 12 - 4.243 - 2.121 = 5.636 \text{ ft}.$$

This is a putt of length and direction

$$P = \sqrt{(2.122^2 + 5.636^2)} = 6.02 \text{ ft};$$

$$\theta = \tan^{-1}\left(\frac{5.636}{2.122}\right) = 69.4° \text{ N of E}.$$

32P

Call the vectors $\vec{a}$ and $\vec{b}$. Since the scalar product of vectors that are perpendicular is zero, the requirement can be restated as

$$(\vec{a} + \vec{b}) \cdot (\vec{a} - \vec{b}) = 0,$$

$$\vec{a} \cdot \vec{a} - \vec{a} \cdot \vec{b} + \vec{b} \cdot \vec{a} - \vec{b} \cdot \vec{b} = 0,$$

$$a^2 - b^2 = 0 \rightarrow a = b.$$

33P

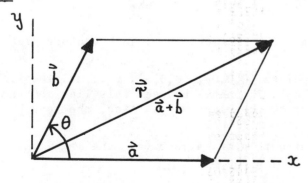

Orient the coordinate axes so that one of the vectors lies along one of the axes; in the figure on the left, for example, the vector $\vec{a}$ lies along the x axis. Then,

$$\vec{a} = a\vec{i}; \quad \vec{b} = (b\cos\theta)\vec{i} + (b\sin\theta)\vec{j},$$

and therefore

$$\vec{a} + \vec{b} = (a + b\cos\theta)\vec{i} + (b\sin\theta)\vec{j}.$$

Hence, the magnitude of this vector is

$$|\vec{a} + \vec{b}| = [(a + b\cos\theta)^2 + (b\sin\theta)^2]^{\frac{1}{2}} = [a^2 + b^2 + 2ab\cos\theta]^{\frac{1}{2}}.$$

34P

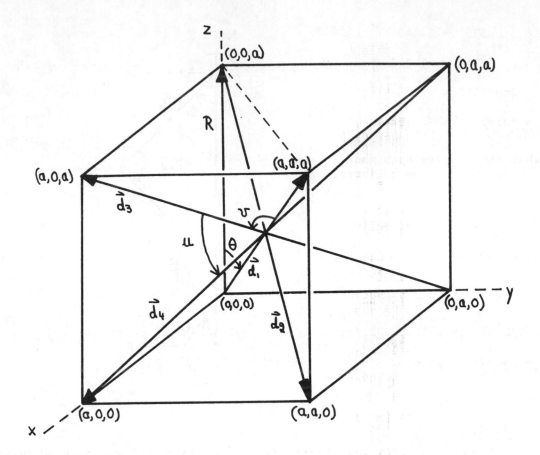

(a) Each diagonal vector can be represented as the difference between vectors drawn from the origin (0,0,0) to its head and from the origin to its tail; that is,

$$\vec{d}_1 = (a - 0)\vec{i} + (a - 0)\vec{j} + (a - 0)\vec{k} = a\vec{i} + a\vec{j} + a\vec{k};$$

$$\vec{d}_2 = (a - 0)\vec{i} + (a - 0)\vec{j} + (0 - a)\vec{k} = a\vec{i} + a\vec{j} - a\vec{k};$$

$$\vec{d}_3 = (a - 0)\vec{i} + (0 - a)\vec{j} + (a - 0)\vec{k} = a\vec{i} - a\vec{j} + a\vec{k};$$

$$\vec{d}_4 = (a - 0)\vec{i} + (0 - a)\vec{j} + (0 - a)\vec{k} = a\vec{i} - a\vec{j} - a\vec{k}.$$

The negative of any or all of the above would also be correct, since such vectors are merely reversed in direction.

(b) Consider the right triangle formed by the points (0,0,0), (0,0,a), (a,a,a). The line connecting (0,0,a) and (a,a,a) has a length of $a\sqrt{2}$. The angle between this line and the edge labelled R is 90°, since R is perpendicular to the plane z = a. Also, the length of R is a. Thus, $\tan\theta = (a\sqrt{2}/a) = \sqrt{2}$, or $\theta = 54.7°$.

(c) Since the components of each diagonal are either +a or −a, see (a), the magnitude (length) of each diagonal vector is $d = \sqrt{(a^2 + a^2 + a^2)} = a\sqrt{3}$.

43E

(a) The scalar product is $\vec{a}\cdot\vec{b} = ab\cos\theta = (10)(6)\cos60° = 30$.

(b) The magnitude of $\vec{a} \times \vec{b}$ is $|\vec{a} \times \vec{b}| = ab\sin\theta = (10)(6)\sin60° = 52.0$. The direction is found from the right-hand rule.

48P

Equating the two expressions for the scalar product yields

$$ab\cos\theta = a_x b_x + a_y b_y + a_z b_z,$$

$$\sqrt{[3^2 + 3^2 + 3^2]}\sqrt{[2^2 + 1^2 + 3^2]}\cos\theta = (3)(2) + (3)(1) + (3)(3),$$

$$(\sqrt{27}\sqrt{14})\cos\theta = 18 \rightarrow \theta = \cos^{-1}(18/\sqrt{378}) = 22.2°.$$

55P

Consider the angle u between the diagonals d_3 and d_4 as an example. Using the method of Problem 48, and examining the figure on p.20,

$$\vec{d}_3 \cdot \vec{d}_4 = (\sqrt{3}a)(\sqrt{3}a)\cos u = a^2 + a^2 - a^2 = a^2,$$

$$3a^2 \cos u = a^2 \rightarrow u = \cos^{-1}(\frac{1}{3}) = 70.5°.$$

The same result will be found for the angle **v** between the diagonals d_1 and d_3.

58P

(a) The vector $\vec{b} \times \vec{a}$ is perpendicular to $\vec{a}$. Since $\cos 90° = 0$,

$$\vec{a} \cdot (\vec{b} \times \vec{a}) = (a)(|\vec{b} \times \vec{a}|)\cos 90° = 0.$$

(b) Call $\vec{b} \times \vec{a} = \vec{c}$. Then $c = ba\sin\phi$. Therefore,

$$|\vec{a} \times (\vec{b} \times \vec{a})| = (a)(|\vec{b} \times \vec{a}|)\sin 90° = (a)(ba\sin\phi) = a^2 b\sin\phi.$$

We have used the result from (a) concerning the direction of $\vec{b} \times \vec{a}$. For the direction of $\vec{a} \times (\vec{b} \times \vec{a})$, see the sketch below.

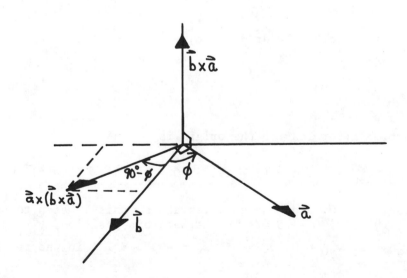

CHAPTER 4

2E

Since 60 km/h = 1 km/min, the successive
distances traveled are 40 km, 20 km, 50 km.
A diagram showing the displacements is drawn
at the right. From this, the total, or net,
displacement is found, by adding the
components of the individual displacements:

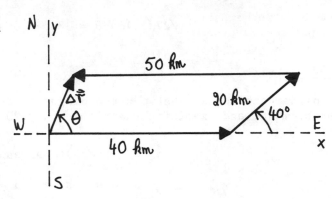

$$\vec{\Delta r} = 40\vec{i} + (20\cos 40°\vec{i} + 20\sin 40°\vec{j}) - 50\vec{i},$$

$$\vec{\Delta r} = 5.321\vec{i} + 12.856\vec{j},$$

in km. The magnitude Δr of the total
displacement is

$$\Delta r = \sqrt{(5.321^2 + 12.856^2)} = 13.914 \text{ km}.$$

The total time Δt of travel is 40 + 20 + 50 = 110 min = 110/60 = 1.833 h. Therefore,

$$\bar{v} = \frac{\Delta r}{\Delta t} = \frac{13.914 \text{ km}}{1.833 \text{ h}} = 7.59 \text{ km/h}.$$

The direction of the average velocity is at an angle θ north of east where

$$\theta = \tan^{-1}(12.856/5.321) = 67.5°,$$

or $90° - 67.5° = 22.5°$ east of north.

4E

(a) By definition of velocity, Eq.2, $\vec{v} = d\vec{r}/dt = 8t\vec{j} + \vec{k}$.
(b) Use Eq.5: $\vec{a} = d\vec{v}/dt = 8\vec{j}$.

8P

(a) The velocity at any time after leaving the origin is

$$\vec{v} = (v_{x0} + a_x t)\vec{i} + (v_{y0} + a_y t)\vec{j} = (3 - t)\vec{i} + (0 - 0.5t)\vec{j} = (3 - t)\vec{i} - 0.5t\vec{j}.$$

At the maximum x position, $v_x = 0$. From the equation above, this is seen to occur at t =
3 s, at which time $v_y = -(0.5)(3) = -1.5$ m/s. Hence, at this position, $\vec{v} = -1.5\vec{j}$, m/s.

(b) With $x_0 = y_0 = 0$, $v_{x0} = 3$, $v_{y0} = 0$, $a_x = -1$, $a_y = -0.5$, Eq.14 in Chapter 2, written
both for x and y, becomes

$$x = 3t - 0.5t^2; \quad y = -0.25t^2.$$

Hence, at t = 3 s, the position of the particle is x = 4.5 m, y = -2.25 m.

10E

(a) The positive x axis is chosen to be horizontal and in the direction in which the electrons enter the region between the plates. The y axis is vertical and up is positive. The origin is at the point at which the electrons enter the field. In view of the large acceleration imparted by the field, the acceleration due to gravity is ignored. The time t required for the electrons to pass through the field is

$$t = \frac{x}{v_{x0}} = \frac{2.0 \text{ cm}}{1.0 \times 10^9 \text{ cm/s}} = 2 \times 10^{-9} \text{ s}.$$

(b) Since the vertical velocity v_{y0} at injection into the field is zero, the vertical displacement is

$$y = \tfrac{1}{2}at^2 = \tfrac{1}{2}(-1.0 \times 10^{17} \text{ cm/s}^2)(2 \times 10^{-9} \text{ s})^2 = -0.2 \text{ cm} = -2 \text{ mm}.$$

(c) The horizontal velocity upon emerging is the same as upon entering, since there is no horizontal acceleration; hence, $v_x = v_{x0} = 1.0 \times 10^9$ cm/s. The vertical velocity upon emerging is

$$v_y = at = (-1.0 \times 10^{17} \text{ cm/s}^2)(2 \times 10^{-9} \text{ s}) = -2 \times 10^{-8} \text{ cm/s}.$$

13E

(a) Since the rifle is fired horizontally, $\theta_0 = 0°$ and $\sin 0° = 0$. In British units, g = 32 ft/s^2. Up is positive. By Eq.10,

$$y - y_0 = (v_0 \sin\theta_0)t - \tfrac{1}{2}gt^2,$$

$$(-0.75 \text{ in})(\frac{1 \text{ ft}}{12 \text{ in}}) = 0 - \tfrac{1}{2}(32 \text{ ft/s}^2)t^2 \rightarrow t = 0.0625 \text{ s}.$$

(b) For the horizontal motion, use Eq.9:

$$x - x_0 = (v_0 \cos\theta_0)t,$$

$$100 \text{ ft} = (v_0 \cos 0°)(0.0625 \text{ s}) \rightarrow v_0 = 1600 \text{ ft/s}.$$

14E

(a) We recognize that this situation is formally similar to Exercise 13. By Eq.10:

$$y - y_0 = (v_0 \sin\theta_0)t - \tfrac{1}{2}gt^2,$$

$$-4 = 0 - \tfrac{1}{2}(32)t^2 \rightarrow t = 0.500 \text{ s}.$$

(b) By Eq.9, $5 = (v_0 \cos 0°)(0.500)$, which gives $v_0 = 10.0$ ft/s.

18E

(a) The displacements are given by Eqs.9 and 10:

$$x - x_0 = (v_0\cos\theta_0)t = (20\cos40°)(1.1) = 16.9 \text{ m},$$

$$y - y_0 = (v_0\sin\theta_0)t - \tfrac{1}{2}gt^2 = (20\sin40°)(1.1) - \tfrac{1}{2}(9.8)(1.1)^2 = 8.21 \text{ m}.$$

(b) Substitute t = 5.0 s into the equations in (a) to get $x - x_0 = 76.6$ m, $y - y_0 = -58.2$ meters; the last answer means 58.2 m below the point from which the stone was thrown, so that the stone must have been thrown from a cliff, or roof of a building, etc.

26P

The motions of the bullet in the x and y directions are given by Eqs.9 and 10:

$$x - x_0 = (v_0\cos\theta_0)t,$$

$$y - y_0 = (v_0\sin\theta_0)t - \tfrac{1}{2}gt^2.$$

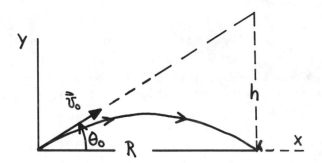

Now let t stand for the time of flight, so that $x(t) - x_0 = R = 150$ ft, and $y(t) - y_0 = 0$ (bullet hits target on the ground). In the sketch, we take $x_0 = y_0 = 0$ (rifle at the origin). Therefore,

$$R = v_0 t\cos\theta_0; \quad 0 = v_0 t\sin\theta_0 - \tfrac{1}{2}gt^2.$$

The first of these equations gives $t = R/v_0\cos\theta_0$; substitute this into the second equation and note that (see Appendix G) $\sin2\theta_0 = 2\sin\theta_0\cos\theta_0$ to get

$$\sin2\theta_0 = gR/v_0^2 = (32)(150)/(1500)^2 = 0.002133 \rightarrow \theta_0 = 0.0611°.$$

Finally, see the sketch,

$$h = R\tan\theta_0 = (150 \text{ ft})(\tan0.0611°) = 0.16 \text{ ft} = 1.92 \text{ in}.$$

27P

At the highest point on the trajectory, $v_y = 0$. Hence, by Eq.12,

$$v_y^2 = 0 = (v_0\sin\theta_0)^2 - 2gy_{max} \rightarrow y_{max} = (v_0\sin\theta_0)^2/2g.$$

35P

If we put the origin at the point the ball was kicked, then $x_0 = y_0 = 0$. Up is positive so the ground is at y = -5 ft (we work the problem in British units). The ball is kicked at t = 0; at t = 4.5 s (the time of flight) the ball hits the ground 50 yd = 150 ft away; i.e., at t = 4.5 s, x = 150 ft, y = -5 ft. By Eq.9,

$$x - x_0 = (v_0\cos\theta_0)t,$$

$$150 = (v_0\cos\theta_0)(4.5) \rightarrow v_0\cos\theta_0 = 33.33.$$

Now use Eq.10:

$$y - y_0 = (v_0\sin\theta_0)t - \tfrac{1}{2}gt^2,$$

$$-5 = (v_0\sin\theta_0)(4.5) - \tfrac{1}{2}(32)(4.5)^2 \rightarrow v_0\sin\theta_0 = 70.89.$$

Therefore,

$$\frac{v_0\sin\theta_0}{v_0\cos\theta_0} = \tan\theta_0 = \frac{70.89}{33.33} \rightarrow \theta_0 = 64.8°.$$

Finally,

$$v_0 = 33.33/\cos\theta_0 = 33.33/\cos 64.8° = 78.3 \text{ ft/s}.$$

43P

(a) The coordinate system used is shown in the sketch: as usual, up is positive; we have put y = 0 at the point on the ground under the plane at the instant the bomb is released. Thus, the initial position of the projectile (probably a bomb) is $x_0 = 0$, $y_0 = +730$ m. At t = 5 s the bomb hits the ground (y = 0); hence, by Eq.10,

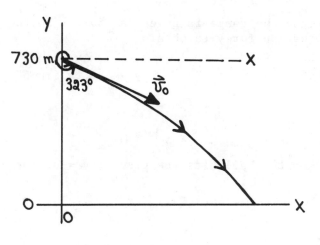

$$y - y_0 = (v_0\sin\theta_0)t - \tfrac{1}{2}gt^2,$$

$$0 - 730 = (v_0\sin 323°)(5) - \tfrac{1}{2}(9.8)(5)^2,$$

$$v_0 = 202 \text{ m/s}.$$

(b) The horizontal distance traveled by the bomb is given by Eq.9:

$$x = (v_0\cos\theta_0)t = (202\cos 323°)(5) = 807 \text{ m}.$$

(c) The horizontal velocity component remains unchanged, so that at impact $v_x = v_0\cos\theta_0 = (202)\cos 323° = 161$ m/s. For the vertical component at impact, use Eq.12:

$$v_y^2 = (v_0\sin\theta_0)^2 - 2g(y - y_0) = (202\sin 323°)^2 - 2(9.8)(0 - 730),$$

$$v_y = -171 \text{ m/s}.$$

The negative sign indicates the expected, that v_y is directed downward.

26

48P

The situation is shown in the sketch. The
ball is launched from a point 4 ft above the
ground so that, with choice of origin as
shown, we have $x_0 = 0$, $y_0 = +4$ ft. By Eqs.9
and 10,

$$x = (v_0\cos45°)t,$$

$$y - 4 = (v_0\sin45°)t - \frac{1}{2}(32)t^2.$$

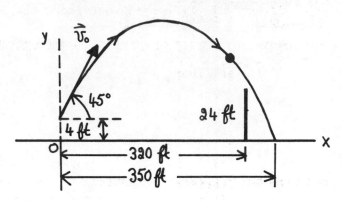

The initial speed of the ball is unknown and
must be found from the range which the ball
would have if there is no fence. (Note that
Eq.14 cannot be used here because the ball
is not projected from ground level.) In the
equations above, then, let t represent, for the moment, the time of flight of the ball
assuming there is no fence to (possibly) interfere with its flight. The first equation
implies that, under these conditions,

$$350 = (v_0\cos45°)t,$$

since the range is given as 350 ft. Solve this equation for t and substitute into the
equation for y to obtain

$$y - 4 = (v_0\sin45°)\frac{350}{v_0\cos45°} - 16(\frac{350}{v_0\cos45°})^2,$$

$$y = 354 - (3.92 \times 10^6)v_0^{-2}.$$

When the ball hits the ground, $y = 0$; the preceding then gives

$$v_0 = \sqrt{(\frac{3.92 \times 10^6}{354})} = 105.23 \text{ ft/s}.$$

Now return to the original equations for x and y, but v_0 is now known. This time set x =
320 ft and solve for t, which now is the time required for the ball to travel 320 ft in
the horizontal direction (i.e., reach the fence). The result is t = 4.30 s. Finally, put
this together with v_0 into the equation for y; it is found that y = 28 ft. Since the
fence is 24 ft high, the ball clears the fence by about 4 ft.

49P

Put the origin at the point on the ground from which the ball is kicked. Then,

$$x = (v_0\cos\theta_0)t,$$

$$y = (v_0\sin\theta_0)t - \frac{1}{2}gt^2.$$

Now let t be the time interval from the kick until the ball reaches a horizontal distance
L, where L = distance from kick to goal posts. Then, $t = L/v_0\cos\theta_0$, from the x equation.

At this same instant t, y = h (ball clears cross-bar); hence, from the y equation,

$$h = -\tfrac{1}{2}g(L/v_0\cos\theta_0)^2 + (v_0\sin\theta_0)(L/v_0\cos\theta_0) = -\tfrac{1}{2}gL^2/(v_0^2\cos^2\theta_0) + L\tan\theta_0.$$

This can be written as

$$ax^2 - Lx + (h + a) = 0,$$

if we let

$$a = \tfrac{1}{2}gL^2/v_0^2, \quad x = \tan\theta_0.$$

Substituting in the numbers gives a = 19.60 m and, since h = 3.44 m, the equation becomes

$$19.6x^2 - 50x + 23.04 = 0 \rightarrow x = 1.9474; \; 0.6036,$$

are the solutions, using the quadratic formula (see Appendix G). With $x = \tan\theta_0$, these two solutions yield angles of 63° and 31° as the required limits.

50P

The origin of the coordinates is put at the antitank gun so that $x_0 = y_0 = 0$. Find the time of flight t of the shell from Eq.10:

$$y - y_0 = (v_0\sin\theta_0)t - \tfrac{1}{2}gt^2,$$

$$-60 - 0 = (240\sin 10°)t - \tfrac{1}{2}(9.8)t^2,$$

$$4.9t^2 - 41.676t - 60 = 0 \rightarrow t = -1.255 \text{ s}, \; 9.760 \text{ s}.$$

Hence t = 9.760 s (the negative solution corresponds to motion before t = 0; the shell is not in free-fall before firing so this solution has no meaning for our situation). Now compute the horizontal range of the shell, by Eq.9:

$$x - x_0 = (v_0\cos\theta_0)t,$$

$$x - 0 = (240\cos 10°)(9.760) \rightarrow x = 2307 \text{ m}.$$

For a hit, the tank must be at x = 2307 m also. Thus, the tank must travel 2307 – 2200 = 107 m if it wants to be hit. The time T needed for the tank to travel this distance is given from

$$x = \tfrac{1}{2}a_{tank}T^2,$$

$$107 = \tfrac{1}{2}(0.9)T^2 \rightarrow T = 15.42 \text{ s}.$$

Hence, the gun crew should wait 15.42 – 9.76 = 5.66 s before firing.

56E

(a) The satellite moves at constant speed so that

$$v = \frac{distance}{time} = \frac{2\pi(R + h)}{t} = \frac{2\pi(6370 + 640)km}{(98\ min)(60\ s/min)} = 7.49\ km/s.$$

(b) See Sample Problem 8:

$$g = \frac{v^2}{R + h} = \frac{(7.49\ km/s)^2}{7010\ km} = 8.00\ X\ 10^{-3}\ km/s^2 = 8.00\ m/s^2.$$

58E

(a) The point moves in a circle, and therefore the distance traveled = $2\pi R = 2\pi(0.15\ m)$ = 0.942 m.

(b) We need the time to complete one revolution in order to get the speed. Since the blade completes 1200 revolutions in one minute, it must take (60 s)/1200 = 0.050 s to complete one revolution. Hence, v = (0.942 m)/(0.050 s) = 18.8 m/s.

(c) The acceleration is $a = v^2/R = (18.8)^2/(0.15) = 2360\ m/s^2$.

61E

(a) Since $a = v^2/R$, we have

$$v = \sqrt{[aR]} = \sqrt{[7gR]} = \sqrt{[(7)(9.8\ m/s^2)(5\ m)]} = 18.5\ m/s.$$

(b) The time required to complete one revolution is

$$t = \frac{2\pi R}{v} = \frac{2\pi(5\ m)}{18.5\ m/s} = 1.698\ s.$$

Therefore, in 1 min = 60 s, the centrifuge will complete (60 s)/(1.698 s/rev) = 35.3 rev which means that it is turning at the rate of 35.3 rev/min.

62P

(a) The acceleration is directed down, toward the axle (center of the circle of motion), and has a magnitude $a = v^2/R$. The time for one revolution is t = 60 s/5 = 12 s. The speed v of the passenger, then, is

$$v = \frac{2\pi R}{t} = \frac{2\pi(15)}{12} = 7.854\ m/s.$$

Hence, the acceleration is $a = v^2/R = (7.854)^2/15 = 4.11\ m/s^2$.

(b) At the lowest point the acceleration is directed up, still toward the center of the circle of motion. The magnitude is the same as in (a).

64P

(a) The acceleration is $a = v^2/R$; since $v = 2\pi R/t$ we have (noting that t = 1 day = 86,400 s),

$$a = (\frac{2\pi R}{t})^2(\frac{1}{R}) = \frac{4\pi^2 R}{t^2} = \frac{4\pi^2(6.37 \times 10^6 \text{ m})}{(8.64 \times 10^4 \text{ s})^2} = 0.0337 \text{ m/s}^2.$$

(b) For this hypothetical situation, solve the equation in (a) for the period t:

$$t = 2\pi\sqrt{(\frac{R}{a})} = 2\pi\sqrt{(\frac{6.37 \times 10^6}{9.8})} = 5066 \text{ s} = 84.4 \text{ min.}$$

68P

The acceleration in uniform circular motion is $a = v^2/R$. To find v, note that when the string breaks the stone is projected in a horizontal direction; i.e., with zero initial vertical speed. Hence, $y = gt^2/2$ and, since the horizontal speed does not change during free-fall, $x = vt$. Combining all these equations yields

$$y = \frac{1}{2}gt^2 = \frac{1}{2}g(\frac{x}{v})^2 \rightarrow v^2 = gx^2/2y;$$

$$a = v^2/R = \frac{gx^2}{2yR} = \frac{(9.8)(10)^2}{2(2)(1.5)} = 163 \text{ m/s}^2.$$

70E

The person can walk at a speed of $(15 \text{ m})/(90 \text{ s}) = 0.167$ m/s. The escalator moves at a speed of $(15 \text{ m})/(60 \text{ s}) = 0.250$ m/s. Thus, the speed of the person walking on the moving escalator, relative to the building, is $0.167 + 0.250 = 0.417$ m/s. Therefore, the time needed to cover the length of the escalator is $(15 \text{ m})/(0.417 \text{ m/s}) = 36.0$ s. To see that the length of the escalator does not really enter, redo the calculation assuming that the escalator is, say, 35 m long.

76E

Let: $\vec{v}_{sc}$ = velocity of snow relative to the car;

   $\vec{v}_{se}$ = velocity of snow relative to the earth;

   $\vec{v}_{ce}$ = velocity of car relative to the earth.

The snow falls vertically relative to the earth and the car moves horizontally, also relative to the earth; thus the vectors $\vec{v}_{se}$ and $\vec{v}_{ce}$ are perpendicular. Hence, the relation between the vectors, Eq.24,

$$\vec{v}_{sc} + \vec{v}_{ce} = \vec{v}_{se},$$

is as shown on the sketch. Since 50 km/h = 13.89 m/s, the angle θ sought is given from tanθ = 13.89/8, or θ = 60°.

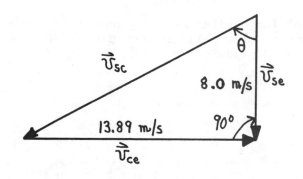

79P

Let: $\vec{v}_{rt}$ = velocity of rain relative to train;

   $\vec{v}_{rg}$ = velocity of rain relative to ground;

   $\vec{v}_{tg}$ = velocity of train relative to ground.

By Eq.24,

$$\vec{v}_{rt} + \vec{v}_{tg} = \vec{v}_{rg},$$

and this is illustrated on the sketch. Since v_{tg} = 30 m/s,

$$v_{rg} = \frac{v_{tg}}{\sin 22°} = \frac{30}{\sin 22°} = 80.1 \text{ m/s}.$$

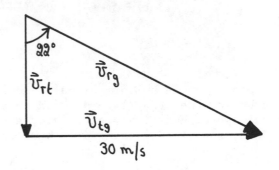

82P

(a) $\vec{v}_{pg}$ = velocity of plane relative to the ground;

   $\vec{v}_{pa}$ = velocity of plane relative to the air;

   $\vec{v}_{ag}$ = velocity of the air relative to the ground.

The speeds are $v_{pa} = v_{pg}$ = 135 mi/h and v_{ag} = 70 mi/h, so that the lengths of all the vectors are known. As far as the directions are concerned, it is given that $\vec{v}_{pg}$ is directed northward, so draw this in first. The other two vectors are arranged to obey the rule for adding vectors by the geometric method. The triangle formed is an isosceles, not a right triangle. (A solution with θ counterclockwise from north is also possible.) The wind direction is given by φ. To find it, draw in the perpendicular bisector (dashed line); this completes a right triangle, so that

$$\phi = \cos^{-1}\left(\frac{35}{135}\right) = 75.0°, \text{ E of S}.$$

(b) The heading of the plane is θ; since the angles of a triangle must total 180°, we have θ = 180° − 2φ = 30°, E of N.

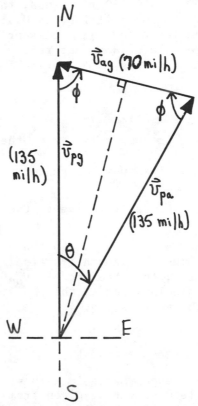

CHAPTER 5

5E

(a) To find the acceleration a, use Eq.2-10. Note that $v_0 = 0$ ("rest" means speed = 0); also, $v = 1600$ km/h $= 1600(0.2778) = 444.5$ m/s (find the conversion factor in Appendix F). Therefore,

$$v = v_0 + at,$$

$$444.5 \text{ m/s} = 0 + a(1.8 \text{ s}),$$

$$a = 247 \text{ m/s}^2.$$

(b) The net force F needed to accelerate the sled is given by Newton's second law:

$$F = ma = (500 \text{ kg})(247 \text{ m/s}^2) = 1.235 \times 10^5 \text{ N}.$$

9P

(a) Draw a free-body diagram of the sphere showing the forces acting on the sphere: F_e due to the electric field, $W = mg$ = weight of the sphere (the weight acts vertically down as always), and T = tension in the string. The sphere stays at rest so its acceleration a = 0. By Newton's second law, this means that the sums of the vertical and horizontal force components must separately add up to zero. Examining the diagram, this condition requires that

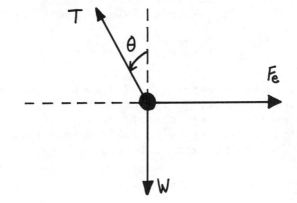

$$T\sin\theta - F_e = 0,$$

$$T\cos\theta - mg = 0.$$

Upon dividing one by the other, these equations imply that

$$\frac{\sin\theta}{\cos\theta} = \tan\theta = \frac{F_e}{mg};$$

see Appendix G for the trig. identity. Substituting the numerical data yields

$$F_e = mg\tan\theta = (3 \times 10^{-4})(9.8)(\tan 37°) = 2.22 \times 10^{-3} \text{ N}.$$

(b) With this result, the tension T can be found from the very first equation:

$$T = \frac{F_e}{\sin\theta} = \frac{2.22 \times 10^{-3}}{\sin 37°} = 3.69 \times 10^{-3} \text{ N}.$$

11P

(a) The acceleration of the sled is $a = F/m = 5.2 \text{ N}/8.4 \text{ kg} = 0.619 \text{ m/s}^2$.

(b) Newton's third law requires that the force exerted by the sled on the girl equals, in magnitude, the force exerted by the girl on the sled. Hence, her acceleration is $A = F/M = 5.2 \text{ N}/40 \text{ kg} = 0.130 \text{ m/s}^2$.

(c) Since they start from rest, the distances traveled by the sled and the girl in the time t required for them to meet are given by $x = \frac{1}{2}at^2$ and $X = \frac{1}{2}At^2$, measured from their starting points. Since they were separated by 15 m to begin with,

$$x + X = 15,$$

$$\frac{1}{2}(0.619)t^2 + \frac{1}{2}(0.130)t^2 = 15,$$

$$t = 6.329 \text{ s.}$$

Therefore, the distance from the girl's original position to the meeting point is just $X = \frac{1}{2}At^2 = \frac{1}{2}(0.130)(6.329)^2 = 2.60 \text{ m}$.

13P

(a) For the two blocks considered as a single entity, Newton's second law gives

$$F = (m_1 + m_2)a.$$

Now focus on block m_1 and draw a free-body diagram for it. F_c is the contact force that is exerted by m_2; this force retards the motion of m_1. Apply Newton's second law to m_1 for the horizontal forces to get

$$F - F_c = m_1 a.$$

Now solve for the acceleration a in the first equation and substitute the result into the second equation to obtain

$$F - F_c = m_1 \left(\frac{F}{m_1 + m_2}\right),$$

$$F_c = \frac{m_2}{m_1 + m_2} F = \frac{1.2}{2.3 + 1.2}(3.2) = 1.10 \text{ N}.$$

(b) Convince yourself that this new situation can be described by simply switching the labels on the blocks; that is, interchange m_1 and m_2 in the formula for F_c above to obtain for the new contact force

$$F_c = \frac{m_1}{m_2 + m_1} F = \frac{2.3}{1.2 + 2.3}(3.2) = 2.10 \text{ N}.$$

14E

(a) Refer to Table 2, p.82 of the text. Weight is a force, so its units are the same as the units of the force column. The lb is a British unit, so W = 1400 lb. The mass is given by m = W/g. In the British system of units g = 32 ft/s^2. Hence, m = (1400 lb)/ (32 ft/s^2) = 43.75 slug, the unit slug being found in the mass column.

(b) The kg is the mass unit in the SI system, so that m = 400 kg. The weight is W = mg = (400 kg)(9.8 m/s^2) = 3920 N, the unit N being the unit of force, and therefore also of weight, in the SI system.

17E

(a) Find the mass of the particle first (this does not change): m = W/g = 20/9.8 = 2.041 kg. The mass of the particle is independent of its location, so m = 2.041 kg at the new point. The weight, however, now becomes W = mg = (2.041 kg)(4.9 m/s^2) = 10 N.

(b) The mass m = 2.041 kg as before. The new weight is W = mg = (2.041)(0) = 0.

19E

The pebble is falling vertically with constant speed; i.e., at constant velocity. Thus, the acceleration of the pebble is zero. By Newton's second law, this indicates that the total (net) force acting on the pebble is zero also. The two forces acting are the weight of the pebble, directed vertically down, and the force due to the water. The only way that two forces can sum to zero is if they are oppositely directed and their magnitudes are equal. Thus the force F_w due to the water, in magnitude, equals the weight W of the pebble:

$$F_w = W = mg = (0.150 \text{ kg})(9.8 \text{ m/s}^2) = 1.47 \text{ N}.$$

Note that the speed of the pebble and the depth in the ocean do not enter.

21E

Since the people pull in opposite directions, the net force (horizontal) is F = 92 – 90 = 2 N. The resulting acceleration is a = F/m = (2 N)/(25 kg) = 0.080 m/s^2 = 8.0 cm/s^2.

28E

By "what strength" is meant: what tension must the fishing line be able to support without breaking in order to stop the fish. This tension is given by T = ma, where m is the mass of the salmon and a its acceleration in being brought to rest. To calculate this acceleration, use Eq.2-15. Set x_0 = 0; x is the stopping distance and this must be converted to feet (the unit of length in the British system of units), so that x = 4.4/12 = 0.3667 ft. Since the fish is brought to a stop v = 0 (final speed = 0). Thus,

$$v^2 = v_0^2 + 2a(x - x_0),$$
$$0^2 = (9.2)^2 + 2a(0.3667),$$
$$a = -115.4 \text{ ft/s}^2.$$

The mass of the fish is $m = W/g = (19 \text{ lb})/(32 \text{ ft/s}^2) = 0.5938$ slug. Hence $T = ma = (0.5938)(115.4) = 68.5$ lb. (The negative sign with the acceleration is not needed to calculate the magnitude of the tension.)

32E

We assume that the meteor is falling vertically. Up is positive. Hence the acceleration and the weight, being directed downward, point in the negative direction. Let f be the frictional force. Apply Newton's second law:

$$F = ma,$$

$$-mg + f = ma,$$

$$-(0.25)(9.8) + f = (0.25)(-9.2),$$

$$f = 0.150 \text{ N}.$$

Since f turns out to be positive, the frictional force is directed up.

38P

First, find the constant acceleration. Up is positive. Hence, $y_0 = 0$, $y = -42$ m, $v = 0$, $v_0 = -12$ m/s. Use Eq.2-15, but write y for x (vertical motion);

$$v^2 = v_0^2 + 2a(y - y_0),$$

$$0^2 = (-12)^2 + 2a(-42),$$

$$a = 1.714 \text{ m/s}^2.$$

This turns out to be positive and therefore the acceleration is directed upward. A free-body diagram of the elevator is shown. Newton's second law yields

$$T - mg = ma,$$

$$T - (1600)(9.8) = (1600)(1.714),$$

$$T = 18,400 \text{ N}.$$

45P

(a) The weight of the person is $w = mg = (80)(9.8) = 784$ N, and the weight of the parachute is $(5)(9.8) = 49$ N. Draw a free-body diagram of the person+parachute system considered as a single object. The total weight is $W = 784 + 49 = 833$ N. Let F be the force exerted by the air on the system. Up is positive. By Newton's second law,

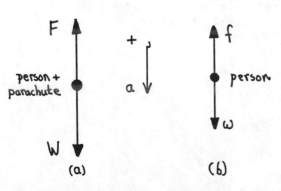

$$F - W = Ma,$$

$$F - 833 = (85)(-2.5),$$

$$F = 620.5 \text{ N.}$$

(b) Now draw a free-body diagram of the person alone. Let f be the force exerted by the parachute on him/her (this will be directed upward); thus, for the person,

$$f - w = ma,$$

$$f - 784 = (80)(-2.5),$$

$$f = 584 \text{ N.}$$

By Newton's third law, this is also the magnitude of the force exerted by the person on the parachute. (The same result is obtained if the forces acting on the parachute are considered instead of the forces acting on the person.)

49P

(a) Descending vertically with constant speed means that the acceleration a = 0; if T = the thrust of the rocket engine, directed upward, we have

$$T - W = ma = 0,$$

$$W = T = 3260 \text{ N.}$$

(b) With the thrust reduced, the acceleration is no longer zero. Up is positive. Hence,

$$T - W = ma,$$

$$2200 - 3260 = m(-0.39),$$

$$m = 2718 \text{ kg.}$$

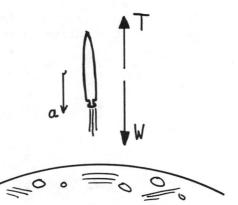

(c) Combining the results of (a) and (b) yields g = W/m = 3260/2718 = 1.20 m/s^2.

51P

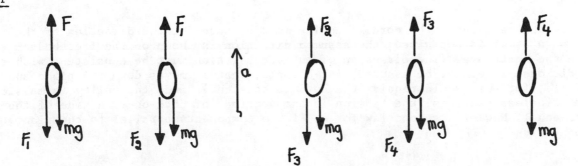

In applying Newton's second law to each link, the following quantities are needed:

$$mg = (0.10 \text{ kg})(9.8 \text{ m/s}^2) = 0.98 \text{ N},$$

$$ma = (0.10 \text{ kg})(2.5 \text{ m/s}^2) = 0.25 \text{ N}.$$

(a) Let F_i = forces between adjacent links, counting from the top pair. Then, using Newton's third law, and since F is the external force lifting the chain (note that F acts on the top link only), Newton's second law yields for each link, starting at the top (up is positive),

$$F - F_1 - mg = ma,$$

$$F_1 - F_2 - mg = ma,$$

$$F_2 - F_3 - mg = ma,$$

$$F_3 - F_4 - mg = ma,$$

$$F_4 - mg = ma.$$

The last equation gives

$$F_4 = mg + ma = 0.98 + 0.25 = 1.23 \text{ N}.$$

Substituting this into the fourth equation gives F_3, which can then be put into the third equation to give F_2, and so on. The results obtained are $F_3 = 2.46$ N, $F_2 = 3.69$ N, and $F_1 = 4.92$ N.

(b) With F_1 known, the first equation gives F = 6.15 N. This result for F can be obtained also by considering the entire chain as a single object of mass 5m; applying Newton's second law,

$$F - W = (5m)a,$$

$$F - (5m)g = (5m)a,$$

$$F = (5m)(g + a) = 6.15 \text{ N}.$$

(c) The net force on each link, by Newton's second law is $F_{net} = ma = 0.25$ N.

52P

(a) Let T be the tension in the cord. A choice must be made for the direction of the acceleration a as it is not given; the assumed direction is shown on the free-body diagram on the next page. (The direction of the acceleration must be consistent with the given conditions; it cannot be assumed, for example, that m_1 moves down the plane and that m_2 falls, for this would require the cord to stretch.) Since the pulley is massless and in frictionless bearings, the tension in the sections of cord on each side of the pulley are equal. Newton's second law for m_1 (force components parallel to the plane) and for m_2 yields

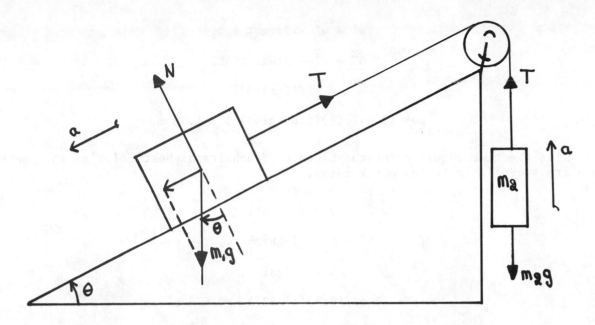

$$m_1 g \sin\theta - T = m_1 a,$$

$$T - m_2 g = m_2 a.$$

Solve for T in the second equation and substitute the resulting expression into the first equation and then solve for a:

$$T = m_2(a + g),$$

$$m_1 g \sin\theta - m_2(a + g) = m_1 a,$$

$$a = \frac{m_1 \sin\theta - m_2}{m_1 + m_2} g = \frac{(3.7)\sin 30° - 2.3}{3.7 + 2.3}(9.8) = -0.735 \text{ m/s}^2.$$

As this is negative, the acceleration actually is in the direction opposite to what was assumed; i.e., m_1 accelerates up the incline.

(b) From the third equation above, using the result for the acceleration found in (a),

$$T = m_2(a + g) = (2.3)(-0.735 + 9.8) = 20.8 \text{ N}.$$

<u>54P</u>

(a) The only force acting parallel to the incline on the block is the component of its weight W parallel to the incline, and this component is $W\sin\theta = mg\sin\theta$. Thus, the acceleration of the block, whether it is moving up or down the incline, is $mg\sin\theta/m = g\sin\theta$, directed down the incline. Let L be the distance that the block moves up the incline. At its highest point on the incline, the speed of the block is zero. The speed at the bottom, $x = 0$, is v_0; if x is positive up the incline, then $a = -g\sin\theta$ and

$$v^2 = v_0^2 + 2a(x - x_0),$$

$$0^2 = v_0^2 + 2(-g\sin\theta)L,$$

$$L = v_0^2/(2g\sin\theta),$$

$$L = (3.5)^2/(2)(9.8)(\sin 32°) = 1.18 \text{ m.}$$

(b) If t is the time required for the block to reach its highest point on the incline (where its speed is instantaneously zero),

$$v = v_0 + at,$$

$$0 = v_0 + (-g\sin\theta)t,$$

$$t = v_0/g\sin\theta,$$

$$t = (3.5)/(9.8)(\sin 32°) = 0.674 \text{ s.}$$

(c) Since there is no friction, the speed of the block when it gets back to the bottom of the incline must be the speed with which it was projected up the incline in the first place; i.e., 3.5 m/s.

60P

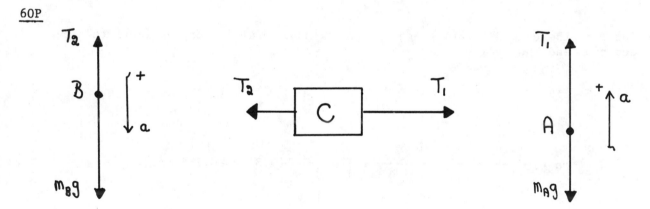

(a) Draw free-body diagrams of the cage (A) and counterweight (B); also shown is the mechanism (C) and the forces exerted on it by the cables. For the cage,

$$T_1 - m_A g = m_A a,$$

$$T_1 = m_A(g + a) = (1100)(9.8 + 2) = 12980 \text{ N.}$$

(b) For the counterweight,

$$T_2 - m_B g = m_B a,$$

$$T_2 = m_B(g + a) = (1000)(9.8 - 2) = 7800 \text{ N.}$$

Note that with the elevator accelerating upwards, the counterweight is accelerating downwards, i.e., in the negative direction, so we put $a = -2$ m/s^2 for the counterweight.

(c) The net force $\vec{F} = \vec{T}_1 + \vec{T}_2$ exerted by the cable on the mechanism is $F = T_1 - T_2 =$ 12980 – 7800 = 5180 N, to the right as pictured. By Newton's third law, the force exerted by the mechanism on the cable is 5180 N to the left, i.e., toward the mechanism.

64P

Let F be the upward lift of the balloon and m the mass of ballast discarded. Up is positive. Newton's second law, written for conditions before and after dropping the ballast are, with a representing the magnitude of the acceleration,

$$F - Mg = M(-a),$$

$$F - (M - m)g = (M - m)a.$$

Subtracting the equations to eliminate F and then solving for m gives $m = 2Ma/(a + g)$.

66P

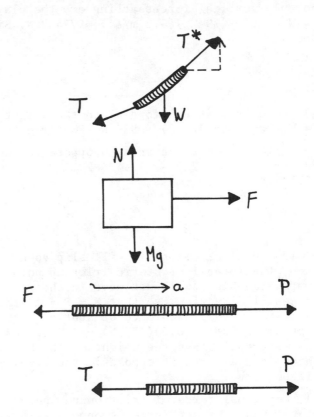

(a) Examine the forces acting on a very small portion of the rope; these forces are the weight W of the portion acting vertically down and the tensions T and T* exerted by particles of the rope immediately to the left and right of the portion considered. If the rope is in equilibrium, the downward acting W must be balanced by upward components of T or T*. If the rope, however, is horizontal, the tensions will have no up or down components. Thus, the rope must sag to some extent, although it may not be apparent.

(b) For the rope, mass m, and the block, mass M, Newton's second law gives

$$P - F = ma,$$

$$F = Ma.$$

F is the force exerted by the rope on the block; by Newton's third law the block exerts an equal force on the rope. Adding these equations gives $a = P/(m + M)$.

(c) Directly from (b),

$$F = Ma = \frac{M}{m + M} P.$$

(d) Let T be the tension in the midpoint of the rope. Draw a free-body diagram of the leading half of the rope. Newton's second law then yields

$$P - T = \frac{m}{2} a = \frac{m}{2} \frac{P}{m + M} \quad \rightarrow \quad T = \frac{P}{2} \frac{m + 2M}{m + M}.$$

CHAPTER 6

1E

(a) The weight W of the bureau is $W = mg = (45 \text{ kg})(9.8 \text{ m/s}^2) = 441$ N. There are no forces acting with components in the **vertical direction** except the weight W and the normal force N; hence, $N = W = 441$ N. The force of static friction that must be overcome is $f = \mu_s N = (0.45)(441) = 198$ N. Thus the applied force F must be at least 198 N to start the bureau moving.

(b) Repeat (a) except that now $W = mg = (45 - 17)(9.8) = 274.4$ N. The normal force $N = 274.4$ N also; $f = \mu_s N = (0.45)(274.4) = 123$ N; $F = f = 123$ N.

2E

The only vertical forces acting are the player's weight W and the normal force N. Hence $N = W = mg = (79 \text{ kg})(9.8 \text{ m/s}^2) = 774.2$ N. Since $f = \mu_k N$, we have $\mu_k = f/N = 470/774.2 = 0.607$.

10E

(a) The only vertical forces are the weight of the crate W and the normal force N. Thus, $N = W = mg = (55)(9.8) = 539$ N. The friction force is $f = \mu_k N = (0.37)(539) = 199.4$ N.

(b) The friction force acts to oppose the motion, so that the net horizontal force F on the crate is $F = 220 - 199.4 = 20.6$ N. Therefore the acceleration is $a = F/m = 20.6/55 = 0.375 \text{ m/s}^2 = 37.5 \text{ cm/s}^2$.

13E

(a) If the block slips, it will slip down; hence, the force f of static friction points upward. The block will not move in the horizontal direction, so that $N = F = 12$ lb. The maximum available force of static friction is $\mu_s N = (0.6)(12 \text{ lb}) = 7.2$ lb. As this is greater than the weight of the block (which is the force it opposes), the block will not move.

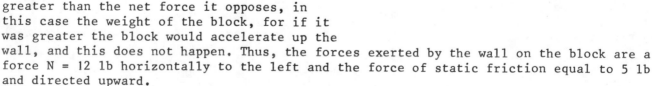

(b) The actual force of friction can be no greater than the net force it opposes, in this case the weight of the block, for if it was greater the block would accelerate up the wall, and this does not happen. Thus, the forces exerted by the wall on the block are a force $N = 12$ lb horizontally to the left and the force of static friction equal to 5 lb and directed upward.

16E

(a) Let v_0 be the initial speed and $v = 0$ the final speed; $x - x_0$ is the distance needed to come to a stop. The only force acting in the horizontal direction as the puck slides to a stop is the force f of kinetic friction. Therefore, $f = ma$. By Eq.15 of Chapter 2,

$$a = \frac{v^2 - v_0^2}{2(x - x_0)} = \frac{0^2 - 6^2}{2(15)} = -1.2 \text{ m/s}^2.$$

Hence the magnitude of f is $f = ma = (0.110 \text{ kg})(1.2 \text{ m/s}^2) = 0.132$ N.

(b) Since $f = \mu_k N$ and $N = mg$ in this situation,

$$f = \mu_k N = \mu_k mg = ma,$$

$$\mu_k = \frac{a}{g} = \frac{1.2}{9.8} = 0.122.$$

<u>23P</u>

(a) Let T be the tension in the rope; f is
the force of friction. With the crate on the
verge of moving, $f = \mu_s N$. Note that $N \neq W$ in
this case since part of the weight W is
balanced by the vertical component of T.
Applying Newton's second law in the x and y
(horizontal and vertical) directions gives

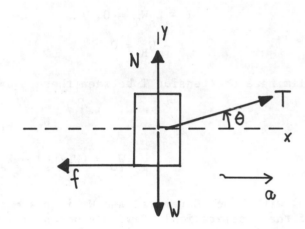

$$T\cos\theta - \mu_s N = 0,$$

$$N + T\sin\theta - W = 0.$$

Eliminate the normal force N and solve for T:

$$T\cos\theta - \mu_s(W - T\sin\theta) = 0,$$

$$T = \frac{\mu_s W}{\cos\theta + \mu_s \sin\theta} = \frac{(0.5)(150)}{\cos 15° + (0.5)\sin 15°} = 68.5 \text{ lb}.$$

(b) With the crate moving the force of friction f becomes $\mu_k N$; the components of the now
nonzero acceleration are $a_x = a$, $a_y = 0$. Therefore, Newton's second law now reads

$$T\cos\theta - \mu_k N = ma,$$

$$N + T\sin\theta - W = 0.$$

Set $m = W/g$ (since the weight, rather than the mass, is given; this is typical in those
problems specified in British units) and solve for the acceleration by eliminating N. The
result can be written

$$\frac{a}{g} = \frac{T}{W}(\cos\theta + \mu_k \sin\theta) - \mu_k \quad \rightarrow \quad a = 4.23 \text{ ft/s}^2,$$

using the value of the tension T found in (a) above (the problem statement implies that
the tension does not change).

25P

(a) With block C placed on block A, the two blocks form a single object with weight $W_T = W_A + W_C$. With C removed, only object A with weight W_A remains on the horizontal surface. The force of friction is f; the free-body diagrams of the two objects are shown on the right. For this part, use W_T, $f = \mu_s N$ and a = 0. In this case $N = W_T$ (f and T have no vertical components), so that Newton's second law applied to the two objects yields

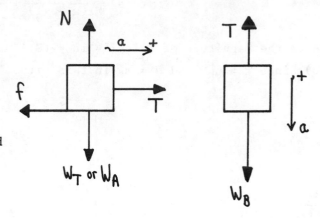

$$T - \mu_s W_T = 0,$$

$$T - W_B = 0.$$

Eliminate the tension T between these equations to obtain

$$W_B - \mu_s(W_A + W_C) = 0,$$

$$W_C = (W_B/\mu_s) - W_A = (22/0.2) - 44 = 66 \text{ N}.$$

(b) With block C removed use W_A in place of W_T; $f = \mu_k N = \mu_k W_A$; also $a \neq 0$ (a = magnitude of the acceleration). Newton's second law gives

$$T - \mu_k W_A = m_A a,$$

$$T - W_B = m_B(-a).$$

Subtracting these equations to remove T yields

$$a = \frac{W_B - \mu_k W_A}{m_B + m_A} = \frac{W_B - \mu_k W_A}{W_B + W_A} g = \frac{22 - (0.15)44}{22 + 44}(9.8) = 2.29 \text{ m/s}^2.$$

33P

(a) The free-body diagrams are drawn for block B moving up the plane, and also accelerating up the plane so that block A accelerates downward. For the opposite direction of acceleration, put −a for a in the equations that follow. Note that there is no correlation between the directions of the velocity and of the acceleration. When friction is present the direction of the velocity must be specified, for the force of friction is directed opposite to the velocity and not necessarily to the acceleration. In fact, reversing the direction of the velocity leads to a different acceleration, so the two directions of velocity must be treated

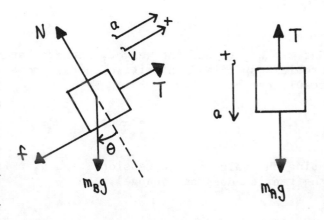

separately. In applying Newton's second law an acceleration up the plane is considered positive, so that with up positive for the hanging mass,

$$T - m_A g = m_A(-a),$$

$$T \pm f - m_B g\sin\theta = m_B a,$$

$$f = \mu N = \mu m_B g\cos\theta.$$

In the middle equation the lower sign applies if B is moving up the plane, as pictured in the sketch, since then f, opposing v, points downward in the negative direction, up the plane being chosen as positive. Substitute for f the quantity indicated in the last equation and then subtract the first two equations, to eliminate T, to obtain

$$a = \frac{m_A g - m_B g\sin\theta \pm \mu m_B g\cos\theta}{(m_A + m_B)g} \, g.$$

For v = 0 use $\mu = \mu_s$, the static coefficient of friction. Numerically, the terms in the numerator are $m_A g$ = 32 lb, $m_B g\sin\theta$ = 70.71 lb (since $\theta - 45°$), and $\mu_s m_B g\cos\theta = 39.60$ lb. For f = 0, a < 0, indicating that the system, if released from rest, will move so that B slides down the plane. But if the maximum possible value of f is used (with the upper sign in the expression for the acceleration) i.e., $\mu_s N$, then a > 0. Thus, in actuality, f < $\mu_s N$ and a = 0, for static friction will not provide a force so large as to reverse the "natural" tendency of B to move down the plane.

(b),(c) For v ≠ 0, $\mu = \mu_k$ = 0.25 so that f = $\mu_k mg\cos\theta$ = 17.68 lb. Substituting the data into the expression for the acceleration gives

$$a = \frac{32 - 70.71 \pm 17.68}{(32 + 100)}(32 \text{ ft/s}^2),$$

$$a = -13.7 \text{ ft/s}^2 \quad \text{(B moving up the plane)};$$

$$a = -5.10 \text{ ft/s}^2 \quad \text{(B moving down the plane)}.$$

The negative signs indicate that the acceleration is directed down the plane in both cases; this means that A is accelerated upward. These directions are opposite to what was assumed for the acceleration in writing the original equations.

36P

(a) Draw a free-body diagram showing the forces acting. Since $f_{16} = \mu_{16} m_{16} g\cos\theta$ and $f_8 = \mu_8 m_8 g\cos\theta$, Newton's second law gives

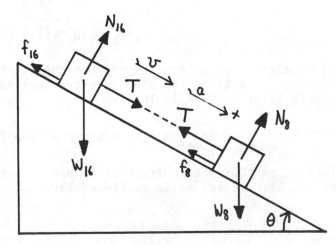

$$m_8 a_8 = m_8 g\sin\theta - T - \mu_8 m_8 g\cos\theta,$$

$$m_{16} a_{16} = m_{16} g\sin\theta + T - \mu_{16} m_{16} g\cos\theta.$$

Now $\mu_{16} = 2\mu_8$ so that if T = 0 initially, then $a_8 > a_{16}$ and the leading block will pull on the following block, with a blocks subsequently moving with a common value of the acceleration, and a tension will be present in the string. Set, then, $a_8 = a_{16} = a$, and add the two equations to get

$$(m_8 + m_{16})a = (m_8 + m_{16})g\sin\theta - (m_8\mu_8 + m_{16}\mu_{16})g\cos\theta \rightarrow a = 11.4 \text{ ft/s}^2,$$

upon insertion of the numerical data.

(b) With the acceleration found, the equation for either block can be used to find T. If the equation for the heavier block is used we have

$$T = m_{16}a - m_{16}g\sin\theta + \mu_{16}m_{16}g\cos\theta = 0.46 \text{ lb.}$$

(c) If the blocks are reversed, interchange the subscripts 16 and 8 in the equations above. If T = 0 initially, $a_8 > a_{16}$, as before. But as the 16 lb block leads to start with, the string remains slack and the blocks move independently, unless the plane is long enough for the 8 lb block to catch up with the other block. The separate pre-collision accelerations can be found from the equations above with T = 0.

39P

(a) Assume, for the moment, that the block slips on the slab. It is essential to note that, by Newton's third law, forces of friction of equal magnitude act on each of the objects; in fact, it is solely a force of friction that causes the slab to accelerate. Let F be the 100 N force acting on the block. As usual, apply Newton's second law using the free-body diagrams as a guide. Since $N_b = m_bg$, the equations are

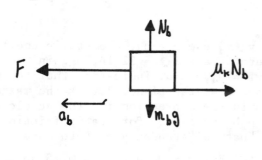

$$F - \mu_k m_b g = m_b a_b,$$

$$100 - (0.4)(10)(9.8) = (10)a_b,$$

$$a_b = 6.08 \text{ m/s}^2.$$

(b) For the slab,

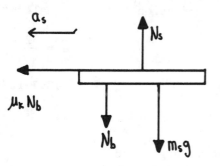

$$\mu_k m_b g = m_s a_s,$$

$$a_s = \mu_k(m_b/m_s)g = (0.4)(10/40)(9.8) = 0.98 \text{ m/s}^2.$$

If remains to be determined whether the block does, in fact, slip on the slab. (If it does not, the block and slab move together as a single object.) The maximum force of static friction "available" is

$$f_{max} = \mu_s N_b = \mu_s m_b g = (0.6)(10)(9.8) = 58.8 \text{ N} < 100 \text{ N,}$$

indicating that static friction cannot hold the block in place and the motion of the system is, indeed, as is outlined above.

41P

The resultant of the two normal forces N, equal in magnitude, is $N_r = 2N\cos45° = \sqrt{2}N$. This acts at angle θ to the vertical. Considering the forces acting perpendicular to the trough,

$$\sqrt{2}N - mg\cos\theta = 0,$$

since there is no acceleration component at a right angle to the trough. Now examine the forces parallel to the trough; the total force of friction is $2(\mu_k N)$ since there are two surfaces of contact. Hence,

$$mg\sin\theta - 2\mu_k N = ma.$$

Now substitute the value of N from the first equation to get

$$mg\sin\theta - 2\mu_k(mg\cos\theta/\sqrt{2}) = ma,$$

$$a = (\sin\theta - \sqrt{2}\mu_k\cos\theta)g.$$

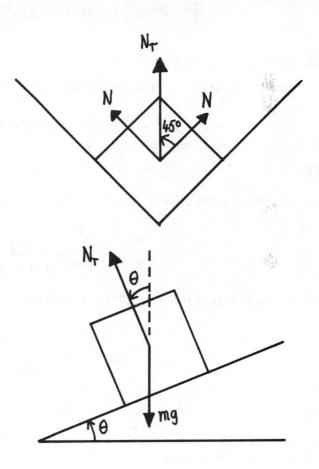

43P

First, note that 90 km/h = 25.0 m/s, 45 km/h = 12.5 m/s. Newton's second law yields

$$f = ma,$$

$$-70v = m\frac{dv}{dt} = (1000)\frac{dv}{dt},$$

$$-0.07dt = dv/v.$$

If T is the time sought,

$$-0.07\int_0^T dt = \int_{25}^{12.5} dv/v,$$

since v(0) = 25 m/s, v(T) = 12.5 m/s. Carrying out the integration gives

$$-0.07T = \ln(\frac{12.5}{25}) = \ln(1/2) = -\ln(2) \quad \rightarrow \quad T = \frac{\ln 2}{0.07} = 9.90 \text{ s.}$$

44E

The force is given by Eq.16. The cross-sectional area is $A = \pi(0.53 \text{ m})^2/4 = 0.2206 \text{ m}^2$. Therefore,

$$D = \tfrac{1}{2}C\rho A v^2 = \tfrac{1}{2}(0.75)(1.2 \text{ kg/m}^3)(0.2206 \text{ m}^2)(250 \text{ m/s})^2 = 6200 \text{ N}.$$

48E

See Sample Problem 7. From Eq.27 we find

$$v = \sqrt{[\mu_s gR]} = \sqrt{[(0.25)(9.8 \text{ m/s}^2)(47.5 \text{ m})]} = 10.79 \text{ m/s} = 38.8 \text{ km/h}.$$

54E

(a) See Sample Problem 2 and Eq.9 (note that 85 km/h = 23.61 m/s):

$$\mu_s = \frac{v_0^2}{2gd} = \frac{(23.61 \text{ m/s})^2}{2(9.8 \text{ m/s}^2)(60 \text{ m})} = 0.474.$$

(b) In this situation, Eq.27 is appropriate:

$$\mu_s = \frac{v^2}{gR} = \frac{(23.61 \text{ m/s})^2}{(9.8 \text{ m/s}^2)(60 \text{ m})} = 0.948.$$

57E

(a) It takes the electron $t = (1 \text{ s})/(6.6 \times 10^{15} \text{ rev}) = 1.515 \times 10^{-16}$ s to complete one revolution. Therefore,

$$v = \frac{2\pi r}{t} = \frac{2\pi(5.3 \times 10^{-11} \text{ m})}{1.515 \times 10^{-16} \text{ s}} = 2.20 \times 10^6 \text{ m/s}.$$

(b) By Eq.18,

$$a = \frac{v^2}{r} = \frac{(2.20 \times 10^6 \text{ m/s})^2}{5.3 \times 10^{-11} \text{ m}} = 9.13 \times 10^{22} \text{ m/s}^2.$$

(c) Eq.19 applies: $F = ma = (9.11 \times 10^{-31} \text{ kg})(9.13 \times 10^{22} \text{ m/s}^2) = 8.32 \times 10^{-8}$ N.

58E

Consider the hanging mass. Since M stays at rest, the tension in the cord must equal its weight: T = Mg. Now examine m. The only centripetal force (force directed toward the center of the circular path) acting on m is the tension T. Therefore,

$$T = mv^2/r = Mg \rightarrow v = \sqrt{(Mgr/m)}.$$

65P

(a) The student's apparent weight W is the force exerted by him on the Ferris wheel. By Newton's third law, this equals in magnitude the force exerted on him by the wheel. Note that W $\neq$ mg here, for W represents his apparent weight, not the force (= mg) that the earth exerts on him (his true weight). Let b refer to the bottom of the wheel and t to the top. The student's acceleration is directed to the center of the wheel. Thus, at the top,

$$W_t - mg = -mv^2/r,$$

$$125 - 150 = -mv^2/r \rightarrow mv^2/r = 25 \text{ lb}.$$

At the bottom,

$$W_b - mg = mv^2/r,$$

$$W_b - 150 = 25 \rightarrow W_b = 175 \text{ lb}.$$

(b) With the new speed V = 2v, $mV^2/r = 4mv^2/r = 100$ lb. Under these conditions, W_t = mg $-$ $mV^2/r = 150 - 100 = 50$ lb.

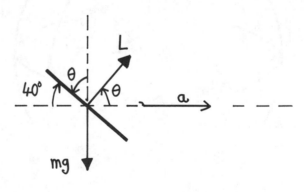

68P

Let r be the radius of the circle. The forces acting on the plane are its weight mg which acts vertically down and the lift L exerted by the air perpendicular to the wings. The acceleration is directed to the center of the circle. Applying Newton's second law in the vertical and horizontal directions gives

$$L\sin\theta = mg; \quad L\cos\theta = mv^2/r.$$

Dividing the equations yields

$$\frac{\sin\theta}{\cos\theta} = \tan\theta = gr/v^2 \rightarrow r = (\frac{v^2}{g})\tan\theta = 2.16 \text{ km},$$

since v = 480 km/h = 133.3 m/s and θ = 90° $-$ 40° = 50°.

73P

(a)

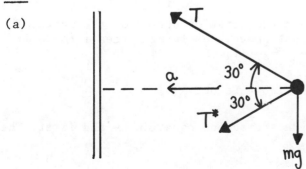

(b) The acceleration of the ball is directed horizontally toward the rod, that is, to the center of the circular path. Hence, the vertical force components must sum to zero; since the angle between the cords and the horizontal is 30°, we have

$$T\sin 30° - T^*\sin 30° - mg = ma_v = 0$$

$$35\sin 30° - T^*\sin 30° - (1.34)(9.8) = 0,$$

48

which gives T* = 8.74 N.

(c) The net force F acts in the direction of the acceleration, and therefore towards the rod. Thus,

$$F = T\cos 30° + T*\cos 30° = (T + T*)\cos 30°,$$

$$F = (35 + 8.74)\cos 30° = 37.9 \text{ N}.$$

(d) $F = ma = mv^2/r$, where r is the perpendicular distance of the ball to the rod (i.e., the radius of the circular path). This distance is $1.7\cos 30° = 1.472$ m and therefore,

$$F = mv^2/r,$$

$$37.9 = (1.34)v^2/(1.472) \quad → \quad v = 6.45 \text{ m/s}.$$

74P

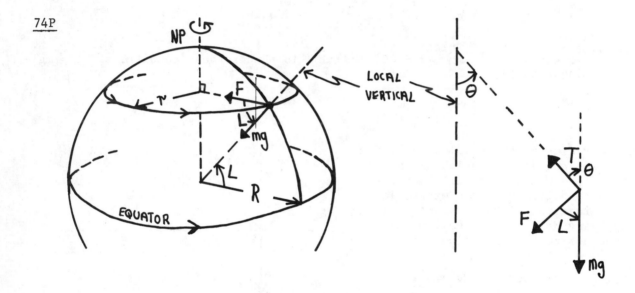

Due to the rotation of the earth, the plumb bob is moving at constant speed in a circle of radius r = RcosL, where R is the radius of the earth and L is the latitude. As a consequence, there must be a resultant force F on the bob directed at the axis of the earth's rotation. This force F makes an angle L with the local vertical and has a magnitude

$$F = m \frac{v^2}{r} = m \frac{(2\pi R\cos L/T)^2}{R\cos L},$$

where m is the mass of the bob and T = 86,400 s is the rotation period of the earth. The local situation is shown in the figure on the right above. Resolving forces into their vertical and horizontal components, we have, since $\vec{mg} + \vec{T} = \vec{F}$,

$$mg - T\cos\theta = F\cos L; \quad T\sin\theta = F\sin L.$$

Transpose mg in the first equation and then divide the second equation by the first. The result is

$$\tan\theta = \frac{F\sin L}{mg - F\cos L}.$$

Now θ is a small angle so that $\tan\theta = \theta$ approximately (see Appendix G). Also, mg is much greater than FcosL. The fact is that the earth does not rotate very rapidly. To a good degree of accuracy, then,

$$\theta = \frac{F\sin L}{mg} = \frac{2\pi^2 R\sin 2L}{gT^2}.$$

(a) With $R = 6.37 \times 10^6$ m, $g = 9.8$ m/s^2, $T = 86{,}400$ s and $L = 40°$, the equation above gives $\theta = 0.0017$ rad $= 5.8'$.

(b),(c) At either $L = 0°$ (equator) or $L = 90°$ (poles), the equation yields $\theta = 0$.

CHAPTER 7

1E

(a) The displacement $\vec{d}$ is in the horizontal, or x, direction. The worker exerts the 200 N force. Hence, the work done by the worker is

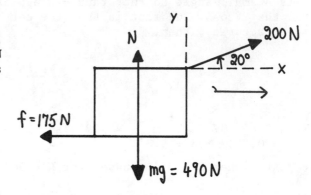

$$W = Fd\cos\phi = (200 \text{ N})(3.0 \text{ m})\cos20° = 564 \text{ J}.$$

(b) The force of sliding friction opposes the motion; i.e., $\phi = 180°$ for sliding friction. The work done by friction is

$$W = fd\cos\phi = (175 \text{ N})(3.0 \text{ m})\cos180° = -525 \text{ J}.$$

(c) The force of gravity is the weight mg, and this always acts vertically down. Since the displacement is horizontal, the angle $\phi = 90°$. But $\cos90° = 0$; this means that the work done by gravity = 0.

(d) The normal force N acts perpendicular (normal) to the surface. But the displacement is parallel to the surface and therefore $\phi = 90°$; the work done by the normal force = 0, again because $\cos90° = 0$.

(e) All of the forces acting on the crate have been accounted for; thus, the total work done is $564 - 525 + 0 + 0 = +39.0$ J.

5P

(a) The force exerted by the cord is the tension T, which can be found by applying Newton's second law:

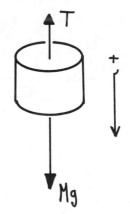

$$T - Mg = M(-\frac{g}{4}),$$

$$T = \frac{3}{4} Mg.$$

The tension pulls up, the displacement is down and therefore $\phi = 180°$. The work done by T is

$$W = Td\cos\phi = (\frac{3}{4} Mg)(d)\cos180° = -\frac{3}{4} Mgd.$$

(b) The weight acts vertically down which is in the same direction as the displacement. Thus, the work done by gravity is

$$W = (Mg)(d)\cos0° = Mgd.$$

9P

(a) The angle between the force F and the displacement d is $\theta = 15°$. That is, the angle ϕ in Eq.2 equals 15°. Hence, the work done by F is

$W = Fd\cos\phi = (7.68)(4.06)\cos 15° = 30.1$ J.

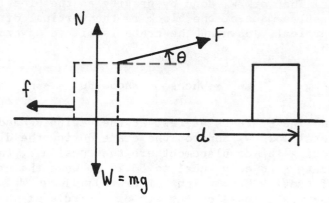

(b) To calculate the work done by friction, we need to find the friction force f. Since the object moves at constant speed and, by assumption, in a straight line, a = 0. Thus, applying Newton's second law in the horizontal direction,

$$F\cos\theta - f = 0.$$

Hence, $f = F\cos\theta$. The work done by friction is

$$W = fd\cos\phi = (F\cos\theta)(d)\cos\phi = (7.68\cos 15°)(4.06)\cos 180° = -30.1 \text{ J.}$$

(c) Gravity, acting vertically down, does no work on a horizontal displacement since $\phi = 90°$. The normal force is perpendicular to the surface along which the displacement takes place, so here too $\phi = 90°$ and the work is zero. All the forces are now accounted for, so the total work done is +30.1 − 30.1 + 0 + 0 = 0.

(d) Since $f = \mu_k N$, calculate the normal force. Applying Newton's second law in the vertical direction,

$$N + F\sin\theta - mg = 0,$$

$$N = mg - F\sin\theta = (3.57)(9.8) - (7.68)\sin 15° = 33.00 \text{ N.}$$

But, from (b), $f = F\cos\theta = (7.68)\cos 15° = 7.418$ N. Finally, then, $\mu_k = f/N = 7.418/33.00 = 0.225$.

10P

(a) In the final position, the sums of the vertical and of the horizontal force components acting on the crate must each be zero, since a = 0; applying this condition to the forces shown on the free-body diagram gives

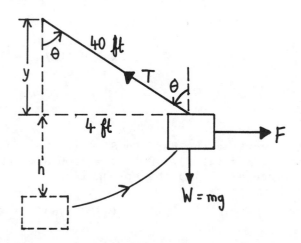

$$T\cos\theta - W = 0,$$

$$T\sin\theta - F = 0.$$

Eliminating T (by dividing the equations) yields $F = W\tan\theta$. But $\tan\theta = 4/y$ and, by the Pythagorean theorem, $y = \sqrt{(40^2 - 4^2)} = 39.80$ feet. Therefore,

$$F = W\tan\theta = (500)(4/39.80) = 50.25 \text{ lb.}$$

(b) Once the block has reached its final position, the work done in holding it there is zero, since there is no additional displacement.

(c) The work W_g done by gravity as the block is moved aside depends only on the displacement of the block in the vertical direction. Since the force W of gravity acts vertically down and the crate is lifted upward, the angle between the vectors is 180°; thus,

$$W_g = Wh\cos\phi = Wh\cos 180° = -Wh = -(500)(40 - 39.80) = -100 \text{ ft·lb.}$$

(d) The crate is at rest at the beginning and end of the displacement. By Eq.16, the total work W_t done on the crate during the displacement is zero. Since the forces acting during the displacement are the tension T, the force F exerted by the agent moving the crate (and only equal to 50.25 lb when the crate is in its final position) and the force of gravity W (= weight of crate) we have $W_t = 0 = W_T + W_F + W_g$. During the displacement, the crate moves on the arc of a circle of which the rope forms a radius. The tension, therefore, is at 90° to the displacement and the work done by the tension must be zero. The work done by the force pushing the crate aside is, then, $W_F = -W_g = +100 \text{ ft·lb.}$

(e) Evidently, this last result is not (50.25 lb)(4 ft), the result expected if F and the displacement vector are parallel and constant. But, even if F stays horizontal, it certainly changes in magnitude in order to bring the crate to rest at the end of the movement and then assume the value needed to hold the crate in place there. Also, the angle between F and the displacement continually changes: F is assumed to act in the horizontal direction and the displacement is tangent to the circular path, so that the angle between the two increases from 0 to θ from the initial to final position. However, regardless of how F varies and the angle changes during the motion, the value of the work done by F is 100 ft·lb.

14P

(a) Assuming $x_0 > 0$ the graph of F vs x is shown at the right. This is a straight line with a slope of F_0/x_0 and intercept of $-F_0$. The work done from $x = 0$ to $x = 2x_0$ is the area "under" the straight line between those limits. This area is zero, since the positive area (that above the $F = 0$ axis) = $F_0 x_0/2$ equals the negative area (that below the $F = 0$ axis) = $-F_0 x_0/2$, giving a total of zero.

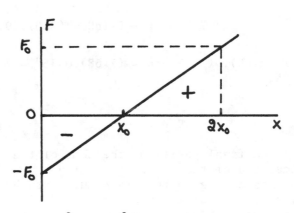

(b) Analytically,

$$W = \int F dx = F_0 \int_0^{2x_0} (x/x_0 - 1)dx = F_0\left(\frac{x^2}{2x_0} - x\right)\Big|_0^{2x_0} = 0.$$

15E

(a) Converting the data to SI units (a consistent set of units), we have for the spring constant $k = (15 \text{ N})/(10^{-2} \text{ m}) = 1500 \text{ N/m}$, and $x = 7.6 \text{ mm} = 7.6 \times 10^{-3} \text{ m}$. The work done by the force extending the spring is

$$W = \tfrac{1}{2}kx^2 = \tfrac{1}{2}(1500 \text{ N/m})(7.6 \times 10^{-3})^2 = 0.0433 \text{ J} = 43.3 \text{ mJ.}$$

(b) Eq.13 gives the work done by the spring. The work done by the agent extending the spring is the negative of this. Also, $x_i = 7.6 \text{ mm}$ and $x_f = 7.6 + 7.6 = 15.2 \text{ mm}$. The work done in extending the spring an additional 7.6 mm is therefore

$$W = \tfrac{1}{2}kx_f^2 - \tfrac{1}{2}kx_i^2 = \tfrac{1}{2}(1500)[(15.2 \times 10^{-3})^2 - (7.6 \times 10^{-3})^2] = 0.130 \text{ J} = 130 \text{ mJ}.$$

17E

Use Eq.15, $K = \tfrac{1}{2}mv^2$. The mass of an electron can be found on the inside front cover of HR ("Some Physical Constants"). The speed of the electron is

$$v = \sqrt{[\tfrac{2K}{m}]} = \sqrt{[\frac{2(6.7 \times 10^{-19} \text{ J})}{9.11 \times 10^{-31} \text{ kg}}]} = 1.21 \times 10^6 \text{ m/s}.$$

22E

(a) By Eq.16, the work done in stopping the bullet is

$$W = \Delta K = \tfrac{1}{2}mv_f^2 - \tfrac{1}{2}mv_i^2 = 0 - \tfrac{1}{2}(30 \times 10^{-3} \text{ kg})(500 \text{ m/s})^2 = -3750 \text{ J}.$$

Often the negative sign is omitted in quoting such results.

(b) Writing $W = Fd\cos\phi = Fd\cos 180° = -Fd$ (to stop the bullet, the force must oppose its motion, so $\phi = 180°$) gives

$$-3750 \text{ J} = -F(0.12 \text{ m}) \rightarrow F = 31,250 \text{ N},$$

recalling that 1 J = 1 N·m.

28E

(a) 60 km/h = 16.67 m/s = v_i. Since $W = K_f - K_i$ (Eq.16), and noting that the brakes do negative work on the car to oppose its motion,

$$-5 \times 10^4 = \tfrac{1}{2}(1000)v_f^2 - \tfrac{1}{2}(1000)(16.67)^2 \rightarrow v_f = 13.3 \text{ m/s} = 48.0 \text{ km/h}.$$

(b) For the additional work required, set $v_f = 0$ (bring the car to a stop) and now $v_i = 13.3$ m/s, so that, again using the work–energy theorem,

$$W = 0 - \tfrac{1}{2}(1000)(13.3)^2 = -88.4 \text{ kJ}.$$

33P

(a) First, it is necessary to find the tension T in the cable. The forces on the astronaut as he (she?) is being lifted are the cable tension T directed up and the weight mg down. Since the acceleration is upward, Newton's second law gives

$$T - mg = ma = m(\tfrac{g}{10}) \rightarrow T = \tfrac{11}{10} mg.$$

Call the upward displacement of the astronaut d. The work W done by the helicopter, by means of the cable, is

$$W = Td\cos\phi = [\tfrac{11}{10} mg]d\cos 0° = [\tfrac{11}{10}(160)](50) = 8800 \text{ ft·lb}.$$

(b) The work W_g done by gravity is

$$W_g = (mg)d\cos\phi = (mg)d\cos 180° = -(mg)d = -(160)(50) = -8000 \text{ ft·lb.}$$

(c) The net work done on the astronaut is the sum of (a) and (b), which is 800 ft·lb. But this is equal to the change ΔK in the kinetic energy of the astronaut. Presumably $K_i = 0$ as the astronaut is plucked from the water, so that at arrival at the helicopter, K = 800 ft·lb.

(d) Set $K = mv^2/2$:

$$800 = mv^2/2 = \frac{1}{2}(\frac{160}{32})v^2 \rightarrow v = 17.9 \text{ ft/s.}$$

36P

(a) Calculate the kinetic energy K directly by Eq.15:

$$K = \frac{1}{2}mv^2 = \frac{1}{2}(8.38 \times 10^{11} \text{ kg})(3 \times 10^4 \text{ m/s})^2 = 3.771 \times 10^{20} \text{ J,}$$

$$K = \frac{3.771 \times 10^{20} \text{ J}}{4.2 \times 10^{15} \text{ J/Mton}} = 89,800 \text{ Mtons.}$$

(b) The energy E and crater diameter D are related by $D = AE^{1/3}$, where A is a constant whose value depends on the units used. We assume that all of the kinetic energy is set free in the explosion, so identify E with K. We are told that D = 1 km if E = 1 Mton. Therefore,

$$A = \frac{D}{E^{1/3}} = \frac{1 \text{ km}}{(1 \text{ Mton})^{1/3}} = 1 \text{ km/Mton}^{1/3}.$$

Hence, for the comet,

$$D = AE^{1/3} = (1 \text{ km/Mton}^{1/3})(8.98 \times 10^4 \text{ Mton})^{1/3} = 44.8 \text{ km.}$$

37P

Since the initial and final speeds are zero, so are the initial and final kinetic energies, and therefore the total work done on the block during its motion is zero, by the work-energy theorem. The forces acting on the block during all or part of its motion are the spring force F_s, the force of kinetic friction f, the weight W and the normal force N. The last two forces do no work as they act in the vertical direction and the block moves horizontally (force and displacement at 90°). Thus, the total work W_t can be written as

$$W_t = 0 = W_s + W_f.$$

In this equation, W_s is the work done by the spring and W_f is the work done by friction. The force exerted by the spring is in the direction of the block's motion, so that W_s is positive; if x is the distance the spring was compressed, $W_s = kx^2/2$. The work done by friction is negative, for the force of sliding friction acts opposite to the direction of motion. If L is the total distance the block moves,

$$W_f = fL\cos 180° = -fL = -\mu_k NL = -\mu_k mgL.$$

Thus, by the work-energy theorem,

$$0 = \tfrac{1}{2}kx^2 - \mu_k mgL,$$

$$\mu_k = \tfrac{1}{2}kx^2/mgL = (\tfrac{1}{2})(200)(0.15)^2/(2)(9.8)(0.6) = 0.191.$$

41E

The woman does work against gravity. In British units, her average power output is

$$\overline{P} = \frac{W}{\Delta t} = \frac{(mg)h}{\Delta t} = \frac{(120)(15)}{3.5} = 514.3 \text{ ft·lb/s} = \frac{514.3}{550} = 0.935 \text{ hp.}$$

In SI units,

$$\overline{P} = \frac{mgh}{\Delta t} = \frac{(55)(9.8)(4.5)}{3.5} = 693 \text{ W.}$$

51P

The kinetic energy K gained by the water equals the work W done on the water by gravity. Since W = mgh, and noting that only 75% (= 0.75) of this energy is converted to electrical energy we have, for the power output of the generator,

$$P = \frac{0.75K}{\Delta t} = \frac{0.75mgh}{\Delta t}.$$

If $\Delta t = 1$ s, then m = $(1200 \text{ m}^3)(10^3 \text{ kg/m}^3) = 1.2 \times 10^6$ kg. Therefore,

$$P = (0.75)(1.2 \times 10^6 \text{ kg})(9.8 \text{ m/s}^2)(100 \text{ m})/(1 \text{ s}) = 8.82 \times 10^8 \text{ W} = 882 \text{ MW.}$$

59P

(a) 1 kW·h corresponds to 3.6×10^6 J. Since gasoline delivers 140 MJ/gal, the number N of gallons required to obtain 1 kW·h is

$$N = \frac{3.6 \times 10^6 \text{ J}}{1.4 \times 10^8 \text{ J/gal}} = 2.571 \times 10^{-2} \text{ gal.}$$

At 30 mi/gal, this amount of gas will carry the car a distance

$$D = (2.571 \times 10^{-2} \text{ gal})(30 \text{ mi/gal}) = 0.771 \text{ mi.}$$

(b) The number of gallons of gas consumed in one hour at 55 mi/h is

$$\frac{55 \text{ mi/h}}{30 \text{ mi/gal}} = 1.833 \text{ gal/h.}$$

Hence, the rate of expenditure of energy is

$$P = (1.833 \text{ gal/h})(140 \text{ MJ/gal}) = 256.6 \text{ MJ/h} = (256.6 \times 10^6 \text{ J})/(3600 \text{ s}) = 71.3 \text{ kW}.$$

60P

With the force f proportional to the speed v we can write $f = Av$, where A is a constant depending on the units used. The constant power $P = fv = Av^2$. Therefore, at two different speeds,

$$P_1 = Av_1^2; \quad P_2 = Av_2^2.$$

Eliminating A by, for example, dividing the equations, we have

$$P_1/P_2 = (v_1/v_2)^2,$$

$$P_1/10 = (7.5/2.5)^2 \rightarrow P_1 = 90 \text{ hp}.$$

62P

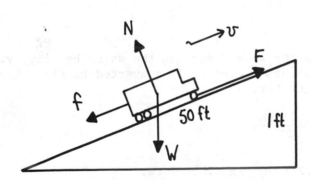
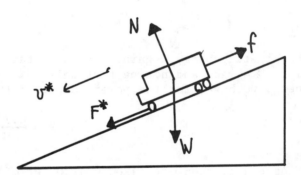

Let F, F* be the forces pulling the truck up the hill with speed v and then down with speed v* (these are friction forces exerted by the road). The resisting force in each direction is $f = W/25$, W = weight of the truck. In moving up and down the hill, the total work done on the truck is zero, for with the speed constant the kinetic energy does not change and $W_t = \Delta K$. Thus, the rate at which work is done is zero also. The forces that do work are F (or F*), the weight W and the resisting force f. Taking note of the directions of these forces, for motion up the hill the condition that the total rate of work done be zero can be written

$$P - W(v/50) - fv = 0,$$

where $P = Fv$ and $v/50$ is the vertical component of the truck's velocity. Since $f = W/25$,

$$P = \frac{3}{50} Wv.$$

Moving down the hill, with $P^* = F^*v^*$,

$$P^* + W(v^*/50) - fv^* = 0 \rightarrow P^* = \frac{1}{50} Wv^*.$$

But $P = P^*$ by supposition, so that,

$$\frac{3}{50} Wv = \frac{1}{50} Wv^* \; \rightarrow \; v^* = 3v = 3(15) = 45 \text{ mph.}$$

63P

(a) Use the work-energy theorem:

$$W = P(\Delta t) = (1.5 \times 10^6)(360) = 5.4 \times 10^8 \text{ J} = \Delta K,$$

$$\Delta K = \frac{1}{2}mv_f^2 - \frac{1}{2}mv_i^2 = \frac{1}{2}m(25^2 - 10^2) \; \rightarrow \; m = 2.057 \times 10^6 \text{ kg.}$$

(b) If v is the speed of the train a time t (in seconds) after the train starts to move, then

$$Pt = \frac{1}{2}mv^2 - \frac{1}{2}mv_i^2,$$

$$v^2 = v_i^2 + \frac{2P}{m} t,$$

$$v = \sqrt{(100 + 1.458t)},$$

with v in m/s and t in s.

(c) From (b) above,

$$\frac{1}{2}mv^2 - \frac{1}{2}mv_i^2 = Pt.$$

Differentiating with respect to time t gives

$$mv\frac{dv}{dt} = mva = P \; \rightarrow \; a = \frac{P}{mv},$$

since v_i and P are constants. The force is F = ma, so that

$$F = \frac{P}{v} = \frac{1.5 \times 10^6}{\sqrt{(100 + 1.458t)}},$$

F in Newtons. (This result also follows directly from Eq.25.)

(d) The total distance x moved is

$$x = \int v(t)dt = \int_0^{360} \sqrt{(100 + 1.458t)}dt.$$

Use Integral 4 in Appendix G to get

$$x = \frac{2}{3}(\frac{1}{1.458})(100 + 1.458t)^{3/2} \Big|_0^{360} = 6685 \text{ m} = 6.685 \text{ km.}$$

66E

(a) The circumference of the earth is $C = 2\pi R = 2\pi(6.37 \times 10^6 \text{ m}) = 4.002 \times 10^7$ m. Since the travel time is 1 s, the speed of the electron is $v = 4.002 \times 10^7$ m/s. In terms of the speed of light, $\beta = v/c = (4.002 \times 10^7 \text{ m/s})/(3 \times 10^8 \text{ m/s}) = 0.133.$

(b) Use Eq.28. The quantity mc^2 is

$$mc^2 = (9.11 \text{ X } 10^{-31} \text{ kg})(3 \text{ X } 10^8 \text{ m/s})^2/(1.6 \text{ X } 10^{-19} \text{ J/eV}) = 5.124 \text{ X } 10^5 \text{ eV}.$$

The other factor in the formula for K is

$$[(1 - \beta^2)^{-\frac{1}{2}} - 1] = 0.008964,$$

using the value of β found in (a). Therefore $K = (5.124 \text{ X } 10^5 \text{ eV})(0.008964) = 4590 \text{ eV} = 4.59 \text{ keV}$.

(c) Classically, we have

$$K = \frac{1}{2}mv^2 = \frac{\frac{1}{2}(9.11 \text{ X } 10^{-31} \text{ kg})(4.002 \text{ X } 10^7 \text{ m/s})^2}{1.6 \text{ X } 10^{-19} \text{ J/eV}} = 4560 \text{ eV} = 4.56 \text{ keV}.$$

The %error is

$$\frac{K_{wrong} - K_{right}}{K_{right}}(100) = \frac{4.56 - 4.59}{4.59}(100) = -0.65\%.$$

CHAPTER 8

1E

Eq.9 applies; note that 1 J = 1 N·m and convert data to SI units;

$$U = \frac{1}{2}kx^2 \rightarrow k = \frac{2U}{x^2} = \frac{2(25\ J)}{(0.075\ m)^2} = 8890\ N/m = 88.9\ N/cm.$$

12E

At the bottom of the ramp put y = 0, so that U = 0 there. The truck's energy initially is all kinetic. At the place on the escape ramp where the truck is brought to rest v = 0 and therefore K = 0. The potential energy there is U = mgy. In terms of the distance along the ramp, we have y = Lsin15°. The mass of the truck is not given; call it m (and hope that it will cancel out). Equating the mechanical energy E, Eq.12, at the bottom of the ramp to the energy where the truck is brought to rest gives

$$0 + \frac{1}{2}mv^2 = mg(Lsin15°) + 0.$$

But v = 80 mi/h = 80(1.467) = 117.4 ft/s. Note that m does drop out. We then have

$$\frac{1}{2}(117.4\ ft/s)^2 = (32\ ft/s^2)(Lsin15°) \rightarrow L = 832\ ft.$$

16E

(a) The kinetic energy is $K = mv^2/2 = (2.4\ kg)(150\ m/s)^2/2 = 27\ kJ.$

(b) The potential energy, by Eq.11, is $U = mgy = (2.4\ kg)(9.8\ m/s^2)(125\ m) = 2.94\ kJ.$

(c) As the projectile strikes the ground U = 0 since y = 0. Thus, the mechanical energy at that moment is all kinetic. The mechanical energy at firing is the sum of (a) and (b), that is, E = 27 kJ + 2.94 kJ = 29.94 kJ. Without air resistance E does not change. Hence, at impact E = 29.94 kJ also, so that, at impact,

$$E = 29.94 \times 10^3\ J = \frac{1}{2}mv^2 = \frac{1}{2}(2.4\ kg)v^2 \rightarrow v = 158\ m/s.$$

Clearly the answers to (a) and (b) do depend on the value of m, but the answer to (c) does not. To convince yourself of this, write out the calculations algebraically, or repeat with a different numerical value for m.

18E

(a) Since U = mgy, the change is $\Delta U = mg(\Delta y) = (0.005\ kg)(9.8\ m/s^2)(+20.08\ m) = 0.984\ J.$

(b) Since the marble "just' reaches its target, put v = 0 at the target. The marble also is at rest just before firing. Therefore $\Delta K = 0$. Since mechanical energy is conserved, $\Delta E = 0$. Hence, the change ΔU in gravitational potential energy must equal the negative of the change in the spring's potential energy. That is,

$$0.984\ J = -(\frac{1}{2}kx_f^2 - \frac{1}{2}kx_i^2) = -(0 - \frac{1}{2}kx_i^2),$$

$$0.984 \text{ J} = \frac{1}{2}k(0.08 \text{ m})^2 \rightarrow k = 308 \text{ N/m} = 3.08 \text{ N/cm}.$$

19E

(a) Put $y = 0$ at the initial position of the ball, so that $U_i = 0$. In the final position $y = L$ so $U_f = mgL$. The change is $\Delta U = U_f - U_i = mgL$.

(b) Since E is constant, $\Delta E = \Delta U + \Delta K = 0$, or $\Delta K = -\Delta U$. Therefore,

$$K_f - K_i = 0 - \frac{1}{2}mv^2 = -mgL \rightarrow v = \sqrt{(2gL)}.$$

25P

(a) First calculate the spring constant k: $k = F/x = (270 \text{ N})/(0.02 \text{ m}) = 13,500 \text{ N/m}$. Let's put $y = 0$ at the "final" position of the block, when it comes to rest with the spring compressed; i.e., $y_f = 0$. In the initial position at the top of the incline $y_i = L\sin 30°$ $= L/2$, where L is the distance sought. The block is at rest in both the initial and final positions, so $K_i = K_f = 0$. Conservation of mechanical energy requires that

$$\frac{1}{2}mv_i^2 + mgy_i + \frac{1}{2}kx_i^2 = \frac{1}{2}mv_f^2 + mgy_f + \frac{1}{2}kx_f^2.$$

Initially the spring is not compressed, so $x_i = 0$. Inserting data in SI units,

$$0 + (12)(9.8)(L/2) + 0 = 0 + 0 + \frac{1}{2}(13,500)(0.055)^2 \rightarrow L = 0.347 \text{ m} = 34.7 \text{ cm}.$$

(b) As the block hits the spring, $y_f = (5.5 \text{ cm})\sin 30° = 2.75 \text{ cm} = 0.0275 \text{ m}$. The spring is not yet compressed, so $x_f = 0$. The initial conditions are unchanged. Therefore, we have

$$0 + (12)(9.8)(0.1735) + 0 = \frac{1}{2}(12)v_f^2 + (12)(9.8)(0.0275) + 0 \rightarrow v_f = 1.69 \text{ m/s}.$$

Note that in this part we left the $y = 0$ position where it was in part (a).

28P

(a) Use SI units throughout. $K_i = mv_i^2/2 = (5)(100)^2/2 = 2.5 \times 10^4$ J.

(b) The shell is not fired vertically. At the top of its path it has a horizontal speed equal to the initial horizontal speed $= (100)\cos 34° = 82.90$ m/s. Hence, its kinetic energy at the top is $K_f = (5)(82.90)^2/2 = 17,180$ J. Since $U_i = 0$, and $U_i + K_i = U_f + K_f$, we have $U_f = 25,000 - 17,180 = 7820$ J.

(c) Since $U_f = mgy_f$, $y_f = U_f/mg = (7820)/(5)(9.8) = 160$ m.

31P

The minimum work W required is that work performed by exerting a force F on the chain just equal to the weight of chain still hanging over the edge of the table. (In this way no kinetic energy is imparted, so that $K = 0$ throughout; actually, of course, the chain must have some kinetic energy, but this can be very small; it will just take a long time to get the chain back on the table.) If at some instant a length x is hanging over, then $F(x) = (m/L)gx$. Since the chain is being moved in a way to decrease x, we introduce a minus sign, so that

$$W = -\int F dx = -\int_{\frac{1}{4}L}^{0} (m/L) gx dx = mgL/32.$$

This result may also be obtained from $W = \Delta U$. To calculate ΔU, note that a length $L/4$ of the chain, with mass $m/4$, must be lifted an average distance of $\frac{1}{2}(L/4) = L/8$. Hence, $\Delta U = (m/4)g(L/8) = mgL/32$.

33P

Choose $y = 0$ at the "final" position of the block, at rest, with the spring compressed by length x. At the initial position of the block $y_i = x + 0.40$ (use SI units throughout). The initial and final kinetic energies both are zero. By conservation of mechanical energy,

$$K_i + mgy_i + \frac{1}{2}kx_i^2 = K_f + mgy_f + \frac{1}{2}kx_f^2,$$

$$0 + (2)(9.8)(x + 0.40) + 0 = 0 + 0 + \frac{1}{2}(1960)x^2,$$

$$19.6x + 7.84 = 980 \, x^2,$$

$$x^2 - 0.02x - 0.008 = 0 \rightarrow x = 0.10 \text{ m} = 10 \text{ cm}.$$

The solution was by the quadratic formula; see Appendix G; the negative root was ignored.

35P

Call the height of the table y, the mass of the ball m, and the horizontal distance to impact L. The horizontal speed v needed to carry the marble a distance L can be found by

$$L = vt \rightarrow t = L/v;$$
$$y = \frac{1}{2}gt^2 = \frac{1}{2}g(L/v)^2 \rightarrow v^2 = L^2 g/2y.$$

If the spring in the gun is compressed a distance x, the speed v with which the marble leaves the gun is given from (see Sample Problem 2)

$$\frac{1}{2}kx^2 = \frac{1}{2}mv^2.$$

Substituting for v^2 from the previous equation leads to

$$kx^2 = mL^2 g/2y.$$

Now call x_1, L_1 and x_2, L_2 the values of x and L for the two shots. The equation above implies that

$$x_1/x_2 = L_1/L_2,$$

$$1.1/x_2 = 1.93/2.2 \rightarrow x_2 = 1.25 \text{ cm}.$$

In the above we used $L_1 = 2.2$ m $- 27$ cm $= 1.93$ m.

38P

If the ball just barely swings around the nail, then the tension in the string must vanish as the ball passes the top of its circular path with speed v, leaving its weight as the only force acting. Noting that the radius of the circular path is r = L – d, Newton's second law for circular motion (see Eq.19 in Ch.6), and the conservation of energy require that

$$mv^2/(L - d) = mg,$$

$$mgL = (mg)[2(L - d)] + \frac{1}{2}mv^2.$$

(In the energy equation, we used the bottom of the swing as y = 0.) Eliminating v^2 from these equations (solve for v^2 in one and substitute into the other) gives d = 3L/5.

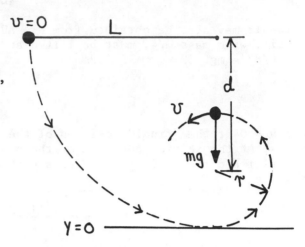

39P

(a) By conservation of energy,

$$U_R + K_R = U_Q + K_Q,$$

$$mg(2L) + 0 = 0 + \frac{1}{2}mv^2 \rightarrow v = 2\sqrt{[gL]}.$$

(b) Consider the forces acting on the mass when at Q. Newton's second law (Eq.19 in Ch.6) yields

$$T - mg = mv^2/L,$$

L being the radius of the circular path. But the second equation in (a) above can be written in the form

$$mv^2/L = 4mg,$$

so that T – mg = 4mg, or T = 5mg.

(c) Now apply Eq.19 in Ch.6 along the direction of the suspension (i.e., along the direction to the center of the circle) to the mass at P, assuming that the angle to the vertical has the value sought; we get

$$T - mg\cos\theta = mu^2/L,$$

u being the speed at P. Conservation of mechanical energy, applied between the points S and P gives

$$mgL + 0 = mgh + \frac{1}{2}mu^2 \rightarrow u^2 = 2gL\cos\theta,$$

since $h = L - L\cos\theta$. Using this result, the first equation in (c) becomes

$$T - mg\cos\theta = 2mg\cos\theta \rightarrow T = 3mg\cos\theta.$$

It follows that if $T = mg$, then $\cos\theta = 1/3$ and $\theta = 71°$.

42P

Let v be the speed with which the boy leaves the ice. Since he is given a very small push at the top, we can safely assume that his initial speed is zero. Energy conservation during the slide on the ice (no friction) gives

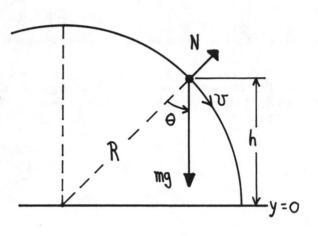

$$0 + mgR = \frac{1}{2}mv^2 + mgR\cos\theta,$$

$$mv^2/R = 2mg(1 - \cos\theta).$$

Now apply Newton's second law (Eq.19 in Ch.6) to get

$$mg\cos\theta - N = mv^2/R.$$

Combining the last two equations yields

$$mg\cos\theta - N = 2mg(1 - \cos\theta).$$

As the boy leaves the ice mound, the normal force N (a force of contact) goes to zero. Putting $N = 0$ in the last equation gives $\cos\theta = 2/3$. But $\cos\theta = h/R$ and therefore, at this point, $h = 2R/3$.

43E

(a) The force F is given by Eq.13:

$$F = -\frac{dU}{dx} = -(\text{slope of U vs x graph}).$$

Near $x = 0$ the slope is about -4 (a straight line tangent to the curve at $x = 0$ passes through $x = 1$, $U = 0$), so that $F(0) = 4$ N. U has zero slope at $x = 2$ m, and therefore $F(2) = 0$. Between $x = 2$ m and $x = 6$ m, the graph of U is almost a straight line with slope 2/4, giving roughly $F = -0.5$ N in this region; the sign indicates that F points in the $-x$ direction here. See the resulting sketch on p.64.

(b) The kinetic energy is $K = E - U = 4 - U$, so the graph of K vs x is just a reflection of U vs x about the line $U = 4$ J.

50E

(a) Using SI units, initially $U = mgy_i = (25)(9.8)(12) = 2940$ J.

(b) At the bottom, $K = mv^2/2 = (25)(5.6)^2/2 = 392$ J.

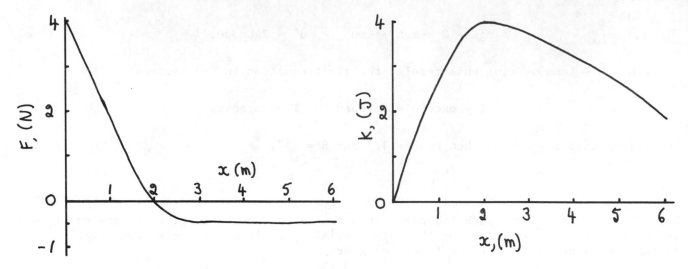

EXERCISE 43

****

(c) At the top, $E_i = K_i + U_i = 0 + 2940 = 2940$ J. At the bottom, $E_f = K_f + U_f = 392 + 0 = 392$ J. The change in mechanical energy is $\Delta E = E_f - E_i = -2548$ J. By Eq.21, this equals the work W_f done by the frictional forces. Since this force opposes the motion, $\phi = 180°$ and therefore

$$W_f = fd\cos 180° = -fd = \Delta E,$$

$$-f(12) = -2548 \rightarrow f = 212 \text{ N}.$$

52E

(a) For this situation, the normal force is $N = mg = (3.5)(9.8) = 34.3$ N. Therefore,

$$W_f = fd\cos 180° = -fd = -\mu_k Nd = -(0.25)(34.3)(7.8) = -66.9 \text{ J}.$$

(b) Over the rough portion of the track, the friction force is the net force acting on the block (N and W cancel). By the work-energy theorem, then,

$$W_f = K_f - K_i,$$

$$-66.9 = 0 - K_i \rightarrow K_i = 66.9 \text{ J}.$$

(c) Sample Problem 2 discusses the horizontal spring gun, and shows that

$$\tfrac{1}{2}kd^2 = \tfrac{1}{2}mv^2 = K_i,$$

$$d = \sqrt{[2K_i/k]} = \sqrt{[\frac{2(66.9)}{640}]} = 0.457 \text{ m} = 45.7 \text{ cm}.$$

54P

(a) With no frictional losses, energy conservation yields (see Sample Problem 4),

$$v = \sqrt{[2gh]} = \sqrt{[2(9.8)(850 - 750)]} = 44.3 \text{ m/s}.$$

(b) Use the work-energy theorem. It is implied that the skier arrives at the second peak with speed $v = 0$. Since the skier started from rest at the upper peak, $\Delta K = 0$. The forces acting on the skier during the run are the skier's weight mg, the normal force N, and the friction force f. The normal force does no work. Since the inclination angles of the two slopes are the same, they can be treated as a single slope of length $L = 3.2$ km. The work-energy theorem requires that $W = \Delta K = W_g + W_f$. But $f = \mu_k N$ and $N = mg\cos\theta$, and therefore,

$$+mgh - \mu_k(mg\cos\theta)L = 0,$$

$$\mu_k = \frac{h}{L\cos\theta} = \frac{100}{3200\cos 30°} = 0.036.$$

<u>59P</u>

(a) The required work, by Eq.10 in Ch.7, is

$$W = \int F dx = \int_{0.5}^{1.0} [52.8x + 38.4x^2]dx = [(52.8)\frac{x^2}{2} + (38.4)\frac{x^3}{3}]\Big|_{0.5}^{1.0} = 39.2 - 8.2 = 31.0 \text{ J}.$$

(b) Use the work-energy theorem (which in this case is the same as the conservation of mechanical energy). The spring does 31 J of work on the mass, which appears as additional kinetic energy. The mass had zero kinetic energy to start with (since it was released from rest). Thus, as it passes $x = 0.50$ m, its kinetic energy is $K = 31$ J. Therefore,

$$K = 31 = \frac{1}{2}mv^2 = \frac{1}{2}(2.17)v^2 \rightarrow v = 5.35 \text{ m/s}.$$

(c) This force is conservative, since there is no friction.

<u>61P</u>

(a) Apply the work-energy theorem. On the way up, gravity and air resistance both do negative work. At its maximum height, the speed and therefore kinetic energy are zero. Hence, $W = \Delta K$ takes the form

$$0 - \frac{1}{2}mv_0^2 = -wh - fh,$$

$$\frac{1}{2}(\frac{w}{g})v_0^2 = (w + f)h = w(1 + f/w)h \rightarrow h = \frac{v_0^2}{2g(1 + f/w)}.$$

(b) Now apply the work-energy theorem over the entire flight. Gravity does negative work as the stone rises, but an equal amount of positive work as it falls, for a total of zero. On the other hand, air resistance does negative work regardless of which way the stone is moving. Therefore, $\Delta K = W$ requires that

$$\frac{1}{2}mu^2 - \frac{1}{2}mv_0^2 = -2fh,$$

$$u^2 = v_0^2 - 4fh/m.$$

Insert the result from (a):

$$u^2 = v_0^2 - \frac{4f}{m} \frac{v_0^2}{2g(1 + f/w)} = v_0^2[1 - \frac{2f}{w} \frac{1}{1 + f/w}] \rightarrow u = v_0\sqrt{[\frac{w - f}{w + f}]}.$$

65P

(a) Let f be the friction force. The mass of the elevator is m = W/g = 4000/32 = 125 slug. If L is the distance traveled before hitting the spring, the work-energy theorem says that

$$\frac{1}{2}mv^2 - 0 = mgL - fL,$$

$$\frac{1}{2}(125)v^2 = (4000 - 1000)(12) \rightarrow v = 24 \text{ ft/s}.$$

(b) At the instant of maximum compression of the spring, the elevator is at rest. Thus, applying the work-energy theorem as the spring is compressed a distance x,

$$0 - \frac{1}{2}mv^2 = mgx - \frac{1}{2}kx^2 - fx,$$

the spring doing negative work in trying to push the elevator back up the shaft; the friction force f always does negative work. Substituting the data, this equation becomes

$$-36,000 = (4000)x - \frac{1}{2}(10,000)x^2 - (1000)x,$$

$$5x^2 - 3x - 36 = 0 \rightarrow x = 3 \text{ ft}.$$

(The negative root is rejected.)

(c) Let s be the distance sought. The elevator is at rest momentarily at the highest point reached and also at the point of maximum compression of the spring. Hence, again by the work-energy theorem (Eq.19),

$$0 = -mgs + \frac{1}{2}kx^2 - fs,$$

the spring force now acting in the same direction the elevator moves; it is assumed that s > x. Using the result from (b),

$$0 = -(4000)s + \frac{1}{2}(10,000)(3)^2 - (1000)s \rightarrow s = 9 \text{ ft}.$$

(d) Gravity and the spring force are conservative forces; all of the energy dissipated is dissipated by friction. If static friction can be ignored (f represents kinetic friction), the spring will be compressed a distance z when the elevator finally comes to rest, where kz = mg, or z = 4000/10,000 = 0.4 ft. The total energy of the system the instant the cable snapped is

$$E_i = mgh = Wh = (4000)(12.4) = 49,600 \text{ ft·lb},$$

taking the final position as the zero level of gravitational potential energy. The final energy is

$$E_f = \frac{1}{2}kz^2 = \frac{1}{2}(10,000)(0.4)^2 = 800 \text{ ft·lb}.$$

The difference was removed by friction doing work on the elevator. If the elevator moved a total distance y, then

$$(49,600 - 800) = fy = 1000y \rightarrow y = 48.8 \text{ ft.}$$

The answer is exact to the extent that static friction can be ignored in determining the elevator's final resting place.

67E

(a) By Eq.26,

$$E = mc^2 = (0.1 \text{ kg})(3 \text{ X } 10^8 \text{ m/s})^2 = 9.0 \text{ X } 10^{15} \text{ J.}$$

(b) In one year, this family consumes an amount of energy E_f given by

$$E_f = Pt = (1 \text{ X } 10^3 \text{ J/s})(3.156 \text{ X } 10^7 \text{ s}) = 3.156 \text{ X } 10^{10} \text{ J.}$$

Hence, the energy in (a) will last for $(9 \text{ X } 10^{15} \text{ J})/(3.156 \text{ X } 10^{10} \text{ J/y}) = 285,000 \text{ y.}$

73P

The energy equivalent E is, by Eq.26,

$$E = mc^2 = (3.2 \text{ X } 10^{-4} \text{ kg})(3 \text{ X } 10^8 \text{ m/s})^2 = 2.88 \text{ X } 10^{13} \text{ J} = 2.88 \text{ X } 10^7 \text{ MJ.}$$

This is equivalent to

$$\frac{2.88 \text{ X } 10^7 \text{ MJ}}{130 \text{ MJ/gal}} = 2.215 \text{ X } 10^5 \text{ gallons of gasoline.}$$

At 30 mi/gal, the car could travel a distance

$$(2.215 \text{ X } 10^5 \text{ gal})(30 \text{ mi/gal}) = 6.645 \text{ X } 10^6 \text{ mi.}$$

This is $(6.645 \text{ X } 10^6)(1.609) = 1.069 \text{ X } 10^7$ km. The earth's circumference is $2\pi R = 2\pi(6370) = 40,024$ km. Hence, the car could be driven $(1.069 \text{ X } 10^7)/(4.0024 \text{ X } 10^4) = 267$ times around the world.

CHAPTER 9

1E

Apply Eq.2 to the earth—moon system:

$$x_{cm} = \frac{m_e x_e + m_m x_m}{m_e + m_m}.$$

Since we want the distance of the center of mass from the center of the earth, put the center of the earth at the origin; i.e., $x_e = 0$. Then x_m = earth—moon distance. Taking needed data from Appendix C,

$$x_{cm} = \frac{0 + (7.36 \times 10^{22} \text{ kg})(3.82 \times 10^8 \text{ m})}{5.98 \times 10^{24} \text{ kg} + 7.36 \times 10^{22} \text{ kg}} = 4.64 \times 10^6 \text{ m} = 4640 \text{ km}.$$

The radius of the earth is 6370 km, so the center of mass of the earth—moon system lies inside the earth.

3E

See Sample Problem 1. Read the x and y coordinates of each particle from the axes of the graph and apply Eq.5:

$$x_{cm} = \frac{(3 \text{ kg})(0) + (8 \text{ kg})(1 \text{ m}) + (4 \text{ kg})(2 \text{ m})}{3 \text{ kg} + 8 \text{ kg} + 4 \text{ kg}} = 1.07 \text{ m},$$

$$y_{cm} = \frac{(3 \text{ kg})(0) + (8 \text{ kg})(2 \text{ m}) + (4 \text{ kg})(1 \text{ m})}{3 \text{ kg} + 8 \text{ kg} + 4 \text{ kg}} = 1.33 \text{ m}.$$

12E

Use Eq.15:

$$(2400 \text{ kg} + 1600 \text{ kg})v_{cm} = (2400 \text{ kg})(80 \text{ km/h}) + (1600 \text{ kg})(60 \text{ km/h}),$$

$$v_{cm} = 72.0 \text{ km/h}.$$

18P

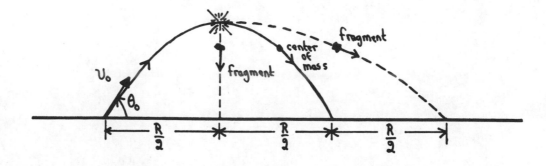

The explosion takes place at the peak of the trajectory. At this point, the vertical component of the momentum of the shell is zero (just before the explosion), and therefore the total vertical component of the momentum of the fragments immediately after the explosion must be zero also. Since one fragment is at rest just after the explosion, the other must have been projected horizontally. Thus the fragments strike the ground at the same time (since both had zero initial vertical speed), and their center of mass, which lies on the line joining them, must strike the ground as they do. But the fragments have equal mass, and therefore the center of mass impacts at a point equidistant from the impact points of the fragments. Since the center of mass moves as though no explosion took place (see p.158 of HR), the situation is as shown on the sketch on p.68, with the fragments each landing a distance R/2 from the center of mass, where R is the range of the gun. The second fragment, then, lands a distance 3R/2 from the gun. The range R is, from Eq.14 in Ch.4,

$$R = v_0^2 \sin 2\theta_0 / g = (1500)^2 \sin 120° / 32 = 60{,}892 \text{ ft.}$$

Thus, the distance from the gun to the impact point of the second fragment is 91,340 ft.

19P

(a) By definition of center of mass, Eq.2, with one of the masses at the origin,

$$x_{cm} = \frac{m(0) + m(x)}{m + m} = \frac{x}{2} = \frac{50}{2} = 25 \text{ mm};$$

i.e., the center of mass lies midway between the bodies, as expected.

(b) Again apply Eq.2; we put x = 0 at the heavier body; then,

$$x_{cm} = \frac{(520)(0) + (480)(50)}{520 + 480} = 24 \text{ mm};$$

i.e., the center of mass is 24 mm from the 520 g body.

(c) In terms of the accelerations of the two bodies, the acceleration a_{cm} of the center of mass is given by Eq.16:

$$a_{cm} = \frac{m_1 a_1 + m_2 a_2}{m_1 + m_2} = \frac{m_1 a + m_2 (-a)}{m_1 + m_2} = \frac{m_1 - m_2}{m_1 + m_2} a,$$

where m_1 = 520 g and a is the acceleration of m_1 (directed down as this is the heavier body). From Sample Problem 9, Chapter 5, we have

$$a = \frac{m_1 - m_2}{m_1 + m_2} g,$$

and therefore

$$a_{cm} = \left(\frac{m_1 - m_2}{m_1 + m_2}\right)^2 g = (0.04)^2 g = 0.0016g,$$

directed down. In the last equation, g = acceleration due to gravity (not grams).

70

20P

Let the shore be at the origin of the x axis, x increasing outward from the shore. Also, let:

x_{cm} = distance of center of mass of dog+boat system from shore;

x_d = distance of dog from shore;

x_b = distance of center of mass of boat from shore.

Then, if W = weight of boat, w = weight of dog, and the star (*) denotes quantities after the dog takes its walk, then from the definition of center of mass, with masses converted to weights by multiplication by g,

$$(W + w)x_{cm} = Wx_b + wx_d,$$

$$(W + w)x^*_{cm} = Wx^*_b + wx^*_d.$$

Since no net external force acts on the system, the conservation of linear momentum requires that $x_{cm} = x^*_{cm}$; set these quantities equal; the equations above imply that

$$W(x^*_b - x_b) = w(x_d - x^*_d).$$

From Eq.17 in Ch.4, (displacement of dog relative to boat) = (displacement of dog relative to shore) + (displacement of shore relative to boat). Since x increases away from the shore, and the dog walks toward the shore (i.e., in the direction of negative x), the equation described in words above becomes

$$-8 = (x^*_d - 20) - (x^*_b - x_b),$$

$$x^*_b - x_b = x^*_d - 12.$$

Substitute this into the third equation above to get

$$W(x^*_d - 12) = w(x_d - x^*_d).$$

But x_d = 20 ft, W = 40 lb and w = 10 lb. The equation above then yields x^*_d = 13.6 ft.

21E

Use Eq.18 in British units. The mass of the car is

$$m = \frac{W}{g} = \frac{3000 \text{ lb}}{32 \text{ ft/s}^2} = 93.75 \text{ slug}.$$

Also, v = 55 mi/h = (55)(1.467) = 80.685 ft/s. Therefore,

$$p = mv = (93.75 \text{ slug})(80.685 \text{ ft/s}) = 7560 \cdot \text{slug ft/s}.$$

23E

(a) By Eq.18, p = mv = (0.004 kg)(950 m/s) = 3.80 kg·m/s.

(b) The kinetic energy is $K = mv^2/2 = (0.004 \text{ kg})(950 \text{ m/s})^2/2 = 1805 \text{ J}$.

(c) Use the result from (a) and apply Eq.18 to the deer; its speed must be

$$v = \frac{p}{m} = \frac{3.8 \text{ kg} \cdot \text{m/s}}{450 \text{ kg}} = 0.00844 \text{ m/s} = 8.44 \text{ mm/s}.$$

32E

The system man+stone has zero momentum initially. To have zero momentum with both in motion after the kick, the magnitudes of their oppositely directed momenta must be equal. Therefore, recalling that $m = W/g$,

$$\left(\frac{200 \text{ lb}}{32 \text{ ft/s}^2}\right)v = \left(\frac{0.15 \text{ lb}}{32 \text{ ft/s}^2}\right)(13 \text{ ft/s}) \rightarrow v = 9.75 \times 10^{-3} \text{ ft/s}.$$

(Note that the $g = 32 \text{ ft/s}^2$ cancels out.)

35E

Apply conservation of momentum to the system man+cart before and after the man jumps:

$$p_{before} = p_{after},$$

$$(75 \text{ kg} + 39 \text{ kg})(2.3 \text{ m/s}) = (75 \text{ kg})(0) + (39 \text{ kg})v \rightarrow v = 6.72 \text{ m/s}.$$

Hence, the speed of the cart increased by $6.72 - 2.3 = 4.42 \text{ m/s}$.

40P

(a) Let V, v be the velocities of the rocket case, mass M, and payload, mass m, after the separation. The velocity of the center of mass of the rocket (which before separation is the same as that of the rocket) is unchanged by the separation; hence, if $u = 7600 \text{ m/s}$ (the speed of the rocket before separation), this speed can be written in terms of the speeds after separation according to Eq.15 (taking account of our different notation),

$$(M + m)u = MV + mv,$$

the initial direction taken as positive. The relative speed after separation is $v - V = 910 \text{ m/s}$ (the less massive component will have the greater speed), and therefore

$$(440)(7600) = 290(v - 910) + 150v \rightarrow v = 8200 \text{ m/s};$$

$$V = 8200 - 910 = 7290 \text{ m/s}.$$

(b) The kinetic energies before and after separation are

$$K_b = \frac{1}{2}(M + m)u^2 = 1.271 \times 10^{10} \text{ J};$$

$$K_a = \frac{1}{2}mv^2 + \frac{1}{2}MV^2 = (0.5043 + 0.7706) \times 10^{10} = 1.275 \times 10^{10} \text{ J}.$$

Evidently $4 \times 10^7 \text{ J}$ (the difference) was stored in the compressed spring.

41P

We are not told the masses of the pieces in kg, that is, we are not given numerical values of these masses, so use letters. We call the mass of each of the smaller pieces m; then the mass of the larger piece is 3m. The diagram at the right shows the momenta of the pieces after the explosion. These must sum to zero, since the vessel was at rest when in a single piece. Setting the sum of the x and of the y components of the momenta each equal to zero gives

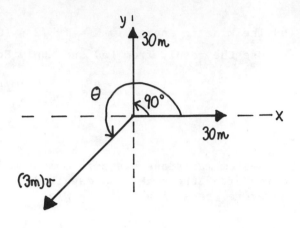

$$(30m)\cos 0° + (30m)\cos 90° + (3mv)\cos\theta = 0,$$

$$v\cos\theta = -10;$$

$$(30m)\sin 0° + (30m)\sin 90° + (3mv)\sin\theta = 0,$$

$$v\sin\theta = -10.$$

Since θ apparently has a negative cosine and a negative sine, it must lie in the third quadrant, as shown. Squaring the two results above and adding gives

$$v^2\cos^2\theta + v^2\sin^2\theta = v^2(\cos^2\theta + \sin^2\theta) = v^2 = (-10)^2 + (-10)^2 \rightarrow v = 14.1 \text{ m/s}.$$

Dividing the same equations yields

$$\frac{\cos\theta}{\sin\theta} = \cot\theta = \frac{-10}{-10} \rightarrow \theta = 45°, 225°.$$

Since 45° is in the first quadrant, not the third, we must have $\theta = 225°$.

50P

First, find the change in speed for only one man on the flatcar, who jumps off in the prescribed manner. Initially, the velocity of the flatcar is v_0 to the right, and after the man has jumped off it is v. Then since the man runs to the left, his velocity with respect to the ground is $v - v_{rel}$ after jumping. The velocity of the center of mass of the system car+man is unaffected by this process and therefore, by Eq.15,

$$v_{cm} = v_0 = \frac{Wv + w(v - v_{rel})}{W + w}.$$

Therefore,

$$\Delta v = v - v_0 = \frac{wv_{rel}}{W + w}.$$

If there are n men and all go at once, use the result above with w replaced with nw:

$$\Delta v = \frac{nwv_{rel}}{W + nw}.$$

If the n men go in succession, calculate the change in speed imparted by each. For the first man,

$$\Delta v_1 = \frac{wv_{rel}}{w + [W + (n - 1)w]},$$

since the original flatcar is now replaced, in essence, by the flatcar plus the $(n - 1)$ men who stay behind as the first jumps. After they have all jumped, the total change in speed is the sum of the individual changes: $\Delta v = \Delta v_1 + \Delta v_2 + \ldots + \Delta v_n$. Replacing $(n - 1)$ in the above equation with $(n - 2)$, $(n - 3) \ldots$, and adding the results gives, after some rearrangement,

$$\Delta v = \frac{wv_{rel}}{W + nw}[1 + \frac{W + nw}{W + (n - 1)w} + \ldots + \frac{W + nw}{W + w}].$$

The term outside the brackets is $1/n$ times the result for all the men jumping together. All of the terms inside the bracket are greater or equal to one, and there are n of these terms. Hence, the second method yields a greater change in speed than the first.

51P

(a) Each hailstone has a volume $= 4\pi r^3/3 = 4\pi(0.5)^3/3 = 0.5236$ cm^3. Therefore the mass m of each hailstone is m $= (0.5236$ cm$^3)(0.92$ g/cm$^3) = 0.482$ g.

(b) The momentum of each hailstone on impact with the roof is

$$p = mv = (0.482 \times 10^{-3} \text{ kg})(25 \text{ m/s}) = 1.205 \times 10^{-2} \text{ kg·m/s}.$$

Now we need to find the number of hailstones striking each second. Since the hail moves at 25 m/s, all hailstones within 25 m of the roof will strike in the next one second. The area of the roof is 200 m^2, so the volume of air containing these hailstones is 5000 m^3 [= $(200$ m$^2)(25$ m$)$]. With 120 hailstones per m^3, this means that $(5000)(120) = 6 \times 10^5$ hailstones strike in one second. The associated force is given by Eq.19. We can replace dP/dt with P/t since we assume that the conditions do not change with time. Therefore,

$$F = \frac{P}{t} = \frac{(6 \times 10^5)(1.205 \times 10^{-2} \text{ kg·m/s})}{1 \text{ s}} = 7230 \text{ N}.$$

55E

We have $M_i = 6000$ kg. The mass of the rocket after firing is $M_f = 6000 - 80 = 5920$ kg. Applying Eq.34,

$$v_f - v_i = u\ln(\frac{M_i}{M_f}),$$

$$v_f - 100 = (250)\ln(\frac{6000}{5920}) \rightarrow v_f = 103 \text{ m/s}.$$

56E

(a) The thrust of the rocket is $T = Ru$ (see top of p.193 of HR), so that

$$T = Ru = (480 \text{ kg/s})(3.27 \times 10^3 \text{ m/s}) = 1.57 \times 10^6 \text{ N}.$$

(b) The final mass is

$$M_f = M_i - Rt = (2.55 \times 10^5 \text{ kg}) - (480 \text{ kg/s})(250 \text{ s}) = 1.35 \times 10^5 \text{ kg}.$$

(c) By Eq.34,

$$v_f = v_i + u\ln(\frac{M_i}{M_f}) = 0 + (3.27 \text{ km/s})\ln(\frac{2.55 \times 10^5}{1.35 \times 10^5}) = 2.08 \text{ km/s}.$$

63P

Assume that the barges pass very close to each other, so that the coal is transported across at virtually zero speed, in m/s. That is, the coal is ejected from the slow barge with a velocity relative to the slow barge of virtually zero. Hence, the force on the slow barge is $F = Ru = 0$. But, although the coal arrives at the fast barge with a transverse (cross stream) speed of almost zero, it arrives at a relative velocity of 20 - 10 = 10 km/h, toward the stern of the barge. The force that is applied to the barge as it is brought to rest on the barge is

$$F = uR = (2.778)(16.67) = 46.3 \text{ N}.$$

We have used the conversions to SI units of 10 km/h = 2.778 m/s and 1000 kg/min = 16.67 kg/s.

64P

(a) We use British units. In injesting the air the plane experiences a resisting force F_1 equal to

$$F_1 = uR = (600 \text{ ft/s})(4.8 \text{ slug/s}) = 2880 \text{ lb}.$$

The mass ejected each second is the 4.8 slug of air plus 0.2 slug of fuel. The thrust imparted to the plane in ejecting this mixture is

$$F_2 = uR = (1600 \text{ ft/s})(5 \text{ slug/s}) = 8000 \text{ lb}.$$

The net thrust is $F_2 - F_1 = 5120$ lb.

(b) The horsepower delivered is, by Eq.26 in Chapter 7,

$$P = (F_2 - F_1)v = \frac{(5120 \text{ lb})(600 \text{ ft/s})}{(550 \text{ ft·lb/s·hp})} = 5585 \text{ hp}.$$

67E

(a) The net external force acting on the woman as she pushes off the floor is $F_{ext} = N - mg$, where N = upward normal force exerted by the floor. Note that $N \neq mg$ since there is a vertical acceleration. Her center of mass moves, while pushing off, a distance $s_{cm} = 90 - 40 = 50$ cm = 0.50 m. By Eq.39,

$$(N - mg)s_{cm} = \tfrac{1}{2}Mv^2 - 0.$$

We are not told v directly, but, by energy conservation, $Mv^2/2 = Mgh$, with $h = 120 - 90 = 30$ cm $= 0.30$ m. Hence,

$$(N - mg)s_{cm} = Mgh,$$

$$[N - (55)(9.8)](0.5) = (55)(9.8)(0.3) \rightarrow N = 862 \text{ N.}$$

(b) From the first equation in (a),

$$[862 - (55)(9.8)](0.5) = \frac{1}{2}(55)v^2 \rightarrow v = 2.42 \text{ m/s.}$$

CHAPTER 10

<u>3E</u>

By Eq.5, the impulse is $J = \overline{F}\Delta t = (50 \text{ N})(10 \text{ X } 10^{-3} \text{ s}) = 0.50 \text{ N} \cdot \text{s}$. Now use Eq.4. In the game of pool, balls are struck by the cue when at rest, so put $v_i = 0$. We then have $J = mv_f$, or

$$v_f = J/m = (0.50 \text{ N} \cdot \text{s})/(0.20 \text{ kg}) = 2.50 \text{ m/s}.$$

<u>7E</u>

Eqs.4 and 5, taken together for one dimensional motion, tell us that

$$\overline{F}\Delta t = mv_f - mv_i = m(v_f - v_i).$$

Take the direction of v_i as positive. We then get

$$(-1000 \text{ N})(27 \text{ X } 10^{-3} \text{ s}) = (0.40 \text{ kg})(v_f - 14 \text{ m/s}) \rightarrow v_f = -53.5 \text{ m/s}.$$

The ball is now moving in the direction opposite to its initial velocity, and its speed is 53.5 m/s.

<u>12P</u>

Since F increases uniformly, its average value is $\overline{F} = (0 + 50)/2 = 25$ N. With $v_i = 0$, we have

$$v_f = \frac{\overline{F}\Delta t}{m} = \frac{(25 \text{ N})(4 \text{ s})}{10 \text{ kg}} = 10.0 \text{ m/s}.$$

<u>19P</u>

Work in the reference frame of the spacecraft before separation. Then $v_i = 0$ for each of the parts into which the spacecraft separates. Their final speeds are

$$v_{1f} = J_1/m_1 = 300 \text{ N} \cdot \text{s}/1200 \text{ kg} = 0.250 \text{ m/s};$$
$$v_{2f} = J_2/m_2 = 300 \text{ N} \cdot \text{s}/1800 \text{ kg} = 0.167 \text{ m/s}.$$

Now, by momentum conservation, the two parts separate in opposite directions. Hence, their relative speed is 0.250 + 0.167 = 0.417 m/s = 41.7 cm/s.

<u>22E</u>

(a) Suppose we choose directions to the right as positive. Momentum conservation dictates that

$$(1.6 \text{ kg})(5.5 \text{ m/s}) + (2.4 \text{ kg})(-2.5 \text{ m/s}) = (1.6 \text{ kg})v + (2.4 \text{ kg})(4.9 \text{ m/s}),$$

which yields v = -5.60 m/s, i.e., a speed of 5.60 m/s to the left.

(b) Calculate the kinetic energies before and after the collision:

$$K_b = \frac{1}{2}(1.6)(5.5)^2 + \frac{1}{2}(2.4)(2.5)^2 = 31.7 \text{ J};$$

$$K_a = \frac{1}{2}(1.6)(5.6)^2 + \frac{1}{2}(2.4)(4.9)^2 = 53.9 \text{ J}.$$

Kinetic energy is not conserved, so the collision is not elastic. (Energy was added; one of the blocks might have had an explosive cap attached to its face.)

26E

(a) We choose the original direction of motion as positive. We are told that m_1 = 340 g, v_{1i} = 1.2 m/s, v_{2i} = 0, v_{1f} = 0.66 m/s. Eq.10 applies:

$$v_{1f} = \frac{m_1 - m_2}{m_1 + m_2}(v_{1i}),$$

$$0.66 = \frac{340 - m_2}{340 + m_2}(1.2) \rightarrow m_2 = 98.7 \text{ g}.$$

(b) Now use Eq.11:

$$v_{2f} = \frac{2m_1}{m_1 + m_2}(v_{1i}) = \frac{2(340)}{340 + 98.7}(1.2) = 1.86 \text{ m/s}.$$

28E

We have m_1 = 2.0 kg. We are not told the initial and final speeds numerically, but we do know that $v_{1f} = v_{1i}/4$. Use Eq.10:

$$\frac{v_{1i}}{4} = \frac{2 - m_2}{2 + m_2}(v_{1i}) \rightarrow m_2 = 1.2 \text{ kg},$$

the unknown initial velocity dropping out.

32P

The given data is: m_1 = 300 g, v_{1f} = 0, $v_{2i} = -v_{1i}$. We seek m_2. Eqs.9 and 10 do not apply, for they were derived under the assumption that v_{2i} = 0. If we work in the reference frame of the problem, we must start from the laws of conservation of momentum and of kinetic energy:

$$m_1 v_{1i} + m_2 v_{2i} = m_1 v_{1f} + m_2 v_{2f},$$

$$\frac{1}{2}m_1 v_{1i}^2 + \frac{1}{2}m_2 v_{2i}^2 = \frac{1}{2}m_1 v_{1f}^2 + \frac{1}{2}m_2 v_{2f}^2.$$

Inserting the given data into these equations gives

$$(300)v_{1i} - m_2 v_{1i} = m_2 v_{2f} \rightarrow v_{1i} = \frac{m_2 v_{2f}}{300 - m_2},$$

$$\frac{1}{2}(300)v_{1i}^2 + \frac{1}{2}m_2 v_{1i}^2 = \frac{1}{2}m_2 v_{2f}^2 \rightarrow v_{1i}^2 = \frac{m_2 v_{2f}^2}{300 + m_2}.$$

Comparing the two resulting equations leads to

$$\left(\frac{m_2 v_{2f}}{300 - m_2}\right)^2 = \frac{m_2 v_{2f}^2}{300 + m_2},$$

$$\frac{m_2}{(300 - m_2)^2} = \frac{1}{300 + m_2} \rightarrow m_2 = 100 \text{ g.}$$

35P

Let the y axis be vertical and positive downwards, with the origin at the top of the shaft. If the ball is dropped at time t = 0, then the subsequent positions of ball y_b and elevator y_e are given by

$$y_b = \frac{1}{2}gt^2,$$

$$y_e = h - vt,$$

where v is the speed of the elevator and h = 60 ft. If ball and elevator collide at time t = t*, then

$$y_b(t*) = y_e(t*),$$

$$h - vt* = \frac{1}{2}gt*^2,$$

$$gt* = -v + \sqrt{(v^2 + 2gh)},$$

choosing, in the last equation, the positive root of the quadratic equation for t*. Relative to the elevator, the velocity of the ball is reversed by the collision (elastic collision): just before the collision the relative velocity is +(v + gt*) and just after it is -(v + gt*). Thus, the velocity of the ball relative to the shaft just after the collision is -(v + gt*) - v = -(2v + gt*). Therefore, the highest point y = H reached by the ball after collision is given by

$$0^2 = (2v + gt*)^2 + 2g[H - y_e(t*)],$$

where $y_e(t*) = h - vt*$. Putting this in and rearranging gives

$$-4v^2 - 6v(gt*) - (gt*)^2 = 2g(H - h).$$

Finally, substitute for gt* from above and solve for H to get

$$H = -2\left(\frac{v}{g}\right)\sqrt{(v^2 + 2gh)}.$$

(a) Put v = +6 ft/s; then H = -23.3 ft (i.e., above the top of the shaft).

(b) Set v = -6 ft/s; in this case H = +23.3 ft (i.e., 23.3 ft below the top of the shaft).

36E

With m,v = mass and speed of the meteor before impact, and M,V the mass and speed of the earth after impact, we have by conservation of momentum,

$$mv = (m + M)V \simeq MV,$$

$$V = (\frac{m}{M})v = (\frac{5 \times 10^{10} \text{ kg}}{5.98 \times 10^{24} \text{ kg}})(7200 \text{ m/s}) = 6.02 \times 10^{-11} \text{ m/s},$$

$$V = (6.02 \times 10^{-8} \text{ mm/s})(3.156 \times 10^{7} \text{ s/y}) = 1.9 \text{ mm/y}.$$

37E

Relative to the ice, the package falls vertically; it has no horizontal velocity component and hence adds no horizontal momentum to the sled. The effect, then, is due to the net increase in mass. Apply conservation of (horizontal) momentum:

$$mv = (m + M)V,$$

$$(6)(9) = (6 + 12)V \rightarrow V = 3 \text{ m/s}.$$

46P

Let v be the speed of the marbles just before impact; since the collisions are completely inelastic, the marbles come to rest upon striking the box. Hence, the momentum given to the box by each marble is $p = m(v_f - v_i) = m(0 - v) = -mv$, or $p = mv$ in magnitude. Since R marbles strike each second, a time interval $t = 1/R$ elapses between impacts. Thus, the average force imparted to the box by the colliding marbles is

$$\overline{F} = \frac{p}{t} = \frac{mv}{1/R} = mRv = mR\sqrt{(2gh)}.$$

But, after time t, Rt marbles with total weight (mg)Rt already reside in the box. Hence, the scale reading SR after time t is

$$SR = mR\sqrt{(2gh)} + mgRt = mgR[\sqrt{(2h/g)} + t],$$

$$SR = (4.5 \times 10^{-3})(9.8)(100)[\sqrt{(15.2/9.8)} + 10] = 49.59 \text{ N}.$$

Metric scales generally display mass, not weight. In this case, the scale reading will be (49.59)/(9.8) = 5.06 kg.

47P

Let f = 0.27 represent the fraction of initial kinetic energy that is dissipated; i.e.,

$$f \cdot \tfrac{1}{2}mv^2 = \tfrac{1}{2}mv^2 - \tfrac{1}{2}(m + M)u^2.$$

Here m = mass of the freight car, M = mass of the caboose; v is the speed of the freight car before collision and u is the speed of the coupled caboose+freight car after impact. The momentum conservation equation is

$$mv = (m + M)u.$$

These two equations can be rewritten, respectively, in the forms

$$(1 - f)mv^2 = (m + M)u^2,$$

$$m^2v^2 = (m + M)^2u^2.$$

Dividing these equations, to eliminate the unknown speeds, gives

$$\frac{1 - f}{m} = \frac{1}{m + M},$$

$$M = \frac{f}{1 - f}\,m,$$

$$W = \frac{f}{1 - f}\,w = \frac{0.27}{0.73}(35) = 12.9 \text{ tons.}$$

48P

(a) Let V be the speed of the gun+ball system at the moment the ball sticks. By the conservation of momentum,

$$mv_i = (m + M)V \;\rightarrow\; V = \frac{m}{m + M}\,v_i.$$

(b) By conservation of total mechanical energy (not just of kinetic energy),

$$\tfrac{1}{2}mv_i^2 = \tfrac{1}{2}(m + M)V^2 + U_s = \tfrac{1}{2}(m + M)V^2 + f\cdot\tfrac{1}{2}mv_i^2.$$

U_s is the potential energy of the ball+spring system when the ball is stuck and f is the desired fraction. Substitute V from (a) into this last equation; a factor $\tfrac{1}{2}mv_i^2$ will then cancel, and the remaining terms yield f = M/(m + M).

56E

(a) Pool balls all have the same mass; in the absence of a numerical value, call the ball mass m. We choose the original direction of motion of the moving cue ball the x axis. After the collision the second ball moves off at an angle θ with respect to this +x-direction. Conservation of momentum requires that (see Eqs.23 and 24)

$$mv\cos 0° + 0 = m(3.5)\cos 65° + m(6.75)\cos\theta,$$

$$mv\sin 0° + 0 = m(3.5)\sin 65° + m(6.75)\sin\theta.$$

Cancelling the mass m and evaluating sin0° and cos0° gives for these equations

$$v = 3.5\cos 65° + 6.75\cos\theta,$$

$$0 = 3.5\sin 65° + 6.75\sin\theta.$$

The second equation yields

$$\theta = \sin^{-1}(-\frac{3.5\sin 65°}{6.75}) = 332°; 208°.$$

If we pick $\theta = 208°$, then the first equation above yields $v < 0$. But v must be positive, for we chose the $+x$ axis in the direction represented by v. Hence, $\theta = 332°$ $(= -28°)$.

(b) Put $\theta = 332°$ into the first equation above for v to find $v = 7.44$ m/s.

<u>59P</u>

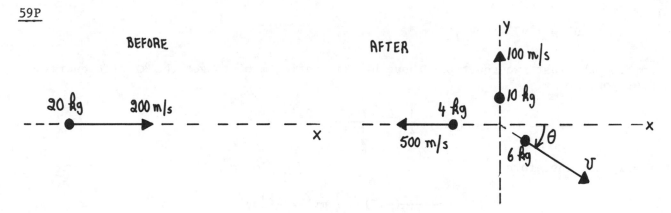

(a) In order that the total momentum after the explosion be entirely in the $+x$ direction, as it is before, the speed v of the third fragment must be directed into the fourth quadrant, as shown above. Conservation of momentum in the x and y directions gives

$$(20)(200) = -(4)(500) + 6v\cos\theta \rightarrow v\cos\theta = 1000,$$

$$0 = (10)(100) - 6v\sin\theta \rightarrow v\sin\theta = 166.7.$$

Squaring the resulting equations and then adding them together gives

$$v^2(\sin^2\theta + \cos^2\theta) = v^2 = 1000^2 + 166.7^2 \rightarrow v = 1014 \text{ m/s}.$$

Using this we have

$$\theta = \sin^{-1}(\frac{166.7}{1014}) = 9.46°.$$

(b) The energy E supplied by the explosive is the excess of the total kinetic energy after the explosion over that present before the explosion:

$$E = \frac{1}{2}(6)(1014)^2 + \frac{1}{2}(4)(500)^2 + \frac{1}{2}(10)(100)^2 - \frac{1}{2}(20)(200)^2 = 3.23 \text{ MJ}.$$

<u>61P</u>

See the sketch, p.82. The conservation of momentum and of kinetic energy requires that

$$mv = MV\cos\theta,$$

$$mu = MV\sin\theta,$$

$$\tfrac{1}{2}mv^2 = \tfrac{1}{2}mu^2 + \tfrac{1}{2}MV^2.$$

The mass of the neutron is m, and the mass of the deuteron is M. Square the first two equations and then add them to get

$$M^2V^2 = m^2v^2 + m^2u^2.$$

From the third equation,

$$m^2u^2 = m^2v^2 - MmV^2.$$

Now combine the last two equations above by eliminating m^2u^2 to get the kinetic energy of the deuteron:

$$\tfrac{1}{2}MV^2 = \frac{m^2}{m + M}\,v^2.$$

But M = 2m and therefore

$$\tfrac{1}{2}MV^2 = \frac{m}{m + M}(mv^2) = \tfrac{1}{3}\,mv^2 = \tfrac{2}{3}(\tfrac{1}{2}mv^2),$$

establishing the result quoted in the problem.

62P

Let m be the mass of each object and v their common initial speed. Use the rule for the addition of vectors by the geometric method to arrange the initial and final velocities so that the vector sum of the initial momenta equals the momentum of the pair after sticking together in the collision. By the conservation of momentum, written for both the x and y directions, we have

$$mv - mv\cos\theta_1 = (2m)(v/2)\cos\theta_2 \;\rightarrow\; 1 - \cos\theta_1 = \cos\theta_2,$$

$$mv\sin\theta_1 = (2m)(v/2)\sin\theta_2 \;\rightarrow\; \sin\theta_1 = \sin\theta_2.$$

The second equation tells us that $\theta_1 = \theta_2$. Using this in the first equation gives $\cos\theta_1 = 1/2$, or $\theta_1 = 60°$. Hence, the angle between the initial velocities (tail to tail) is 120°. (This result could be anticipated by noting that the three momentum vectors, being of equal length, form an equilateral triangle.)

65P

The geometry at impact is shown on the
sketch. From this, we have

$$\sin\theta = \frac{R}{2R} = \frac{1}{2} \rightarrow \theta = 30°.$$

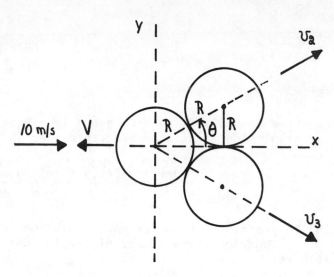

Also, from the symmetry of the collision, v_2
$= v_3 = v$. This also follows from examining
the momentum in the y direction. Also, since
the momentum in the y direction before the
impact is zero, it must be zero just after
the collision. It follows that the velocity
of the first ball after collision must be in
the x direction (not necessarily the +x
direction); call this velocity V. By the
conservation of momentum,

$$m(10) = 2mv\cos30° + mV \rightarrow v = \frac{1}{\sqrt{3}}(10 - V).$$

By conservation of kinetic energy,

$$\frac{1}{2}m(10)^2 = 2(\frac{1}{2}mv^2) + \frac{1}{2}mV^2 \rightarrow 100 = 2v^2 + V^2.$$

Squaring the momentum equation yields

$$v^2 = \frac{1}{3}(10 - V)^2 = \frac{1}{3}(100 - 20V + V^2).$$

Substitute this into the equation preceding it to find that

$$100 = \frac{2}{3}(100 - 20V + V^2) + V^2,$$

$$(5V + 10)(V - 10) = 0 \rightarrow V = -2 \text{ m/s}.$$

That is, after the collision the ball originally in motion moves off in the direction
from whence it came. The other solution of the quadratic equation for V reproduces the
conditions before the collision, as expected. Hence, with this result for V, the second
equation gives

$$v = (\frac{1}{\sqrt{3}})(10 + 2) = 6.93 \text{ m/s}.$$

67E

In terms of the rest masses, Q is defined by $Q = (m_i - m_f)c^2$, where m_i = sum of initial
masses, and m_f = sum of final masses. For our reaction, then, we have

$$Q = (m_p + m_F - m_\alpha - m_0)c^2.$$

Substitute the given masses, and use the mass-energy conversion to get

$$Q = (0.008712 \text{ u})c^2 = (0.008712)(uc^2) = (0.008712)(932 \text{ MeV}) = 8.12 \text{ MeV}.$$

CHAPTER 11

4E

(a) From Appendix F, we see that 1 ly = 9.46 X 10^{12} km. Hence, the time needed to make one revolution about the galactic center is

$$t = \frac{2\pi r}{v} = \frac{2\pi[(2.3 \text{ X } 10^4)(9.46 \text{ X } 10^{12} \text{ km})]}{(250 \text{ km/s})} = 5.47 \text{ X } 10^{15} \text{ s}.$$

(b) In years, the time for one revolution is (5.47 X 10^{15} s)/(3.16 X 10^7 s/y) = 1.731 X 10^8 y. Therefore, the number of revolutions made is (4.5 X 10^9 y)/(1.731 X 10^8 y/rev) = 26.0 rev.

6E

(a) The angular position at t = 0 is $\theta(t = 0) = 2 + 4(0)^2 + 2(0)^3 = 2$ rad. The angular velocity at any time is found from Eq.5:

$$\omega = \frac{d\theta}{dt} = 8t + 6t^2.$$

Hence, at t = 0, $\omega(t = 0) = 8(0) + 6(0)^2 = 0$.

(b) At t = 4 s, $\omega(t = 4) = 8(4) + 6(4)^2 = 128$ rad/s.

(c) The angular acceleration at any time is found from Eq.7, using the result in (a) for ω at any time:

$$\alpha = \frac{d\omega}{dt} = 8 + 12t.$$

Therefore, at t = 2 s, $\alpha(t = 2) = 8 + 12(2) = 32$ rad/s^2. The angular acceleration depends on t (see above) and therefore cannot be constant.

13E

(a) We are given ω_0 = 78 rev/min, ω = 0, t = 30 s = 0.5 min. To find α, use Eq.8:

$$\omega = \omega_0 + \alpha t,$$

$$0 = 78 \text{ rev/min} + \alpha(0.5 \text{ min}) \rightarrow \alpha = -156 \text{ rev/min}^2.$$

(b) To find θ using only the original data given in the problem statement (to avoid relying on the answer for α being correct), use Eq.11:

$$\theta = \frac{1}{2}(\omega_0 + \omega)t = \frac{1}{2}(78 \text{ rev/min} + 0)(0.5 \text{ min}) = 19.5 \text{ rev}.$$

20P

Let t = t* at the start of the 4 s interval. Since α = 3 rad/s^2, we have, by Eq.9, the angle turned through up to time t*:

$$\theta(t*) = \tfrac{1}{2}\alpha t*^2 = 1.5t*^2,$$

since $\omega_0 = 0$ at $t = 0$. At $t = t* + 4$, $\theta = \theta(t*) + 120$; again, by Eq.9,

$$\theta(t*) + 120 = 1.5(t* + 4)^2 = 1.5t*^2 + 12t* + 24,$$

$$120 = 12t* + 24 \rightarrow t* = 8.00 \text{ s.}$$

25P

(a) Use Eq.9. The rate of change of the rotation speed is evidently very small, so put $\alpha = 0$. Then, after time T (one period), θ increases by 2π rad, so that

$$\theta = \omega_0 t + \tfrac{1}{2}\alpha t^2,$$

$$2\pi = \omega_0(T) + 0,$$

$$\omega_0 = 2\pi/T = \omega.$$

The last step ($\omega = \omega_0$) is allowed if $\alpha \simeq 0$.

(b) In what follows the angular acceleration is not ignored. It is given that, at present,

$$T = 0.033 \text{ s}; \quad \frac{dT}{dt} = 1.26 \times 10^{-5} \text{ s/y.}$$

By definition (see Eq.7), $\alpha = d\omega/dt$ and, from (a), $\omega = 2\pi/T$. Therefore,

$$\alpha = \frac{d}{dt}\left(\frac{2\pi}{T}\right) = -\frac{2\pi}{T^2}\frac{dT}{dt},$$

$$\alpha = -\frac{2\pi}{(0.033 \text{ s})^2}(1.26 \times 10^{-5} \text{ s/y}) = -7.2698 \times 10^{-2} \text{ (s} \cdot \text{y)}^{-1},$$

$$\alpha = -\frac{7.2698 \times 10^{-2} \text{ s}^{-1}}{3.16 \times 10^7 \text{ s}} = -2.30 \times 10^{-9} \text{ rad/s}^2.$$

(c) Call the present value of the angular speed ω_0; its numerical value, from (a), is

$$\omega_0 = \frac{2\pi}{T_0} = \frac{2\pi}{0.033} = 190.4 \text{ rad/s.}$$

When the pulsar stops rotating, $\omega = 0$. Hence, by Eq.8,

$$\omega = \omega_0 + \alpha t,$$

$$0 = 190.4 + (-2.30 \times 10^{-9})t,$$

$$t = 8.278 \times 10^{10} \text{ s} = 2620 \text{ y.}$$

If the present is 1988, then the pulsar will stop in the year 1988 + 2620 = 4608 A.D.

(d) Now call the value of the angular velocity when the pulsar was formed ω_0, and the present value ω. The pulsar's age is $1988 - 1054 = 934$ y $= 2.951 \times 10^{10}$ s. Thus

$$\omega_0 = \omega - \alpha t = 190.4 - (-2.30 \times 10^{-9})(2.951 \times 10^{10}) = 258.3 \text{ rad/s}.$$

This corresponds to a period of rotation

$$T_0 = 2\pi/\omega_0 = 2\pi/258.3 = 0.024 \text{ s} = 24 \text{ ms}.$$

26E

With the angular speed constant, we have $\alpha = 0$. Hence $a_t = 0$ (Eq.17). The acceleration is purely radial. By Exercise 2,

$$\omega = (33\tfrac{1}{3})(0.1047 \text{ rad/s}) = 3.49 \text{ rad/s}.$$

Noting that the radius r is one-half the diameter, Eq.18 gives

$$a_r = \omega^2 r = (3.49 \text{ rad/s})^2(0.15 \text{ m}) = 1.83 \text{ m/s}^2.$$

28E

We are given $r = 110$ m and $v = 50$ km/h $= 50(0.2778) = 13.89$ m/s. By Eq.15,

$$\omega = \frac{v}{r} = \frac{13.89 \text{ m/s}}{110 \text{ m}} = 0.126 \text{ rad/s}.$$

35P

(a) The earth turns through an angle of 2π rad every day (86,400 s). By Eq.4, the corresponding angular speed is

$$\omega = \frac{\Delta\theta}{\Delta t} = \frac{2\pi \text{ rad}}{86,400 \text{ s}} = 7.27 \times 10^{-5} \text{ rad/s},$$

and is the same everywhere on earth since the earth rotates like a rigid body.

(b) The radius r of the daily circular path traced out by a point on the earth's surface in the course of one day is given by $r = R\cos L$, where R = radius of the earth and L = latitude of the point. Hence, the linear speed v due to the rotation of the earth is

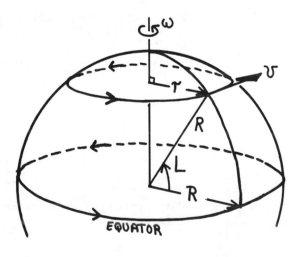

$$v = r\omega = R\omega\cos L = (6370 \text{ km})(7.27 \times 10^{-5} \text{ rad/s})\cos 40° = 0.355 \text{ km/s} = 355 \text{ m/s}.$$

(c) The angular speed ω of rotation is the same everywhere on the earth. The linear speed v depends on the latitude L. At the equator, put $L = 0°$ in (b) to obtain

$$v = R\omega = (6370 \text{ km})(7.27 \times 10^{-5} \text{ rad/s}) = 0.463 \text{ km/s} = 463 \text{ m/s}.$$

39P

(a) A wheel with 500 teeth has 500 slots also, so that the angular separation between adjacent slots is $2\pi/500$ rad. Thus, the wheel turns through an angle $\pi/250$ rad in the time t it takes for light to travel the distance 2ℓ. This time t is

$$t = \frac{2\ell}{c} = \frac{2(0.5\ \text{km})}{3\ \text{X}\ 10^5\ \text{km/s}} = 3.333\ \text{X}\ 10^{-6}\ \text{s}.$$

Hence,

$$t = \frac{\theta}{\omega} = \frac{\pi/250}{\omega} = 3.333\ \text{X}\ 10^{-6}\ \text{s} \rightarrow \omega = 3770\ \text{rad/s}.$$

(b) The linear speed v at the rim is

$$v = r\omega = (0.05\ \text{m})(3770\ \text{rad/s}) = 188.5\ \text{m/s}.$$

41P

Since the belt does not slip, the linear speeds at the rims of the two wheels must be equal; that is,

$$r_A \omega_A = r_C \omega_C.$$

Differentiating with respect to time gives

$$\alpha_C = (r_A/r_C)\alpha_A.$$

Now $\omega_C = \alpha_C t + \omega_{OC}$ and $\omega_{OC} = 0$; therefore,

$$\omega_C = (r_A/r_C)\alpha_A t.$$

Since $\omega_C = 100$ rev/min $= (100)(2\pi)/(60) = 10.47$ rad/s, the equation above gives

$$10.47 = (\frac{10}{25})(1.6)t \rightarrow t = 16.4\ \text{s}.$$

45E

Use Eq.22. As noted there, ω must be in radian measure:

$$\omega = 600\ \text{rev/min} = 600(2\pi\ \text{rad})/(60\ \text{s}) = 62.83\ \text{rad/s}.$$

We can now solve for I:

$$I = 2K/\omega^2 = 2(24,400\ \text{J})/(62.83\ \text{rad/s})^2 = 12.4\ \text{kg}\cdot\text{m}^2.$$

47E

The rotational inertia of a uniform cylinder is found in Table 2: $I = MR^2/2$. Combining this with Eq.22 for the kinetic energy gives

$$K = \frac{1}{2}I\omega^2 = \frac{1}{2}(MR^2/2)\omega^2 = \frac{1}{4}MR^2\omega^2.$$

For the first cylinder,

$$K = \frac{1}{4}(1.25 \text{ kg})(0.25 \text{ m})^2(235 \text{ rad/s})^2 = 1080 \text{ J}.$$

The second cylinder, with three times the radius of the first, has $K = 9708$ J.

57P

The center of mass of the meter stick is at the 50 cm mark. The 20 cm mark is 30 cm away. By the parallel-axis theorem,

$$I = I_{cm} + Mh^2.$$

We have $h = 30$ cm $= 0.30$ m. Find I_{cm} in Table 2. The desired rotational inertia becomes

$$I = \frac{1}{12}ML^2 + Mh^2 = \frac{1}{12}(0.56 \text{ kg})(1 \text{ m})^2 + (0.56 \text{ kg})(0.30 \text{ m})^2 = 0.0971 \text{ kg·m}^2.$$

59E

The torque about the pivot P is, by Eq.28,

$$\tau = rF\sin\phi,$$

where $F = W = mg$. Therefore,

$$\tau = rmg\sin\phi,$$

$$\tau = (1.25 \text{ m})(0.75 \text{ kg})(9.8 \text{ m/s}^2)\sin 30°,$$

$$\tau = 4.59 \text{ N·m}.$$

65E

(a) By Eq.8,

$$\alpha = \frac{\omega - \omega_0}{t} = \frac{6.2 \text{ rad/s} - 0}{0.220 \text{ s}} = 28.2 \text{ rad/s}^2.$$

(b) The torque follows from Eq.31:

$$\tau = I\alpha = (12 \text{ kg·m}^2)(28.2 \text{ rad/s}^2) = 338 \text{ N·m}.$$

68E

(a) By Eq.31,

$$I = \tau/\alpha = (960 \text{ N·m})/(6.2 \text{ rad/s}^2) = 155 \text{ kg·m}^2.$$

(b) Figure (f) in Table 2 tells us that

$$M = \frac{3I}{2R^2} = \frac{3(155 \text{ kg} \cdot \text{m}^2)}{2(1.9 \text{ m})^2} = 64.4 \text{ kg}.$$

72P

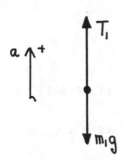 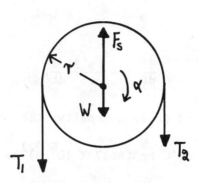 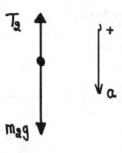

(a) By Eq.14 of Chapter 2, using CGS units,

$$x - x_0 = v_0 t + \frac{1}{2}at^2,$$

$$75 = 0 + \frac{1}{2}a(5)^2 \rightarrow a = 6.00 \text{ cm/s}^2.$$

(b) Draw free-body diagrams of the blocks and pulley. Let m_2 be the mass of the heavier block; by Newton's second law,

$$T_1 - m_1 g = m_1 a,$$

$$T_1 = m_1(g + a) = (460)(980 + 6) = 4.54 \times 10^5 \text{ dyne}.$$

For m_2,

$$T_2 - m_2 g = m_2(-a),$$

$$T_2 = m_2(g - a) = (500)(980 - 6) = 4.87 \times 10^5 \text{ dyne}.$$

Note that the tensions on either side of the pulley are not equal; if they were equal, they could not exert a net torque on the pulley.

(c) If the cord does not slip, then the linear speed v of the cord must equal the linear speed $r\omega$ of a point on the rim of the pulley; i.e., $v = r\omega$. Differentiating gives

$$\alpha = \frac{a}{r} = \frac{6.00 \text{ cm/s}^2}{5.0 \text{ cm}} = 1.20 \text{ rad/s}^2.$$

(d) Neither the weight W of the pulley, nor the force F_s exerted by the support exert any torque on the pulley, for they act through the axle. Apply Eq.31:

$$I = \frac{\tau}{\alpha} = \frac{T_2 r - T_1 r}{\alpha} = \frac{(T_2 - T_1)r}{\alpha},$$

$$I = (4.87 \times 10^5 \text{ dyne} - 4.54 \times 10^5 \text{ dyne})(5 \text{ cm})/(1.2 \text{ rad/s}^2) = 1.375 \times 10^5 \text{ g} \cdot \text{cm}^2.$$

75P

(a) The rotational kinetic energy of the earth, in terms of the period T of rotation, is

$$K = \frac{1}{2}I\omega^2 = \frac{1}{2}I(\frac{2\pi}{T})^2 = \frac{2\pi^2 I}{T^2}.$$

Therefore,

$$\frac{dK}{dt} = -\frac{4\pi^2 I}{T^3}\frac{dT}{dt}.$$

Picturing the earth as a sphere of uniform density, the rotational inertia I will be

$$I = \frac{2}{5}MR^2 = \frac{2}{5}(5.98 \times 10^{24} \text{ kg})(6.37 \times 10^6 \text{ m})^2 = 9.706 \times 10^{37} \text{ kg} \cdot \text{m}^2.$$

The rate of change of the earth's rotation, in SI units, is

$$\frac{dT}{dt} = \frac{1 \text{ ms}}{\text{cent}} = \frac{10^{-3} \text{ s}}{(100)(3.156 \times 10^7 \text{ s})} = 3.169 \times 10^{-13},$$

and the current period of rotation is T = 24 h = 86,400 s. Therefore,

$$\frac{dK}{dt} = -4\pi^2 \frac{9.706 \times 10^{37} \text{ kg} \cdot \text{m}^2}{(8.64 \times 10^4 \text{ s})^3}(3.169 \times 10^{-13}) = -1.88 \times 10^{12} \text{ J/s}.$$

(b) First, find the angular acceleration. Since $\omega = 2\pi/T$, it follows that

$$\alpha = \frac{d\omega}{dt} = -\frac{2\pi}{T^2}\frac{dT}{dt}.$$

Dropping the sign to solve for the magnitude of the acceleration yields numerically

$$\alpha = \frac{2\pi}{(8.64 \times 10^4 \text{ s})^2}(3.169 \times 10^{-13}) = 2.667 \times 10^{-22} \text{ rad/s}^2.$$

(c) The torque slowing the earth therefore must be

$$\tau = I\alpha = (9.706 \times 10^{37} \text{ kg} \cdot \text{m}^2)(2.667 \times 10^{-22} \text{ rad/s}^2) = 2.589 \times 10^{16} \text{ N} \cdot \text{m}.$$

The perpendicular distance of the seabeds, where the forces are applied, to the axis of the earth's rotation, is (writing L for latitude)

$$r = R\cos L = (6.37 \times 10^6 \text{ m})\cos 60° = 3.185 \times 10^6 \text{ m}.$$

The forces are applied in both the northern and southern hemispheres, so that $\tau = 2Fr$, or

$$F = \tau/2r = (2.589 \times 10^{16} \text{ N} \cdot \text{m})/2(3.185 \times 10^6 \text{ m}) = 4.06 \times 10^9 \text{ N}.$$

<u>76P</u>

(a) The radial acceleration of a point at
the top of the chimney, height h, is, by Eq.
18, $a_r = h\omega^2$. By energy conservation,

$$mg(\tfrac{1}{2}h) = \tfrac{1}{2}I\omega^2 + mg(\tfrac{1}{2}h\cos\theta),$$

where $I = mh^2/3$ since the chimney, in effect,
is rotating about an axis through its base;
see Table 2. Substituting this expression for
I into the preceding equation gives

$$mgh(1 - \cos\theta) = \tfrac{1}{3}(mh^2\omega^2),$$

and therefore, by Eq.18,

$$a_r = h\omega^2 = 3g(1 - \cos\theta).$$

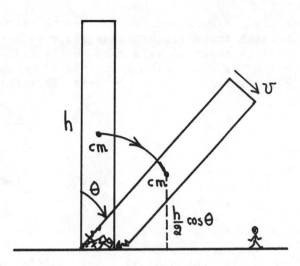

(b) The tangential acceleration, by Eq.17, is $a_t = h\alpha$. Use the result from (a),

$$\omega^2 = \frac{3g}{h}(1 - \cos\theta),$$

to find that

$$2\omega \frac{d\omega}{dt} = (\frac{3g}{h})\sin\theta(\frac{d\theta}{dt}),$$

$$\frac{d\omega}{dt} = \alpha = (\frac{3g}{2h})\sin\theta,$$

since $\omega = d\theta/dt$. Thus, the tangential acceleration is

$$a_t = (\frac{3g}{2})\sin\theta.$$

(c) The total linear acceleration is

$$a = \sqrt{[a_r^2 + a_t^2]} = \frac{3g}{2}\sqrt{[(1 - \cos\theta)(5 - 3\cos\theta)]}.$$

This is zero at $\theta = 0°$, and increases as θ increases, passing g near $\theta = 34.5°$. If the
chimney does not break up, then as the top hits the ground, we have $\theta = 90°$ and $a = 3.35g$
so that the answer to the question is YES.

(d) As the chimney tips over, the weight of the upper part acts transverse to (across)
the column. Chimneys standing upright are not subject to this force (although they are
subject to transverse forces due to the wind), and are not specifically designed to
withstand it.

<u>82P</u>

Let I be the rotational inertia of the body about the axis (leg) of rotation. The center
of mass of the body lies at the midpoint of the cross-piece and falls through a vertical
distance of $\ell/2$. By the conservation of energy,

$$(3m)g(\tfrac{1}{2}\ell) = \tfrac{1}{2}I\omega^2,$$

m the mass of each rod. It remains to calculate I. The rotational inertia of the leg about which the rotation takes place is zero, for the moment arm is zero. The cross-piece is rotating about an axis through its end, so that its rotational inertia is $m\ell^2/3$. The other leg has a rotational inertia of $m\ell^2$ about the axis since each part of the leg is at the same distance ℓ from the axis. Therefore,

$$I = \tfrac{1}{3}m\ell^2 + m\ell^2 = \tfrac{4}{3}m\ell^2.$$

Hence,

$$(3m)g(\tfrac{1}{2}\ell) = \tfrac{1}{2}(\tfrac{4}{3}m\ell^2)\omega^2 \;\rightarrow\; \omega = \tfrac{3}{2}\sqrt{(\tfrac{g}{\ell})}.$$

86P

(a) An angular speed of 39 rev/s = $39(2\pi)$ = 245 rad/s. For a constant angular acceleration Eq.8 yields

$$\alpha = \frac{\omega - \omega_0}{t} = \frac{0 - 245}{32} = -7.66 \text{ rad/s}^2.$$

(b) The torque is given by $\tau = I\alpha$. The rotational inertia is

$$I = I_{rod} + 2I_{ball},$$

$$I = (\tfrac{1}{12})ML^2 + (2)m(\tfrac{L}{2})^2 = \tfrac{1}{12}(6.4)(1.2)^2 + (2)(1.06)(0.6)^2 = 1.5312 \text{ kg·m}^2.$$

Therefore, using the result from (a), $\tau = I\alpha = (1.5312)(-7.66) = -11.7$ N·m.

(c) The work W done equals the change in kinetic energy, by the work-energy theorem. Since the final kinetic energy is zero, the work is

$$W = -\tfrac{1}{2}I\omega_0^2 = -\tfrac{1}{2}(1.5312)(245)^2 = -45955 \text{ J}.$$

(d) The number N of revolutions is $\theta/2\pi$, where θ is the angle in radians turned through in being brought to rest. This angle is found from Eq.10:

$$\theta = \frac{\omega^2 - \omega_0^2}{2\alpha} = \frac{0^2 - (245)^2}{2(-7.66)} = 3918 \text{ rad},$$

$$N = \theta/2\pi = 624.$$

(e) The calculation in (a) assumes constant angular acceleration and therefore also a constant torque. If the torque varies, then a single value such as found in (b) cannot be assigned to it. The equation in (d) for the angle turned through likewise assumed a constant angular acceleration. Only (c), which relied solely on the work-energy theorem, survives dropping the assumption of a constant torque.

87P

(a) Apply conservation of energy:

$$Mgh = \frac{1}{2}I\omega^2 + \frac{1}{2}Mv^2.$$

The angular speed ω of the flywheel is proportional to the speed v of the car; let c be the proportionality constant; i.e.,

$$\omega = cv.$$

Using this to eliminate ω from the energy equation gives

$$Mgh = \frac{1}{2}I(cv)^2 + \frac{1}{2}Mv^2 = \frac{1}{2}v^2[c^2I + M].$$

To evaluate the constant c, use the data given that $v = 80$ km/h must yield $\omega = 240$ rev/s. Convert to SI units to find that

$$c = \frac{\omega}{v} = \frac{1508 \text{ rad/s}}{22.22 \text{ m/s}} = 67.87 \text{ m}^{-1}.$$

The rotational inertia I of the flywheel is

$$I = \frac{1}{2}mR^2 = \frac{1}{2}(\frac{200}{9.8})(0.55)^2 = 3.087 \text{ kg·m}^2.$$

The vertical distance the car descends is $h = L\sin\theta = 1500\sin5° = 130.7$ m. The mass M of the car is 800 kg. Putting all of this data, which is now in SI units, into the energy equation above gives

$$(800)(9.8)(130.7) = \frac{1}{2}v^2[(67.87)^2(3.087) + 800] \rightarrow v = 11.68 \text{ m/s} = 42.1 \text{ km/h}.$$

(b) Return to the first energy equation; this time replace v with ω/c to obtain

$$Mgh = \frac{1}{2}\omega^2[I + \frac{M}{c^2}].$$

Differentiate with respect to time t, noting that

$$\frac{d}{dt}(\omega^2) = 2\omega\frac{d\omega}{dt} = 2\omega\alpha.$$

Hence,

$$Mg\frac{dh}{dt} = (\omega\alpha)[I + \frac{M}{c^2}].$$

But,

$$\frac{dh}{dt} = v\sin\theta = \frac{\omega}{c}\sin\theta,$$

so that

$$\frac{Mg\omega\sin\theta}{c} = (\omega\alpha)[I + \frac{M}{c^2}].$$

Now solve this for the desired angular acceleration:

$$\alpha = \frac{Mg\sin\theta}{c[I + M/c^2]} = \frac{cv^2}{2L} = \frac{(67.87 \text{ m}^{-1})(11.68 \text{ m/s})^2}{2(1500 \text{ m})} = 3.09 \text{ rad/s}^2.$$

(The near numerical equality in the values of I and α is a coincidence.)

(c) The power is

$$P = \tau\omega = (I\alpha)\omega = (I\alpha)(cv),$$

$$P = (3.087 \text{ kg} \cdot \text{m}^2)(3.09 \text{ rad/s}^2)(67.87 \text{ m}^{-1})(11.7 \text{ m/s}) = 7570 \text{ W} = 7.57 \text{ kW}.$$

CHAPTER 12

3E

The work W that must be done on the hoop is, by the work-energy theorem,

$$W = \Delta K = K_f - K_i = 0 - K_i = -K_i.$$

The initial kinetic energy is the sum of the translational and rotational kinetic energies (see Eq.5):

$$K_i = \tfrac{1}{2}Mv_{cm}^2 + \tfrac{1}{2}I\omega^2.$$

By Eq.2 this can be written

$$K_i = \tfrac{1}{2}Mv_{cm}^2 + \tfrac{1}{2}I(v_{cm}/R)^2.$$

Now refer to Table 2, Chapter 11, frame (a) to find $I = MR^2$. Putting everything together gives

$$W = -[\tfrac{1}{2}Mv_{cm}^2 + \tfrac{1}{2}(MR^2)(v_{cm}/R)^2] = -Mv_{cm}^2 = -(140 \text{ kg})(0.15 \text{ m/s})^2 = -3.15 \text{ J}.$$

(Often, the negative sign is omitted in quoting such a result.)

5E

Let v be the speed of the car; its total translational kinetic energy is $K_c = (1000)v^2/2 = 500v^2$. The kinetic energy of rotation of the wheels about their axles is

$$K_w = 4[\tfrac{1}{2}I\omega^2] = 2I(v/r)^2.$$

We are told to use $I = mr^2/2$ (see Table 2 in Chapter 11); hence,

$$K_w = 2(mr^2/2)(v/r)^2 = mv^2 = 10v^2.$$

Hence, the desired fraction is

$$K_w/K_c = 10v^2/500v^2 = 0.02.$$

10P

If we know the speed v with which the ball leaves the right-hand end of the track, then the desired distance x can be found by the methods of Section 4-6:

$$h = \tfrac{1}{2}gt^2; \quad x = vt \;\rightarrow\; x = v\sqrt{[\tfrac{2h}{g}]}.$$

To find v, apply energy conservation to the motion of the ball on the track. The mass and radius of the ball are not given numerically, so call them m and r. Apply Eqs.5 and 2, and pick I from Table 2 in Chapter 11 to get

$$K = \frac{1}{2}(2mr^2/5)(v/r)^2 + \frac{1}{2}mv^2 = \frac{7}{10}mv^2.$$

Thus, energy conservation gives

$$mgH + 0 = mgh + \frac{7}{10}mv^2,$$

$$v = \sqrt{[\frac{10}{7}g(H - h)]} = \sqrt{[(\frac{10}{7})(9.8)(60 - 20)]} = 23.66 \text{ m/s}.$$

With v determined, we now have

$$x = (23.66 \text{ m/s})\sqrt{[\frac{2(20 \text{ m})}{9.8 \text{ m/s}^2}]} = 47.8 \text{ m}.$$

15E

(a) Use Eq.16. Employing CGS units,

$$a = g \frac{1}{1 + I/MR_0^2} = (980) \frac{1}{1 + 950/(120)(0.32)^2} = 12.5 \text{ cm/s}^2.$$

(b) Use Eq.4 in Chapter 2 with $x_0 = 0$, $v_0 = 0$ and $x = h$:

$$t = \sqrt{[2h/a]} = \sqrt{[2(120)/(12.5)]} = 4.38 \text{ s}.$$

(c) The potential energy gained in falling a distance h is mgh. If the yo-yo is in only rotational motion, then $K = I\omega^2/2$. Applying energy conservation (still in CGS units),

$$\omega = \sqrt{[2mgh/I]} = \sqrt{[2(120)(980)(120)/(950)]} = 172.4 \text{ rad/s} = \frac{172.4}{2\pi} = 27.4 \text{ rev/s}.$$

23E

(a) Adopting the hint, we have from Exercise 22:

$$\ell_x = (0.25)[(0)(5) - (-2)(0)] = 0,$$

$$\ell_y = (0.25)[(-2)(-5) - (2)(5)] = 0,$$

$$\ell_z = (0.25)[(2)(0) - (0)(-5)] = 0.$$

Therefore $\vec{\ell} = 0$.

(b) Use Exercise 17:

$$\vec{\tau} = \vec{i}[(0)(0) - (-2)(4)] + \vec{j}[(-2)(0) - (2)(0)] + \vec{k}[(2)(4) - (0)(0)],$$

$$\vec{\tau} = +8\vec{i} + 8\vec{k}, \text{ N·m}.$$

26P

Suppose the origin is at O. The angular momentum of the particles are in opposite directions (one into, the other out of, the page). Hence, the total angular momentum is

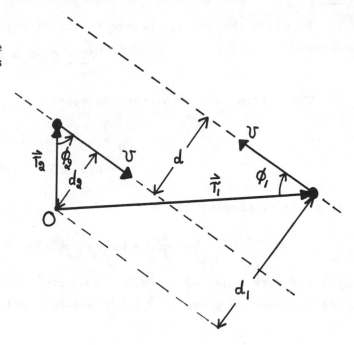

$$L = mvr_1\sin\phi_1 - mvr_2\sin\phi_2,$$

$$L = mv(r_1\sin\phi_1 - r_2\sin\phi_2).$$

From the sketch, however, it can be seen that

$$r_1\sin\phi_1 = d_1; \quad r_2\sin\phi_2 = d_2,$$

where d_1 and d_2 are the perpendicular distances from O to the lines of motion of the particles. Thus,

$$L = mv(d_1 - d_2).$$

But $d_1 - d_2 = d$, the perpendicular distance between the lines of motion, so that

$$L = mvd.$$

The quantities on the right-hand side of this equation are independent of the location of the origin O, provided it is outside the lines of motion. It remains to examine the case where O lies between the paths of the particles, and it is left to the reader to show that in this case also we get $L = mvd$.

30E

(a) We are given that m = 3 kg, $\vec{r} = 3\vec{i} + 8\vec{j}$, $\vec{v} = 5\vec{i} - 6\vec{j}$, $\vec{F} = -7\vec{i}$, all in SI units. The angular momentum is, by Exercise 22,

$$\ell_x = (3)[(8)(0) - (0)(-6)] = 0,$$

$$\ell_y = (3)[(0)(5) - (3)(0)] = 0,$$

$$\ell_z = (3)[(3)(-6) - (8)(5)] = -174;$$

i.e., $\vec{\ell} = -174\vec{k}$, in SI units (kg·m^2/s).

(b) Use Exercise 17; since $F_y = F_z = 0$ and z = 0 we get

$$\vec{\tau} = \vec{k}[(3)(0) - (8)(-7)] = 56\vec{k}.$$

(c) By Eq.23, $d\vec{\ell}/dt = \vec{\tau} = 56\vec{k}$.

32E

(a) By Eq.30 the average torque is (in SI units),

$$\bar{\tau} = \frac{\Delta L}{\Delta t} = \frac{0.80 - 3}{1.5} = -1.47 \text{ N·m.}$$

(b) The initial and final angular speeds are

$$\omega_0 = L_0/I = 3/0.14 = 21.43 \text{ rad/s,}$$

$$\omega = L/I = 0.8/0.14 = 5.714 \text{ rad/s.}$$

By Eq.11 of Chapter 11,

$$\theta = \frac{1}{2}(\omega + \omega_0)t = \frac{1}{2}(5.714 + 21.43)(1.5) = 20.4 \text{ rad.}$$

(c) The work done is, by Eq.35 in Chapter 11, $W = \tau\theta = (-1.47)(20.4) = -30.0$ J.

(d) The average power is $\bar{P} = W/t = (-30.0)/(1.5) = -20.0$ W.

35P

By Newton's third law, the friction forces f exerted by each wheel on the other are equal in magnitude and opposite in direction. Let ω_1 be the angular speed of the larger wheel when slipping ceases. The frictional forces are exerted over a finite time, during which they may change in magnitude. However, by Problem 34,

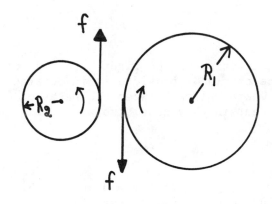

$$\int f dt = I(\omega_f - \omega_i),$$

the integral extending over the time that the frictional forces act. Applying this integral to each of the wheels gives

$$\int f R_2 dt = I_2\omega_2 - I_2(0),$$

$$-\int f R_1 dt = I_1\omega_1 - I_1\omega_0,$$

the torques being in opposite directions. These equations can be rewritten in the form

$$R_2\int f dt = I_2\omega_2,$$

$$-R_1\int f dt = I_1(\omega_1 - \omega_0).$$

Now eliminate the unknown integral (by dividing the equations, for example) to get

$$-\frac{R_2}{R_1} = \frac{I_2\omega_2}{I_1(\omega_1 - \omega_0)}.$$

When slipping ceases, the linear speeds of the rims of the wheels must be equal:

$$v = R_1\omega_1 = R_2\omega_2.$$

Now eliminate ω_1 from the second preceding equation using the equation directly above to obtain

$$\omega_2 = \omega_0[\frac{R_1I_2}{R_2I_1} + \frac{R_2}{R_1}]^{-1}.$$

37E

(a) Using Eq.21 of Chapter 11,

$$I = \Sigma m_i r_i^2 = m\ell^2 + m(2\ell)^2 + m(3\ell)^2 = 14m\ell^2.$$

(b) For the middle particle,

$$L_m = I_m\omega = [m(2\ell)^2]\omega = 4m\ell^2\omega.$$

(c) Use the result in (a):

$$L = I\omega = 14m\ell^2\omega.$$

41E

(a) Apply conservation of angular momentum:

$$I_1\omega_1 = I_2\omega_2,$$
$$(6)(1.2) = (2)\omega_2 \rightarrow \omega_2 = 3.6 \text{ rev/s.}$$

(b) To calculate kinetic energy, the angular speeds must be in rad/s:

$$\omega_1 = 1.2 \text{ rev/s} = (1.2)(2\pi) = 7.540 \text{ rad/s,}$$
$$\omega_2 = 3.6 \text{ rev/s} = (3.6)(2\pi) = 22.62 \text{ rad/s.}$$

Hence,

$$K_1 = \frac{1}{2}I_1\omega_1^2 = \frac{1}{2}(6)(7.540)^2 = 170.6 \text{ J,}$$
$$K_2 = \frac{1}{2}I_2\omega_2^2 = \frac{1}{2}(2)(22.62)^2 = 511.7 \text{ J.}$$

Therefore, the desired ratio is

$$K_2/K_1 = (511.7 \text{ J})/(170.6 \text{ J}) = 3.00.$$

43E

(a) Apply conservation of angular momentum to the two wheels:

$$I_1\omega_1 + I_2\omega_2 = (I_1 + I_2)\omega_f,$$

$$I_1(800 \text{ rev/min}) + 0 = (I_1 + 2I_1)\omega_f \rightarrow \omega_f = 267 \text{ rev/min}.$$

(b) Since we are asked to compute a ratio, we can continue to use rev/min as a unit for angular speed. The original kinetic energy is

$$K_i = \frac{1}{2}I_1\omega_1^2 = \frac{1}{2}I_1(800)^2 = 32 \times 10^4 I_1.$$

(We cannot get a numerical value since we do not have a numerical value for I_1.) The kinetic energy of the final combination is

$$K_f = \frac{1}{2}(I_1 + I_2)\omega_f^2 = \frac{1}{2}(I_1 + 2I_1)(267)^2 = 10.69 \times 10^4 I_1.$$

Hence, the desired ratio is

$$\frac{\Delta K}{K_i} = \frac{(32 - 10.69) \times 10^4}{32 \times 10^4} = 0.666.$$

46E

(a) Use Problem 56 in Chapter 11 to get

$$I = Mk^2 = (180 \text{ kg})(0.91 \text{ m})^2 = 149 \text{ kg·m}^2.$$

(b) Since the child runs tangent to the rim, we have $r\sin\phi = d = R$, where R is the radius of the merry-go-round (see Fig.11). Hence,

$$\ell = mvR = (44 \text{ kg})(3.0 \text{ m/s})(1.2 \text{ m}) = 158 \text{ kg·m}^2/\text{s}.$$

(c) Apply conservation of angular momentum. The angular momentum of the merry-go-round is zero initially. The rotational inertia of the child after jumping on is $(44 \text{ kg})(1.2 \text{ m})^2 = 63.36 \text{ kg·m}^2$. Therefore, we have

$$158 \text{ kg·m}^2/\text{s} = [(149 + 63.36) \text{ kg·m}^2]\omega \rightarrow \omega = 0.744 \text{ rad/s}.$$

50P

The initial angular momentum of the system train+wheel is zero; by the conservation of angular momentum, the angular momentum with the power on must be zero also. Thus, the wheel and the train rotate in opposite directions relative to the earth. The magnitudes of their angular momenta must be equal:

$$L_{wheel} = L_{train}.$$

Since the rotational inertia of the wheel is MR^2, we have

$$L_{wheel} = MR^2\omega.$$

The speed V of the train relative to the earth is

$$V = v - R\omega,$$

since $R\omega$ is the speed of the track relative to the earth. Thus,

$$MR^2\omega = mVR = m(v - R\omega)R,$$

$$\omega = \frac{m}{M + m}\left(\frac{v}{R}\right).$$

55P

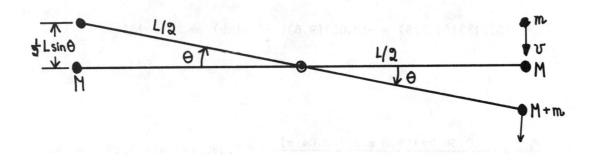

(a) Apply conservation of angular momentum about the axis of the rod of length L; if the balls can be represented as point masses, then

$$mv\left(\frac{L}{2}\right) = \left[M\left(\frac{L}{2}\right)^2 + (M + m)\left(\frac{L}{2}\right)^2\right]\omega.$$

We substitute the data in SI units:

$$\omega = \frac{2mv}{(2M + m)L} = \frac{2(0.05)(3)}{(4.05)(0.5)} = 0.148 \text{ rad/s}.$$

(b) The desired ratio is

$$\frac{K_{aft}}{K_{bef}} = \frac{\frac{1}{2}I\omega^2}{\frac{1}{2}mv^2} = \frac{\frac{1}{2}[(2M + m)(L/2)^2]\omega^2}{\frac{1}{2}mv^2} = \left(\frac{2M + m}{m}\right)\left(\frac{L\omega}{2v}\right)^2.$$

From (a),

$$\frac{L\omega}{2v} = \frac{m}{2M + m},$$

so that

$$\frac{K_{aft}}{K_{bef}} = \frac{m}{2M + m} = \frac{0.05}{4.05} = 0.0123.$$

(c) Conservation of energy can be applied to the motion of the system subsequent to the completely inelastic collision between the putty and the stick. It is assumed that the rod has not had a chance to move during the impact, so that it is still in the original horizontal position immediately after impact. In this position, set the gravitational potential energy U = 0. At the extreme position of its swing, the kinetic energy is zero instantaneously. Since part of the rod is below the U = 0 position at this time, the potential energy U will contain a negative part. By conservation of energy, then,

$$K_i = U_f,$$

$$(0.0123)(\tfrac{1}{2}mv^2) = [M - (M + m)]g[\tfrac{L}{2}\sin\theta],$$

using the result from (b). But

$$\tfrac{1}{2}mv^2 = \tfrac{1}{2}(0.05)(3)^2 = 0.225 \text{ J.}$$

Hence,

$$(0.0123)(0.225) = -(0.05)(9.8)(\tfrac{0.5}{2}\sin\theta) \;\rightarrow\; \theta = 181°.$$

56E

Apply Eq.39:

$$\Omega = \frac{Mgr}{I\omega} = \frac{(0.50 \text{ kg})(9.8 \text{ m/s}^2)(0.04 \text{ m})}{(5 \times 10^{-4} \text{ kg·m}^2)[(30)(2\pi)\text{rad/s}]} = 2.080 \text{ rad/s} = 0.331 \text{ rev/s.}$$

The direction of precession is the same as that of the rotation; in this case, clockwise as seen from above.

CHAPTER 13

5E

(a) Applying Eq.3, we must have $\vec{F}_1 + \vec{F}_2 + \vec{F}_3 = 0$, so that

$$\vec{F}_3 = -[\vec{F}_1 + \vec{F}_2] = -[(10\vec{i} - 4\vec{j}) + (17\vec{i} + 2\vec{j})] = -[27\vec{i} - 2\vec{j}] = -27\vec{i} + 2\vec{j}.$$

(b) The direction of F_3, counterclockwise from the +x axis, is at an angle

$$\theta = \cos^{-1}(F_{3x}/F_3).$$

But $F_3 = \sqrt{(27^2 + 2^2)} = 27.07$ N. Hence,

$$\theta = \cos^{-1}(-27/27.07) = 176°, \ 184°.$$

Since $F_{3y} > 0$, we have $\theta = 176°$.

10E

(a) Resolve the forces acting into their horizontal and vertical components; these must each sum to zero;

$$N - T\cos\theta = 0,$$

$$w - T\sin\theta = 0.$$

The second equation gives for the tension T,

$$T = \frac{w}{\sin\theta}.$$

It remains to find the angle θ. From the geometry of the arrangement,

$$\sin\theta = \frac{L}{\sqrt{(L^2 + r^2)}},$$

and therefore

$$T = w \frac{\sqrt{(L^2 + r^2)}}{L}.$$

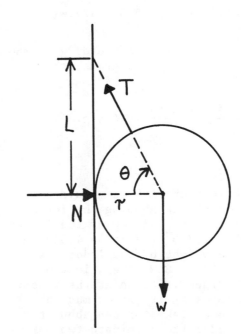

(b) Divide the original pair of equations to get $N = w\cot\theta = wr/L$.

13E

Follow the method of Sample Problem 1. Let's assume that the forces exerted by the two pedestals each are directed up, as shown in the sketch on the next page. For equilibrium,

$$\Sigma F = F_1 + F_2 - 580 = 0,$$

and, taking torques about the left end of the board, counting counterclockwise rotations as positive,

$$\Sigma \tau = F_1(0) + F_2(1.5) - 580(4.5) = 0,$$

$$F_2 = +1740 \text{ N}.$$

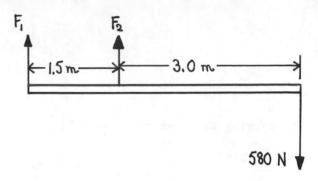

This being positive, F_2 actually is directed as was assumed in the sketch. By Newton's third law, the force exerted by the board on the pedestal is down, i.e., is a force of compression on the pedestal. Now put this value of F_2 into the first equation above to get $F_1 = 580 - 1740 = -1160$ N. This means, by the negative sign, that the force exerted by the pedestal on the board is opposite to what was assumed, i.e., is actually down. The force exerted by the board on the pedestal is therefore up, a force of tension on the pedestal.

14E

Take torques about the pivot P (to eliminate the force N exerted by the pivot),

$$Mg(50 - 45.5) - 2mg(45.5 - 12) = 0;$$

we are using CGS units. The factor g (= 980 cm/s^2) cancels; since m = 5 grams, this equation yields

$$M = \frac{(2)(5)(33.5)}{4.5} = 74.4 \text{ g}.$$

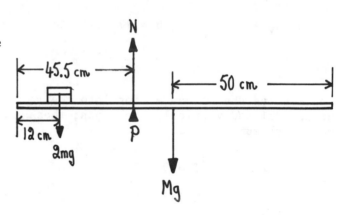

16E

(a) Using Sample Problem 3 as a guide, we draw a labelled sketch showing the forces acting on the ladder. The top of the ladder is at a height of $\sqrt{(5^2 - 2.5^2)} = 4.33$ m above the ground. The weight of the ladder is (10 kg)(9.8 m/s^2) = 98 N; the weight of the man is 735 N. Since friction at the window is ignored, the force F_w it exerts must be at 90° to the window. Take torques about the base of the ladder (to eliminate two unknown forces); counterclockwise rotation is positive so we have

$$-(98)(2.5\cos\theta) - 735(3\cos\theta) + F_w(4.33) = 0.$$

But $\cos\theta = 2.5/5 = 0.5$ (using the triangle formed by ladder, wall and ground). With this substitution, the equation gives $F_w = 283$ N.

(b) The only horizontal force components acting are F_{gx} to the right and F_w to the

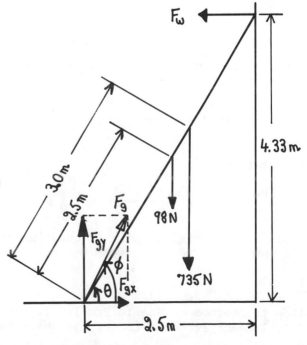

left; hence F_{gx} = 283 N also, in order that these two oppositely directed horizontal force components sum to zero. In order that the vertical force components also sum to zero, we must have F_{gy} = 98 + 735 = 833 N. Therefore, the force exerted by the ground, in magnitude and direction, is

$$F_g = \sqrt{[283^2 + 833^2]} = 880 \text{ N}; \quad \phi = \tan^{-1}(\frac{833}{283}) = 71.2°,$$

the angle being with respect to the ground. Note that F_g does not point along the ladder, which is at an angle θ = 60° with the ground.

20P

Adopting the hint, put the x axis along the right-hand cable. The angle between the cables must be 180° - 51° - 66° = 63°. The angle between the 1800 lb force (the tension in the short vertical cable directly supporting the bucket), which is directed vertically down, and the x axis is 270° - 66° = 204°. With all angles measured counter-clockwise from the x axis, the equilibrium condition, Eq.3, applied to the knot at O, can be written as

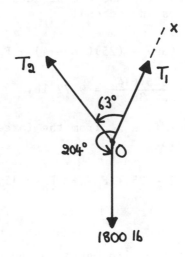

$T_1\cos0° + T_2\cos63° + 1800\cos204° = 0,$

$T_1\sin0° + T_2\sin63° + 1800\sin204° = 0.$

But $\cos0°$ = 1 and $\sin0°$ = 0. Hence, the second equation can be solved immediately for T_2:

$$T_2 = -1800 \frac{\sin204°}{\sin63°} = 821.7 \text{ lb.}$$

Now substitute this result into the x equation to get

$$T_1 + 821.7\cos63° + 1800\cos204° = 0 \rightarrow T_1 = 1271 \text{ lb.}$$

Note that there is no need to solve "two equations in two unknowns" in the form that would be obtained if conventional horizontal and vertical x and y axes had been used.

26P

The normal force vanishes as the wheel lifts. Taking torques about O (to eliminate the force exerted by the step at O) as the wheel lifts gives

$$Wx = F(r - h).$$

But,

$$x^2 = r^2 - (r - h)^2,$$

$$x = \sqrt{(2rh - h^2)},$$

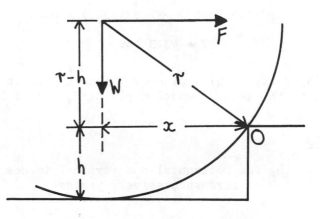

so that

$$F = W \frac{\sqrt{(2rh - h^2)}}{r - h}.$$

28P

(a) We assume that the forces are directed as shown on the sketch. There are two forces to be found: F_c and F_h. We can eliminate F_h, and so find F_c first, by taking torques about the hinge (we use the BE data and note the moment arms in inches):

$$\Sigma\tau = 0 = F_c(36) - (25)(18 - 4) + F_h(0),$$

$$F_c = \frac{(25)(14)}{36} = 9.72 \text{ lb.}$$

(b) We can now find F_h from the force condition, Eq.3:

$$\Sigma F = 0 = 9.72 - 25 + F_h \rightarrow F_h = 15.3 \text{ lb.}$$

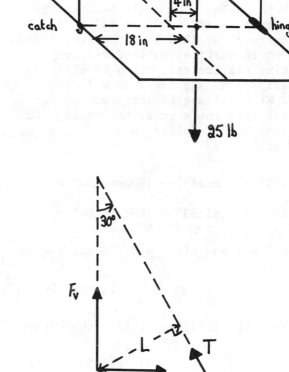

30P

(a) Note that the wire makes an angle of 60° with the horizontal. Summing the horizontal and vertical force components separately to zero yields the following equations:

$$F_h - T\cos60° = 0,$$

$$F_v + T\sin60° - 50 = 0.$$

Take torques about the hinge: the moment arms of both F_h and F_v are zero. Hence, with ℓ and L representing the moment arms of the weight and of the tension in the wire,

$$50(\ell) - T(L) = 0,$$

$$50(1.5\sin60°) - T(3\sin30°) = 0,$$

$$T = 43.3 \text{ lb.}$$

(b) Putting this value of T into the first equation gives F_h = 21.65 lb; substituting T into the second equation yields F_v = 12.5 lb.

33P

Summing the horizontal and vertical forces components to zero, and summing the torques about 0 to zero successively yields

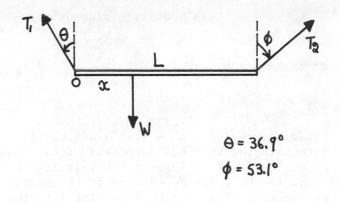

$$T_2\sin\phi - T_1\sin\theta = 0,$$

$$T_2\cos\phi + T_1\cos\theta - W = 0,$$

$$-Wx + (T_2\cos\phi)L = 0.$$

Multiply the first equation by $\cos\theta$, the
second equation by $\sin\theta$ and subtract the
resulting equations (to eliminate T_1) to
obtain for T_2,

$$T_2 = \frac{W\sin\theta}{\sin(\theta + \phi)};$$

$$\theta = 36.9°$$
$$\phi = 53.1°$$

we used a trigonometric identity; see App.G.
Note from their numerical values, however,
that $\theta + \phi = 90°$ so that $T_2 = W\sin\theta$. Now substitute this expression for T_2 into the third
equation above to get

$$x = L\sin\theta\cos\phi = L\sin^2\theta = (6.1)\sin^2 36.9° = 2.20 \text{ m.}$$

38P

See the sketch at the right. The force f of
static friction opposes the tendency of the
planck to slip back to the left and therefore
points to the right. The roller, being free
of friction, exerts only a normal force
(called R to distinguish it from the normal
force exerted by the floor) on the planck.
Examine the situation at $\theta = 70°$, where f =
$\mu_s N$, its limiting value. Summing the forces
and torques to zero (the latter taken about
the bottom of the planck) gives

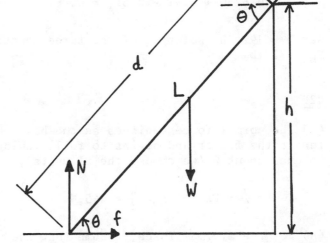

$$R\cos\theta + N - W = 0,$$

$$R\sin\theta - \mu_s N = 0,$$

$$Rd - W(\frac{L}{2})\cos\theta = 0.$$

To eliminate N, multiply the first equation
by μ_s and add to the second equation to get

$$R(\sin\theta + \mu_s\cos\theta) = \mu_s W.$$

The third equation above gives R = (WL/2d)$\cos\theta$; put this into the equation directly above
to find that

$$\sin\theta\cos\theta + \mu_s\cos^2\theta = \frac{2d}{L}\mu_s.$$

But $d = h/\sin\theta$, and therefore

$$\sin^2\theta\cos\theta + \mu_s\sin\theta\cos^2\theta = \frac{2h}{L}\mu_s \;\rightarrow\; \mu_s = \frac{\sin^2\theta\cos\theta}{2h - L\sin\theta\cos^2\theta}\,L.$$

Numerically, L = 2h and θ = 70°, giving μ_s = 0.339.

39P

The distance $d = \sqrt{(4^2 - 1.25^2)} = 3.80$ ft.
Note that the vertical component of the force
F_H exerted by the hinge on the left leg is
equal to the force F_E exerted by the ground
on the right leg, for Eq.3 for the left leg
and for the ladder as a whole are identical:

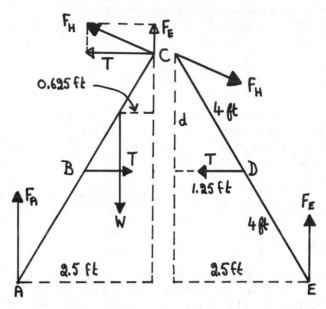

$$F_A + F_E - W = 0.$$

Taking torques about C gives, first for the
right leg

$$2.5F_E - 3.8T = 0,$$

and for the left leg,

$$0.625W + 3.8T - 2.5F_A = 0.$$

Set W = 192 lb; solution of the three equations gives (a) T = 47 lb, and (b) F_A = 120 lb,
F_E = 72 lb.

42P

(a) The normal force vanishes as the box
leaves the flooor and begins to roll. Taking
torques about O (we choose the SI units),

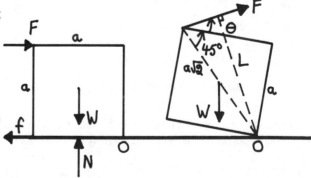

$$W\frac{a}{2} = Fa \;\rightarrow\; F = \frac{W}{2} = 445 \text{ N.}$$

(b) For the minimum force, assume that the
box was on the verge of sliding, so that f =
$\mu_s N$. Translational equilibrium demands that
f = F; thus

$$F = \mu_s N = \mu_s W \;\rightarrow\; \mu_s = \frac{F}{W} = 0.50.$$

(c) If F is applied at an angle θ above the horizontal at the top of the box (for the
longest moment arm L), then again taking torques about O,

$$F(a\sqrt{2})\sin(\theta + 45°) = (890)\frac{a}{2} \;\rightarrow\; F\sin(\theta + 45°) = 315.$$

For F to be a minimum, set sin(θ + 45°) = 1, its maximum possible value. Hence, θ = 45°
and F = 315/sin90° = 315 N.

46E

(a) The strain is $\Delta L/L = (2.8 \text{ cm})/(1500 \text{ cm}) = 1.87 \times 10^{-3}$.

(b) The stress is $F/A = W/A = mg/A$, where A is the cross-sectional area of the rope. We have, then,

$$\text{stress} = \frac{mg}{\pi d^2/4} = \frac{(95 \text{ kg})(9.8 \text{ m/s}^2)}{\pi(0.0096 \text{ m})^2/4} = 1.29 \times 10^7 \text{ N/m}^2.$$

(c) From Eq.25,

$$\text{modulus} = E = \frac{\text{stress}}{\text{strain}} = \frac{1.29 \times 10^7 \text{ N/m}^2}{1.87 \times 10^{-3}} = 6.90 \times 10^9 \text{ N/m}^2.$$

49E

(a) The shear stress is $F/A = T/A = Mg/A$, and therefore

$$\text{stress} = \frac{Mg}{A} = \frac{(1200 \text{ kg})(9.8 \text{ m/s}^2)}{\pi(0.048 \text{ m})^2/4},$$

$$\text{stress} = 6.50 \times 10^6 \text{ N/m}^2.$$

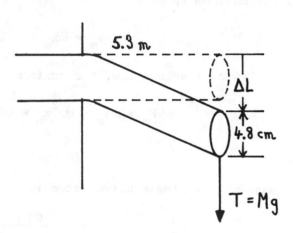

(b) Use Eq.26; identify E with the shear modulus and use the value of F/A (= Mg/A) found in (a) above to get

$$\Delta L = (\frac{F}{A})(\frac{L}{E}) = (6.50 \times 10^6 \text{ N/m}^2)(\frac{5.3 \text{ m}}{3 \times 10^{10} \text{ N/m}^2}) = 1.15 \times 10^{-3} \text{ m} = 1.15 \text{ mm}.$$

CHAPTER 14

4E

(a) See Sample Problem 1. We work in SI units. The period T is the time required for the motion to repeat itself, so that $T = 0.50$ s.

(b) By Eq.2 the frequency $\nu = 1/T = 1/0.50 = 2$ Hz.

(c) The angular frequency, by Eq.4, is $\omega = 2\pi\nu = 2\pi(2) = 12.6$ rad/s.

(d) The spring constant is given by Eq.11:

$$k = m\omega^2 = (0.50)(12.6)^2 = 79.4 \text{ N/m}.$$

(e) The maximum speed is

$$v_m = \omega x_m = (12.6)(0.35) = 4.41 \text{ m/s}.$$

(f) By Newton's second law, the maximum force is

$$F_m = ma_m = m\omega^2 x_m = (0.50)(12.6)^2(0.35) = 27.8 \text{ N}.$$

5E

The mass m of a single silver atom is

$$m = \frac{108 \text{ g}}{6.02 \text{ X } 10^{23}} = 1.794 \text{ X } 10^{-22} \text{ g}.$$

Combining Eqs.4 and 11 gives

$$2\pi\nu = \sqrt{\left(\frac{k}{m}\right)},$$

$$k = 4\pi^2\nu^2 m = 4\pi^2(10^{13} \text{ s}^{-1})^2(1.794 \text{ X } 10^{-25} \text{ kg}) = 708 \text{ N/m}.$$

7E

From Eq.6 the acceleration amplitude is $a_m = \omega^2 x_m$. But, from Eq.4, $\omega = 2\pi\nu$; therefore,

$$a_m = (2\pi\nu)^2 x_m,$$

$$\nu = \frac{1}{2\pi}\sqrt{[a_m/x_m]} = \frac{1}{2\pi}\sqrt{[(9.8 \text{ m/s}^2)/(10^{-6} \text{ m})]} = 498 \text{ Hz}.$$

This is the frequency for which $a_m = g$. From the first equation, we see that for $\nu > 498$ Hz, $a_m > g$.

11E

(a) From Eq.4, $\omega = 2\pi/T = 2\pi/(10^{-5} \text{ s}) = 6.28 \text{ X } 10^5$ rad/s.

(b) By Eq.5, $x_m = v_m/\omega = (1000 \text{ m/s})/(6.28 \text{ X } 10^5 \text{ rad/s}) = 1.59 \text{ X } 10^{-3} \text{ m} = 1.59 \text{ mm}.$

13E

(a) See Fig.2; the amplitude is $x_m = (2 \text{ mm})/2 = 1 \text{ mm}.$
(b) By Eqs.4 and 5,

$$v_m = \omega x_m = 2\pi\nu x_m = 2\pi(120 \text{ s}^{-1})(10^{-3} \text{ m}) = 0.754 \text{ m/s}.$$

(c) Use Eqs.4 and 6:

$$a_m = \omega^2 x_m = (2\pi\nu)^2 x_m = 4\pi^2\nu^2 x_m = 4\pi^2(120 \text{ s}^{-1})^2(10^{-3} \text{ m}) = 568 \text{ m/s}^2.$$

19P

The frequency ν of the motion of the piston is

$$\nu = \frac{180 \text{ osc}}{60 \text{ s}} = 3 \text{ Hz}.$$

Noting that the amplitude is $(0.76 \text{ m})/2 = 0.38 \text{ m}$, the maximum speed is

$$v_m = 2\pi\nu x_m = 2\pi(3 \text{ s}^{-1})(0.38 \text{ m}) = 7.16 \text{ m/s}.$$

26P

If it does not slip on the shake table, then the block also executes simple harmonic motion. Under this condition, the maximum force on the block in the horizontal direction will be

$$F_m = ma_m = m(4\pi^2\nu^2 x_m).$$

This force is supplied by static friction. If the amplitude x_m increases, F_m increases also. But the force of static friction cannot increase beyond $\mu_s N$, where $N = mg$ in this case. Hence, the maximum amplitude can be found from

$$\mu_s mg = 4\pi^2 m\nu^2 x_{m,max},$$

$$x_{m,max} = \mu_s g/[4\pi^2\nu^2] = (0.5)(9.8 \text{ m/s}^2)/[4\pi^2(2 \text{ s}^{-1})^2] = 0.031 \text{ m} = 3.1 \text{ cm}.$$

30P

(a) The amplitude x_m of oscillation is $(10 \text{ cm})/2 = 5 \text{ cm}$. As the object passes through the point 5 cm below its initial position, the resultant force on it must be zero, since the acceleration midway between the turning points is zero. The two forces acting are the weight mg of the object directed down and the spring force kx_m directed up. Therefore, at this point,

$$mg = kx_m \rightarrow k/m = g/x_m.$$

The frequency of oscillation is

$$\nu = \frac{1}{2\pi}\sqrt{(k/m)} = \frac{1}{2\pi}\sqrt{(g/x_m)} = \frac{1}{2\pi}\sqrt{(9.8/0.05)} = 2.23 \text{ Hz.}$$

(b) Let the position x of the object be given by

$$x = x_m \sin 2\pi\nu t,$$

so that the object passes through its equilibrium position $x = 0$ at time $t = 0$. From (a), $2\pi\nu = 14$ Hz, $x_m = 5$ cm so that, in cm,

$$x = 5\sin 14t.$$

When the object is 8 cm below its original position, it is at $x = 8 - 5 = 3$ cm. This occurs at a time t* given by

$$3 = 5\sin 14t* \rightarrow \sin 14t* = 0.60.$$

The velocity v at any time is

$$v = \frac{dx}{dt} = 70\cos 14t,$$

so that at $t = t*$,

$$v = 70\cos 14t* = 70(0.8) = 56 \text{ cm/s,}$$

where we used the identity $\cos^2\theta + \sin^2\theta = 1$, $\theta = 14t*$.

(c) Let the added mass be M. The new frequency is

$$\nu = \frac{1}{2\pi}\sqrt{\left(\frac{k}{M + m}\right)}.$$

This is half the original frequency, so that

$$\frac{1}{2\pi}\sqrt{\left(\frac{k}{M + m}\right)} = \frac{1}{2}\frac{1}{2\pi}\sqrt{\left(\frac{k}{m}\right)}.$$

Square this equation to get

$$4m = M + m \rightarrow m = \frac{M}{3} = 100 \text{ g.}$$

(d) At the new equilibrium position the resultant force is zero [see (a)]:

$$(m + M)g = kx_m^*,$$

with x_m^* the new amplitude. From (a), this result becomes (since $k = mg/x_m$),

$$x_m^* = \frac{m + M}{m} x_m = \frac{100 + 300}{100}(5) = 20 \text{ cm,}$$

that is, 20 cm below the original position of the first mass in (a).

31P

(a) The amplitude of each motion is A/2 and their periods of motion are T = 1.5 s. If their displacements are called x_1 and x_2, then

$$x_1 = \frac{A}{2}\cos(\frac{2\pi t}{T}),$$

$$x_2 = \frac{A}{2}\cos(\frac{2\pi t}{T} - \frac{\pi}{6}),$$

so that particle 1 is at one end of the line at time t = 0 and particle 2 is at this same point at time

$$t = \frac{\pi/6}{2\pi/T} = \frac{1}{8} \text{ s;}$$

we expressed the phase difference in radian measure. Hence, particle 1 leads particle 2. It is required to find their distance apart at time t = 0.125 + 0.5 = 0.625 s. Using the equations for displacements, it is found that at this time,

$$x_1 = \frac{A}{2}\cos(5\pi/6) = -0.433A,$$

$$x_2 = \frac{A}{2}\cos(4\pi/6) = -0.250A.$$

Hence, their distance apart is $d = x_2 - x_1 = 0.183A$.

(b) To establish the directions of motion, examine the velocities at t = 0.625 s. Since v = dx/dt, these velocities involve the sines of the angles encountered in (a). But $\sin(5\pi/6)$ and $\sin(4\pi/6)$ are both positive, indicating that the particles are moving in the same direction at this time.

35P

Assume that the spring constants are not the same, as shown on the sketch. As the mass oscillates, the spring of force constant k_1 is stretched (or compressed) a distance x_1; the corresponding distance for the other spring is called x_2. By Newton's third law, the forces exerted by the springs on each other are equal and given by

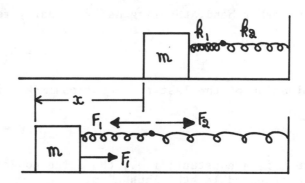

$$F_1 = k_1 x_1; \quad F_2 = k_2 x_2,$$

so that

$$k_1 x_1 = k_2 x_2.$$

Now, the displacement of the mass itself is $x = x_1 + x_2$ and therefore

$$x_1 = x - x_2 = x - \frac{k_1}{k_2}x_1 \rightarrow x_1 = \frac{k_2}{k_1 + k_2}x.$$

The force acting on the mass m is $F_1 = k_1 x_1$ (F_2 does not act on m); in terms of the displacement x of the mass, the force is

$$F_1 = \frac{k_1 k_2}{k_1 + k_2} \, x.$$

Thus, the springs joined in this way act like a single spring of force constant k_{eff} that is given by

$$k_{eff} = \frac{k_1 k_2}{k_1 + k_2}.$$

For the case where $k_1 = k_2 = k$, $k_{eff} = k/2$, so that

$$\nu = \frac{1}{2\pi}\sqrt{(k_{eff}/m)} = \frac{1}{2\pi}\sqrt{(k/2m)}.$$

37P

See Problem 35, where it is shown that for two springs joined in tandem, the effective spring constant is given by

$$k_{eff} = \frac{k_1 k_2}{k_1 + k_2}.$$

If the two springs originally were part of one spring, let $k_{eff} = k$, where k is the spring constant of the uncut spring. Making this substitution, the equation above can be manipulated into the form

$$\frac{1}{k} = \frac{1}{k_1} + \frac{1}{k_2}.$$

On the other hand, the lengths are clearly related by

$$L = L_1 + L_2.$$

Examination of the last two equations makes it reasonable to suppose that

$$k = \frac{C}{L},$$

where C is a constant; i.e., the force constant of a spring is inversely proportional to its unstressed length. Thus also,

$$k_1 = \frac{C}{L_1}, \quad k_2 = \frac{C}{L_2},$$

in which C = kL. With $L_1 = nL_2$ we have

$$L_1 + L_2 = nL_2 + L_2 = L,$$

$$L_2 = \frac{L}{n+1}, \quad L_1 = \frac{n}{n+1} L.$$

Therefore,

$$k_1 = \frac{n+1}{n} k, \quad k_2 = (n+1)k.$$

If $n = 1$, these give $k_1 = 2k = k_2$, or $k = k_1/2 = k_2/2$, agreeing with the results of Problem 35.

39E

Use Eq.19. In SI units, we are given $k = 130$ N/m and $x_m = 0.024$ m. Therefore,

$$E = \tfrac{1}{2}kx_m^2 = \tfrac{1}{2}(130 \text{ N/m})(0.024 \text{ m})^2 = 0.0374 \text{ J}.$$

41E

(a) At equilibrium, the weight mg of the block is balanced by the upward force exerted by the spring, so that

$$k = \frac{mg}{y} = \frac{(1.3 \text{ kg})(9.8 \text{ m/s}^2)}{0.096 \text{ m}} = 133 \text{ N/m} = 1.33 \text{ N/cm}.$$

(b) The period is

$$T = \frac{2\pi}{\omega} = 2\pi\sqrt{[\tfrac{m}{k}]} = 2\pi\sqrt{[\tfrac{1.3 \text{ kg}}{133 \text{ N/m}}]} = 0.621 \text{ s}.$$

(c) The frequency is $\nu = 1/T = 1/0.621 = 1.61$ Hz.

(d) The block is at rest at the turning points (ends) of its line of motion; hence $x_m = 5$ cm.

(e) By Eq.19,

$$E = \tfrac{1}{2}kx_m^2 = \tfrac{1}{2}(133 \text{ N/m})(0.05 \text{ m})^2 = 0.166 \text{ J}.$$

(f) The maximum speed is $v_m = \omega x_m = 2\pi\nu x_m = 2\pi(1.61 \text{ Hz})(5 \text{ cm}) = 50.6$ cm/s.

44E

(a) If $x = x_m/2$, then by Eqs.16 and 19,

$$U = \tfrac{1}{2}kx^2 = \tfrac{1}{2}k(x_m/2)^2 = \tfrac{1}{8}kx_m^2 = \tfrac{1}{4}(\tfrac{1}{2}kx_m^2) = \tfrac{1}{4}E.$$

Since $E = U + K$, we have $K = E - \tfrac{1}{4}E = \tfrac{3}{4}E$.

(b) If $U = E/2$, then again by Eqs.16 and 19,

$$\tfrac{1}{2}kx^2 = \tfrac{1}{2}(\tfrac{1}{2}kx_m^2) \rightarrow x = x_m/\sqrt{2}.$$

50P

(a) Momentum conservation gives

$$mv = (m + M)V \rightarrow V = \frac{m}{m + M} v.$$

After the bullet has come to rest, energy conservation requires that

$$\tfrac{1}{2}(m + M)V^2 + \tfrac{1}{2}kx^2 = \tfrac{1}{2}k(x^* - x)^2 + (m + M)gx^*.$$

The equilibrium position of the block before the bullet was fired is taken as the zero level of gravitational potential energy; x is the distance the block stretches the spring before the bullet impacts. The difference in equilibrium positions will be accounted for later. The initial equilibrium position is given from

$$Mg = kx.$$

Substituting this and the result for V into the energy equation above gives

$$kx^{*2} + 2mgx^* - \frac{m^2v^2}{m + M} = 0,$$

$$500x^{*2} + 0.98x^* - 13.89 = 0 \rightarrow x^* = 0.166 \text{ m} = 16.6 \text{ cm}.$$

The bullet+block will oscillate about the new equilibrium position x_e found from

$$(m + M)g = kx_e.$$

Hence, the new equilibrium position is lower than the initial equilibrium position by

$$\Delta x = x_e - x = \frac{mg}{k} = \frac{(0.05)(9.8)}{500} = 9.8 \times 10^{-4} \text{ m} = 0.098 \text{ cm}.$$

Therefore, the amplitude of oscillation will be 16.6 + 0.1 = 16.7 cm = x_m.

(b) The desired fraction is

$$\tfrac{1}{2}kx_m^2 / \tfrac{1}{2}mv^2 = \tfrac{1}{2}(500) \ (0.167)^2 / \tfrac{1}{2}(0.05)(150)^2 = 1.24 \times 10^{-2}.$$

51P

The translational kinetic energy is

$$K_t = \frac{1}{2}Mv^2.$$

Since the rotational inertia of a cylinder about its axis is $MR^2/2$, and the cylinder is rolling without slipping so that $v = R\omega$, the kinetic energy of rotation is

$$K_r = \frac{1}{2}I\omega^2 = \frac{1}{2}(\frac{1}{2}MR^2)(\frac{v}{R})^2 = \frac{1}{4}Mv^2.$$

Finally, the potential energy is $kx^2/2$. Hence, the total energy E is

$$E = \frac{3}{4}Mv^2 + \frac{1}{2}kx^2.$$

Evaluate E at the moment of release; since the cylinder was released from rest, E in SI units is

$$E = \frac{1}{2}kx_i^2 = \frac{1}{2}(3)(0.25)^2 = \frac{3}{32}.$$

At equilibrium $x = 0$ and $v = v_{max}$; therefore, energy conservation yields

$$E = \frac{3}{4}Mv_{max}^2 = \frac{3}{32} \rightarrow Mv_{max}^2 = \frac{1}{8}.$$

Hence, (a) $K_t = 1/16$ J, and (b) $K_r = 1/32$ J.
(c) As before,

$$E = \frac{3}{4}Mv^2 + \frac{1}{2}kx^2.$$

Differentiate with respect to time t; note that $dE/dt = 0$ and then factor out $2dx/dt = 2v$ to get

$$\frac{d^2x}{dt^2} + (\frac{2k}{3M})x = 0.$$

But $d^2x/dt^2 = a = $ acceleration, so we have

$$F = Ma = -(\frac{2k}{3})x.$$

This meets the requirement for SHM; see Eq.8. Hence we have

$$\frac{2k}{3} = M\omega^2,$$

$$T = \frac{2\pi}{\omega} = 2\pi\sqrt{[\frac{3M}{2k}]}.$$

54P

(a) Since the period is T = 0.5 s, the frequency is ν = 1/T = 2 Hz; combining this with the given amplitude of π rad gives

$$\theta = \pi\cos(4\pi t + \phi),$$

for the angular position as a function of time; ϕ is undetermined since we are not told the position of the wheel at t = 0; also sine could be used instead of cosine. The angular velocity is

$$\frac{d\theta}{dt} = -4\pi^2\sin(4\pi t + \phi).$$

(Note that $d\theta/dt \neq \omega$.) Since the sine varies between +1 and -1, the maximum value of the angular velocity is $4\pi^2$ = 39.5 rad/s.

(b) When the displacement is $\theta = \pi/2$, then $\cos(4\pi t + \phi)$ = 0.5, by the first equation. Hence $\sin(4\pi t + \phi) = \pm\sqrt{3}/2$, so that $d\theta/dt = \pm 2\pi^2\sqrt{3} = \pm34.2$ rad/s at this instant. In either case, the angular speed is 34.2 rad/s.

(c) When $\theta = \pi/4$, $\cos(4\pi t + \phi)$ = 0.25. But the angular acceleration is

$$\frac{d^2\theta}{dt^2} = -16\pi^3\cos(4\pi t + \phi),$$

giving, for the position $\theta = \pi/4$, an angular acceleration of $-4\pi^3$ = -124 rad/s^2.

58E

Reaarange Eq.25 to solve for the length L:

$$L = gT^2/4\pi^2 = (9.8 \text{ m/s}^2)(2 \text{ s})^2/4\pi^2 = 0.993 \text{ m} = 99.3 \text{ cm}.$$

63E

This is a physical pendulum to which Eq.27 and Fig.10 apply. The distance d corresponds to the distance h in Fig.10. For the rotational inertia I about the pivot, use the parallel-axis theorem and find the rotational inertia about the center of mass from Fig. 2 in Chapter 11. That is,

$$I = I_{cm} + md^2 = mL^2/12 + md^2.$$

Therefore, by Eq.27,

$$T = 2\pi\sqrt{[\frac{mL^2/12 + md^2}{mgd}]} = 2\pi\sqrt{[\frac{L^2 + 12d^2}{12gd}]}.$$

67E

(a) The rotational inertia of the pendulum is the sum of the rotational inertias of the disk and rod. For the disk, apply the parallel-axis theorem and refer to Fig.2 in Chapter 11; we get

$$I_{disk} = \frac{1}{2}(500 \text{ g})(10 \text{ cm})^2 + (500 \text{ g})(60 \text{ cm})^2 = 1.825 \times 10^6 \text{ g} \cdot \text{cm}^2.$$

For the rod, Fig.2 in Chapter 11 gives directly

$$I_{rod} = \frac{1}{3}(270 \text{ g})(50 \text{ cm})^2 = 2.25 \times 10^5 \text{ g} \cdot \text{cm}^2.$$

Hence, the rotational inertia of the pendulum is 2.05×10^6 g·cm^2.

(b) Apply Eq.4 of Chapter 9 to the rod, considered localized at its center of mass, and to the disk, also considered as localized at its center of mass. Measuring distances from the pivot, we have for the location of the center of mass of the pendulum,

$$h = \frac{(500 \text{ g})(60 \text{ cm}) + (270 \text{ g})(25 \text{ cm})}{500 \text{ g} + 270 \text{ g}} = 47.7 \text{ cm}.$$

(c) Use Eq.27 with the results of (a) and (b):

$$T = 2\pi\sqrt{[\frac{I}{mgh}]} = 2\pi\sqrt{[\frac{2.05 \times 10^6 \text{ g} \cdot \text{cm}^2}{(770 \text{ g})(980 \text{ cm/s}^2)(47.7 \text{ cm})}]} = 1.50 \text{ s}.$$

73P

(a) Apply Eq.25; using SI units, we have

$$\nu = \frac{1}{T} = \frac{1}{2\pi}\sqrt{[\frac{g}{L}]} = \frac{1}{2\pi}\sqrt{[\frac{9.8}{2}]} = 0.352 \text{ s}^{-1}.$$

(b) See Sample Problem 10 in Chapter 5. The effect of the acceleration of the elevator is to replace g with g + a, so that now

$$\nu = \frac{1}{2\pi}\sqrt{[\frac{g + a}{L}]} = \frac{1}{2\pi}\sqrt{[\frac{9.8 + 2}{2}]} = 0.387 \text{ s}^{-1}.$$

(c) For free-fall a = -g; (b) above now gives $\nu = 0$ (no tendency to oscillate).

76P

The torque exerted by the spring force F about the pivot P is

$$\tau = F(\frac{L}{2} \cos\theta).$$

If θ is very small, then $\cos\theta \simeq 1$; thus

$$\tau = F(\frac{L}{2}) = (-kx)(\frac{L}{2}).$$

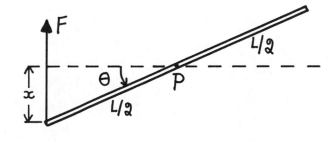

But $x = (L/2)\sin\theta = (L/2)\theta$ for very small θ. Therefore

$$\tau = -k(\frac{L}{2} \theta)(\frac{L}{2}) = -\frac{1}{4}kL^2\theta = -\kappa\theta,$$

where $\kappa = kL^2/4$. This equation agrees in form with Eq.20, for which the corresponding period is given in Eq.21. From Fig.2, Chapter 11 we find $I = mL^2/12$ so that

$$T = 2\pi\sqrt{[\frac{I}{\kappa}]} = 2\pi\sqrt{[\frac{mL^2/12}{kL^2/4}]} = 2\pi\sqrt{[\frac{m}{3k}]}.$$

81E

Set $t = 20T = 20(0.34 \text{ s}) = 6.8 \text{ s}$, using the result for T from Sample Problem 8. The quantity $bt/2m$ in Eq.31 will have the value

$$\frac{bt}{2m} = \frac{(70 \text{ g/s})(6.8 \text{ s})}{2(250 \text{ g})} = 0.952.$$

Hence the desired ratio is

$$(x_m e^{-bt/2m})/x_m = e^{-0.952} = 0.386.$$

82E

(a) Use the result of Sample Problem 8(a). We have $b/2m = (0.230 \text{ kg/s})/2(1.5 \text{ kg}) = 0.07667 \text{ s}^{-1}$. We want to find t such that

$$x_m e^{-bt/2m} = \frac{1}{3} x_m,$$

$$e^{-0.07667t} = \frac{1}{3} \rightarrow t = 14.3 \text{ s}.$$

Note that the numerical value of x_m does not enter, since the amplitude ratio was given.

(b) Assume that the frequency of oscillation equals the undamped frequency, so that the period of oscillation is

$$T = \frac{2\pi}{\omega} = 2\pi\sqrt{[\frac{m}{k}]} = 2\pi\sqrt{[\frac{1.5}{8}]} = 2.721 \text{ s}.$$

Therefore, the number of oscillations completed in 14.3 s is $(14.3 \text{ s})/(2.721 \text{ s/osc}) = 5.26$.

CHAPTER 15

2E

Rearrange Eq.1 to solve for the separation r:

$$r = \sqrt{\left[\frac{Gm_1m_2}{F}\right]} = \sqrt{\left[\frac{(6.67 \times 10^{-11} \text{ N} \cdot \text{m}^2/\text{kg}^2)(5.2 \text{ kg})(2.4 \text{ kg})}{2.3 \times 10^{-12} \text{ N}}\right]} = 19.0 \text{ m}.$$

5P

Call the mass of the spaceprobe m. Apply Eq.1 to the force between space probe and earth, with x the distance between them, and to the force between space probe and sun, with $d - x$ their separation, where d = earth-sun distance. Setting the forces equal gives

$$\frac{GM_{earth}m}{x^2} = \frac{GM_{sun}m}{(d-x)^2}.$$

Taking the square root gives

$$\frac{\sqrt{M_{earth}}}{x} = \frac{\sqrt{M_{sun}}}{d-x},$$

$$x = \frac{\sqrt{M_{earth}}}{\sqrt{M_{earth}} + \sqrt{M_{sun}}}(d) = \frac{\sqrt{(5.98 \times 10^{24} \text{ kg})}}{\sqrt{(5.98 \times 10^{24} \text{ kg})} + \sqrt{(1.99 \times 10^{30} \text{ kg})}}(1.50 \times 10^{11} \text{ m}),$$

$$x = 2.60 \times 10^8 \text{ m} = 2.60 \times 10^5 \text{ km}.$$

8E

The satellite is a thin spherical shell; the gravitational force it exerts on the meteor is given by Eq.1, where r is the distance between the meteor and the center of the satellite at closest approach. This distance is r = 15 + 3 = 18 m. Hence, in SI units,

$$F = Gm_1m_2/r^2 = (6.67 \times 10^{-11})(20)(7)/(18)^2 = 2.88 \times 10^{-11} \text{ N}.$$

15P

The gravitational force on m exerted by the hollowed sphere plus the force on m that was exerted by the portion that was subsequently removed must equal the force on m exerted by a uniform sphere, mass M, before hollowing (principle of superposition). Hence, the difference between the last two forces gives the desired attraction of the hollowed sphere. The force F* exerted on m by M before the hollowing is

$$F* = \frac{GMm}{d^2}.$$

The mass M_h of the portion hollowed out is given from

$$M_h/M = V_h/V = \frac{4\pi}{3}(\frac{R}{2})^3 / \frac{4\pi}{3}R^3 \rightarrow M_h = M/8.$$

The force produced by the portion subsequently removed is

$$F_h = \frac{GmM_h}{(d - R/2)^2} = \frac{GmM_h/d^2}{(1 - R/2d)^2} = \frac{1}{8}\frac{GmM/d^2}{(1 - R/2d)^2}.$$

Thus, the force exerted by the remaining portion of the original sphere is

$$F = F^* - F_h = \frac{GMm}{d^2}[1 - \frac{1}{8(1 - R/2d)^2}].$$

<u>16P</u>

(a) The component of the gravitational force F on the object along the chord is $F\cos\theta = F(x/R)$, towards $x = 0$; r is the distance of the object from the center of the earth. If M* is the mass of that part of the earth that lies interior to a sphere of radius r, and ρ is the assumed constant density of the earth then

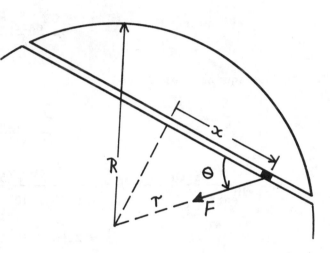

$$F = GM^*m/r^2 = G(\frac{4}{3}\pi r^3 \rho)m/r^2 = \frac{4}{3}\pi G\rho mr.$$

Therefore,

$$F\cos\theta = (\frac{4}{3}\pi G\rho m)x = -m\frac{d^2x}{dt^2},$$

$$a = -(\frac{4}{3}\pi G\rho)x = -\omega^2 x = -(4\pi^2 \nu^2)x,$$

with a minus sign inserted since the force is always directed back toward $x = 0$. This is an equation describing SHM (see Eq.7 in Chapter 14) with a frequency ν given by

$$\nu = \sqrt{(\frac{G\rho}{3\pi})}.$$

(b) By comparison with the case of a chute dug along a diameter of the earth, the frequencies are seen to be equal (see Sample Problem 3). Hence, the periods $T = 1/\nu = 84.2$ min are the same also.

(c) In simple harmonic motion, $v_{max} = \omega x_m = 2\pi\nu x_m = 2\pi x_m/T$. In the present case the amplitude $x_m = $ (chord length)$/2 < R$, the latter being the amplitude for a chute dug along a diameter. Since the periods T are the same along chord and diameter, v_m will be smaller along the chord.

<u>18E</u>

From Fig.11 we obtain $g = 9.780$ m/s^2 at the equator and $g = 9.835$ m/s^2 at the pole. Use Eq.25 of Chapter 11 for the pendulum at the equator to find the length L of the pendulum:

$$L = gT^2/4\pi^2 = (9.780 \text{ m/s}^2)(1 \text{ s})^2/4\pi^2 = 0.2477 \text{ m}.$$

Now apply the same equation to get the period at the pole:

$$T = 2\pi\sqrt{\left[\frac{L}{g}\right]} = 2\pi\sqrt{\left[\frac{0.2477 \text{ m}}{9.835 \text{ m/s}^2}\right]} = 0.997 \text{ s.}$$

In reality, the length of the pendulum does not affect the final result; to see this, take the ratio of Eq.25 of Chapter 11 applied at the equator and at the pole.

20E

Use Eq.21:

$$r = \sqrt{\left[\frac{GM}{g_0}\right]} = \sqrt{\left[\frac{(6.67 \times 10^{-11} \text{ m}^3/\text{kg}\cdot\text{s}^2)(5.98 \times 10^{24} \text{ kg})}{4.9 \text{ m/s}^2}\right]} = 9.02 \times 10^6 \text{ m.}$$

This is the distance from the center of the earth; the altitude of this point is 9020 km − 6370 km = 2650 km.

33P

(a) Let the suspended object have mass m, the spring a force constant k. The forces on m in suspension are its true weight mg directed down and the spring force kx directed up. This latter force is equal in magnitude to the force exerted by m on the spring and is equal to the scale reading. The mass m is moving on the equator (a circle) with linear speed V relative to space; hence, by Newton's second law (see Eq.19 in Chapter 6),

$$mg - kx = \frac{mV^2}{R};$$

R is the radius of the earth. The speed V is made up of the speed $R\omega$ due to the earth's rotation and the speed v of the ship relative to the earth:

$$V = R\omega \pm v.$$

The + sign holds if the ship is sailing in the direction of the earth's rotation, i.e., to the east, and the − sign if it is sailing in the opposite direction. With this, the scale reading becomes

$$kx = mg - \frac{m(R\omega \pm v)^2}{R} = mg - mR\omega^2 \pm 2m\omega v - \frac{mv^2}{R},$$

the + sign now for a ship sailing west. Now, v << $R\omega$, i.e., the speed of the ship relative to the earth is very much smaller than the speed due to the earth's rotation (about 1000 mph at the equator). This means that the last term in the equation above is numerically very much smaller than the others and can be dropped with little error. When the ship is at rest (dead in the water), the scale reading W_0 is given by

$$W_0 = mg - mR\omega^2,$$

so that

$$W = W_0 \pm 2m\omega v = W_0\left(1 \pm \frac{2m\omega v}{W_0}\right),$$

In the last term on the right in the last equation, apply the binomial theorem (see Appendix G) to get

$$\frac{2m\omega v}{W_0} = \frac{2m\omega v}{mg - mR\omega^2} = \frac{2\omega v}{g}(1 + \frac{R\omega^2}{g} + \ldots).$$

$R\omega^2$ is the acceleration imparted by the rotation of the earth and is much less than g, the acceleration due to gravity. Thus, to a good approximation,

$$W = W_0(1 \pm 2\omega v/g).$$

(b) Since the term mV^2/R was transposed during the analysis, in the final equation above the − sign applies when sailing east, and + sign when sailing west along the equator.

34E

(a) Consider the force components parallel to the line joining the particles; these two components are oppositely directed, so that the acceleration they impart is

$$g_p = G[\frac{m_1}{r_1^2}(\cos\theta_1) - \frac{m_2}{r_2^2}(\cos\theta_2)].$$

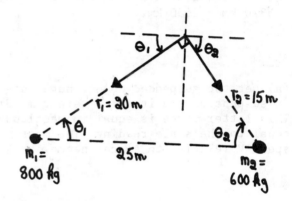

Noting that the triangle is a right triangle (this is not stated explicitly in the Exercise, but left for the reader to notice) we have, in SI units,

$$g_p = G[\frac{800}{20^2}(\frac{20}{25}) - \frac{600}{15^2}(\frac{15}{25})] = 0.$$

The perpendicular components are in the same direction and add, giving in SI units,

$$g = G[\frac{m_1}{r_1^2}(\sin\theta_1) + \frac{m_2}{r_2^2}(\sin\theta_2)],$$

$$g = (6.67 \times 10^{-11})[\frac{800}{20^2}(\frac{15}{25}) + \frac{600}{15^2}(\frac{20}{25})] = 2.22 \times 10^{-10} \text{ N.}$$

(b) The potential energies are scalars, so there are no "components":

$$U = -\frac{Gm_1m_3}{r_1} - \frac{Gm_2m_3}{r_2}.$$

Using $m_3 = 1$ kg, as directed, we get

$$U = -(6.67 \times 10^{-11})[\frac{800}{20} + \frac{600}{15}] = -5.34 \times 10^{-9} \text{ J/kg.}$$

40E

The mass M of the Milky Way, or at least that portion of it that lies inside the sun's orbit, is given from Eq.33:

$$M = \frac{4\pi^2 r^3}{GT^2}.$$

Since $T = (2.5 \times 10^8 \text{ y})(3.156 \times 10^7 \text{ s/y}) = 7.89 \times 10^{15}$ s, we have in SI units,

$$M = \frac{4\pi^2 (2.2 \times 10^{20})^3}{(6.67 \times 10^{-11})(7.89 \times 10^{15})^2} = 1.012 \times 10^{41} \text{ kg.}$$

Assuming that the stars in the Milky Way have an average mass equal to the mass of the sun, then the number # of stars * must be

$$\# = \frac{1.012 \times 10^{41} \text{ kg}}{2 \times 10^{30} \text{ kg/*}} = 5.06 \times 10^{10} \text{ *.}$$

42P

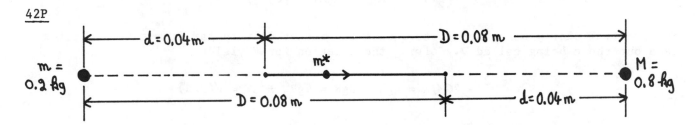

(a) We work entirely in SI units, since only the value of G in SI units is provided in Eq.2. Refer to the sketch above for the notation. The force F due to M and force f due to m on m* are

$$F = \frac{GMm*}{D^2}; \quad f = \frac{Gmm*}{d^2};$$

these are oppositely directed. But D = 2d, so that $D^2 = 4d^2$; also M = 4m; hence, F = f and the resultant vector sum of F and f is zero, regardless of the value of m*.

(b) The potential energies add as scalars (no components):

$$U = -\frac{Gmm*}{d} - \frac{GMm*}{D} = -Gm*\left(\frac{m}{d} + \frac{M}{D}\right).$$

For m* = 1 kg we get

$$U = -(6.67 \times 10^{-11})(1)\left[\frac{0.2}{0.04} + \frac{0.8}{0.08}\right] = -1.00 \times 10^{-9} \text{ J,}$$

or −1.00 nJ/kg = −1.00 pJ/g.

(c) In the final position the distances d and D are interchanged, so that now

$$U = -Gm*\left[\frac{m}{D} + \frac{M}{d}\right] = -(6.67 \times 10^{-11})(10^{-3})\left[\frac{0.2}{0.08} + \frac{0.8}{0.04}\right] = -1.50 \times 10^{-12} \text{ J} = -1.50 \text{ pJ.}$$

The work required to move m* (very slowly so as to impart no kinetic energy) is the difference in potential energies at the two positions; using the result from (b) we have

$W = U_f - U_i = -1.50 \text{ pJ} - (-1.00 \text{ pJ}) = -0.50 \text{ pJ}.$

43P

The potential energies simply add as scalars:

$$U = -GM_e m/R - GM_m m/r.$$

44P

(a) The escape speed at the earth's surface, by Eq.29, is $v_e = \sqrt{(2GM/R)}$. Using $g = GM/R^2$, this becomes $v_e = \sqrt{(2gR)}$. Since the initial speed of the rocket is $v = 2\sqrt{(gR)} > v_e$, the rocket escapes.

(b) By conservation of energy,

$$K_i + U_i = K_f + U_f,$$

$$\tfrac{1}{2}mv^2 - GMm/R = \tfrac{1}{2}mV^2 - 0,$$

the potential energy being nearly zero when the rocket is very far from the earth, its speed out there being called V. Solving the equation for V yields

$$v^2 = v^2 - 2GM/R = 4gR - 2gR = 2gR \;\to\; V = \sqrt{(2gR)}.$$

45P

(a) Let M, R be the mass and radius of the planet. The speed of escape, from Eq.29, is $v_e = \sqrt{(2GM/R)}$. Since, by Eq.21, $g = GM/R^2$, the escape speed in terms of the surface gravity is

$$v_e = \sqrt{[2gR]} = \sqrt{[2(3 \text{ m/s}^2)(5 \times 10^5 \text{ m})]} = 1732 \text{ m/s}.$$

(b) Let h be the height reached above the surface of the planet. The potential energy is $U = -GMm/r$, where r = distance to the center of the planet and m = mass of the particle. From (a),

$$M = gR^2/G,$$

so that

$$U = -\frac{G(gR^2/G)m}{R + h} = -mg\,\frac{R^2}{R + h}.$$

Hence, if v = launch speed from h = 0 and V = the speed at height h, conservation of energy requires that

$$\tfrac{1}{2}mv^2 - mgR = \tfrac{1}{2}mV^2 - mg\,\frac{R^2}{R + h},$$

$$v^2 - v^2 = 2g(R - \frac{R^2}{R + h}) = 2gR(\frac{h}{R + h}).$$

At the top of the path V = 0 and therefore

$$v^2 = 2gR \frac{h}{R + h}.$$

Substituting the numbers gives h = 250 km.

(c) Apart from the numbers, the motion in (c) is just the reverse of the motion in (b), so that the last equation in (b) above still holds. Hence, with h = 1000 km = 2R,

$$v^2 = \frac{4}{3}gR = \frac{4}{3}(3 \text{ m/s}^2)(5 \text{ X } 10^5 \text{ m}) \rightarrow v = 1414 \text{ m/s}.$$

54E

Rearrange Eq.33 to solve for the central mass M:

$$M = \frac{4\pi^2 r^3}{GT^2} = \frac{4\pi^2(9.4 \text{ X } 10^6 \text{ m})^3}{(6.67 \text{ X } 10^{-11} \text{ m}^3/\text{kg}\cdot\text{s}^2)(27540 \text{ s})^2} = 6.48 \text{ X } 10^{23} \text{ kg}.$$

58E

(a) The necessary orbital speed follows from Eq.39. Note that r = R + h = 6370 + 160 = 6530 km. Working in SI units, we have

$$v = \sqrt{[\frac{GM}{r}]} = \sqrt{[\frac{(6.67 \text{ X } 10^{-11})(5.98 \text{ X } 10^{24})}{6.53 \text{ X } 10^6}]} = 7.82 \text{ X } 10^3 \text{ m/s} = 7.82 \text{ km/s}.$$

(b) Eq.33 gives the period T:

$$T^2 = \frac{4\pi^2 r^3}{GM} = \frac{4\pi^2(6.53 \text{ X } 10^6)^3}{(6.67 \text{ X } 10^{-11})(5.98 \text{ X } 10^{24})} \rightarrow T = 5250 \text{ s} = 87.5 \text{ min}.$$

60E

(a) The perigee and apogee distances from the center of the earth are R_p = 180 + 6370 = 6550 km and R_a = 360 + 6370 = 6730 km. By Fig.19,

$$a = \frac{1}{2}(R_a + R_p) = \frac{1}{2}(6730 \text{ km} + 6550 \text{ km}) = 6640 \text{ km}.$$

(b) By Eq.36,

$$e = 1 - R_p/a = 1 - (6550 \text{ km})/(6640 \text{ km}) = 0.0136.$$

67P

The object spends most of its journey under the gravitational influence of the sun. Only near the earth does the earth's field dominate. An approximate answer can be obtained by neglecting the sun's field when the object is very close to the earth, and neglecting the earth's field when the object is very far from the earth ("patched-conic approximation"). The transition occurs at about 260,000 km from the earth (see Problem 5), this being taken as the radius of the earth's "sphere of influence". The object must leave the earth's sphere of influence with a speed u, relative to the sun, sufficient to project it onto a parabolic escape trajectory. This parabolic escape speed is u = $\sqrt{2}v_0$, where v_0 is the earth's orbital speed; see Exercise 39. Thus, as the object leaves the earth's sphere of influence, its speed V relative to the sun will be

$$V = u - v_0 = (\sqrt{2} - 1)v_0.$$

Apply conservation of energy within the earth's sphere of influence so that, if v_{esc} is the speed sought,

$$\frac{1}{2}mv_{esc}^2 - \frac{GMm}{R} = \frac{1}{2}mV^2 - 0,$$

ignoring the potential energy very far from the earth; R is the radius of the earth. Substituting for V gives

$$v_{esc}^2 = \frac{2GM}{R} + [(\sqrt{2} - 1)v_0]^2.$$

Now, $2GM/R$ = (speed of escape from earth)^2 = (11.2 km/s)^2, by Sample Problem 6. From Appendix C, the earth's orbital speed is 29.8 km/s = v_0. Hence,

$$v_{esc}^2 = (11.2)^2 + [(\sqrt{2} - 1)29.8]^2 \rightarrow v_{esc} = 16.7 \text{ km/s}.$$

<u>68P</u>

The motion of the masses is about the center of mass (cm) of the system, and therefore the center of mass must be located. It lies at the intersection of the symmetrical dividing lines; see Fig.4 in Chapter 9. (Only two of these lines are needed). The lengths of the lines is $L\sin60° = L\sqrt{3}/2$. Hence, the center of mass is at a distance x from each side, as shown, where x is found from Eq.4 in Chapter 9:

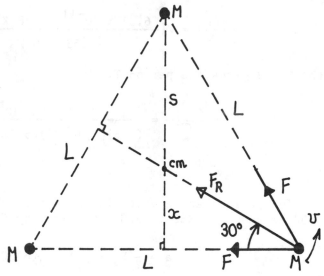

$$(3M)x = M(L\sqrt{3}/2) + 2M(0) \rightarrow x = L\sqrt{3}/6.$$

The distance s of any of the three masses from the center of mass is given by

$$s = L\sqrt{3}/2 - x = L\sqrt{3}/3.$$

Now s is the radius of the circular orbit of each mass; for uniform circular motion,

$$F_R = Mv^2/s,$$

F_R the resultant force on each mass. But,

$$F_R = 2F\cos30° = F\sqrt{3},$$

F the gravitational force between any pair of masses. By the law of gravitation,

$$F = GM^2/L^2.$$

Putting the last four equations together yields

$$\sqrt{3}GM^2/L^2 = Mv^2/(L\sqrt{3}/3) \;\rightarrow\; v = \sqrt{(GM/L)}.$$

71E

(a) For the asteroid and for the earth we have

$$T_a^2 = (\frac{4\pi^2}{GM})r_a^3; \quad T_e^2 = (\frac{4\pi^2}{GM})r_e^3.$$

Dividing the equations (to eliminate the common factor $4\pi^2/GM$) gives

$$T_a = T_e(r_a/r_e)^{3/2} = (1 \; year)(2)^{3/2} = 2.83 \; y.$$

(b) Eq.39 gives the kinetic energies:

$$K_a = \frac{GMm_a}{2r_a}; \quad K_e = \frac{GMm_e}{2r_e}.$$

Dividing to obtain the desired ratio yields

$$K_a/K_e = (m_a/m_e)(r_e/r_a) = (2 \times 10^{-4})(1/2) = 10^{-4}.$$

74E

Applying the quoted relations leads to

$$K + U = E,$$

$$\frac{1}{2}mv^2 - \frac{GMm}{r} = -\frac{GMm}{2a},$$

$$v^2 = GM(\frac{2}{r} - \frac{1}{a}).$$

77P

Let r be the radius of the orbit, R the radius of the earth; M, m are the masses of the earth and of the satellite. The minimum energy K needed to lift the satellite from the earth's surface to a distance r from the earth's center, but giving it no kinetic energy, is given from energy considerations:

$$K_i + U_i = K_f + U_f,$$

$$K - GMm/R = 0 - GMm/r,$$

$$K = GMm(\frac{1}{R} - \frac{1}{r}).$$

The kinetic energy K* required to put the satellite into a circular orbit once the object is at the desired height is $mv^2/2$, with

$$K^* = \frac{GMm}{2r},$$

by Eq.39.

(a) With r = 1.25R, the equations above give K = 0.2GMm/R, K* = 0.4GMm/R, so that the answer is NO.

(b) For r = 1.5R, K = K* = GMm/3R, and the energies are equal.

(c) Finally, with r = 1.75R, K = 3GMm/7R, K* = 2GMm/7R; evidently in this case the answer is YES.

81P

(a) Use the notation R, h, M, m for the radius of the earth, altitude of the circular orbit above the surface of the earth (so that the radius of the orbit is r = R + h), mass of the earth and mass of the satellite. For circular orbits, from Eq.39,

$$v = \sqrt{\left(\frac{GM}{R + h}\right)} = 7.54 \text{ km/s},$$

using M = 5.98 X 10^{24} kg and R = 6.37 X 10^6 m.

(b) The period T of revolution is found from

$$T = \frac{2\pi(R + h)}{v} = \frac{2\pi(7.01 \times 10^3)}{7.54} = 5.84 \times 10^3 \text{ s} = 97.4 \text{ min.}$$

(c) The total mechanical energy, by Eq.40, is

$$E = -\frac{GMm}{2(R + h)}.$$

In the initial orbit, this energy numerically is, using SI units,

$$E_i = -\frac{(6.67 \times 10^{-11})(5.98 \times 10^{24})(220)}{2[(6370 + 640)(1000)]} = -6.259 \times 10^9 \text{ J.}$$

After 1500 revolutions, the mechanical energy is

$$E_f = E_i - (1500)(1.4 \times 10^5) = -6.469 \times 10^9 \text{ J.}$$

The new altitude above the earth's surface can therefore be found from Eq.40:

$$E_f = -6.469 \times 10^9 = -\frac{GMm}{2(R + h^*)}.$$

Substituting numerical data on the right gives h* = 412 km. The equations in (a) and (b) now yield v* = 7.67 km/s and T* = 92.6 min.

(d) Let F be the force. The rate of energy loss (power) equals the product of torque and angular speed (see Eq.36 in Chapter 11); the angular speed is $2\pi/T$ so that

$$P = \tau\omega,$$

$$\frac{dE}{dt} = F(R + h)\left(\frac{2\pi}{T}\right).$$

By Eq.33 the period is

$$T = 2\pi \frac{(R + h)^{3/2}}{\sqrt{(GM)}},$$

so that

$$F = \sqrt{\left(\frac{R + h}{GM}\right)}\left(\frac{dE}{dt}\right).$$

For the highest (initial) orbit,

$$\frac{dE}{dt} = \frac{1.4 \times 10^5 \text{ J/rev}}{5.84 \times 10^3 \text{ s/rev}} = 24 \text{ J/s},$$

in absolute value. The previous equation now yields for the force $F = 3.18 \times 10^{-3}$ N.

(e) The resistive force exerts a torque on the satellite and its orbital angular momentum diminishes. If all influences originating outside of the earth-satellite system are ignored, the system is isolated and its total angular momentum must remain constant. This implies that the spin angular momentum of the earth must increase.

84P

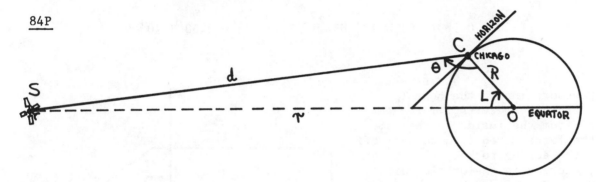

From Sample Problem 9, the radius r of the synchronous orbit is r = 42,200 km. The triangle COS has no right angle, so use the cosine law (see Appendix G) to find d, the straight-line distance between the satellite and the city of Chicago:

$$d^2 = r^2 + R^2 - 2rR\cos L,$$

$$d^2 = (42,200)^2 + (6370)^2 - 2(42,200)(6370)\cos 47.5° \rightarrow d = 38,190 \text{ km.}$$

Now use the law of sines (also in Appendix G):

$$\frac{d}{\sin L} = \frac{r}{\sin\theta},$$

$$\frac{38,190}{\sin 47.5°} = \frac{42,200}{\sin\theta} \rightarrow \theta = 54.6°; \ 125.4°.$$

The angle 54.6° cannot apply, since it would require aiming the antenna below the horizon. Thus, the antenna should be pointed at an angle 125.4° - 90° = 35.4° above the horizon, toward the south (i.e., toward the equatorial plane of the satellite's orbit).

CHAPTER 16

3E

Apply Eq.2. Working in SI units we have

$$p = \frac{\Delta F}{\Delta A} = \frac{42\ N}{\pi(0.011\ m)^2} = 1.10 \times 10^5\ Pa.$$

5E

The inside air pushes out and the outside air pushes in (i.e., in opposite directions), so that the net force exerted by the air is

$$\Delta F = p_{in}\Delta A - p_{out}\Delta A = (p_{in} - p_{out})\Delta A,$$

and is directed outward since $p_{in} > p_{out}$. The window area is $\Delta A = (3.4)(2.1) = 7.14\ m^2$. Also, 1 atm = 1.01 X 10^5 Pa. Therefore, the net force outward on the window is

$$\Delta F = [(1.0 - 0.96)(1.01 \times 10^5\ N/m^2)][7.14\ m^2] = 2.88 \times 10^4\ N.$$

7P

Let p be the pressure inside the box. The air remaining in the box exerts a force pA against the lid from the inside in the same direction as the force F required to pull off the lid. The force acting to keep the lid on the box is PA, where P is the outside air pressure. If F is the minimum force needed to remove the lid, then

$$pA + F = PA,$$

$$p = P - \frac{F}{A} = 15 - \frac{108}{12} = 6\ lb/in^2.$$

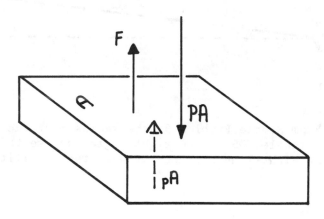

9P

(a) Each horse must exert a force F equal in magnitude to the resultant of the forces due to the air inside and outside the hemisphere to which it is hitched. The vertical forces due to the air sum to zero; it is the horizontal components that must be added. Thus, if θ is the angle made by the forces due to the air with the horizontal, then we have

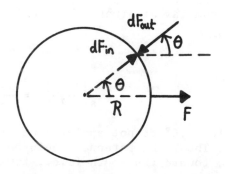

$$F = \int(dF_{out} - dF_{in})\cos\theta = \int(p_{out} - p_{in})(\cos\theta)dA = \int(p\cos\theta)dA = p\int(\cos\theta)dA.$$

In spherical coordinates $dA = 2\pi R^2 \sin\theta d\theta$ (see your calculus text); using this, the force formula becomes

$$F = 2\pi p R^2 \int_0^{\pi/2} \sin\theta\cos\theta d\theta = \pi R^2 p.$$

(b) Since atmospheric pressure is $p_{out} = 14.7$ lb/in^2, we have numerically

$$p = p_{out} - p_{in} = (0.9)(14.7) = 13.23 \text{ lb/in}^2,$$

giving

$$F = \pi(1 \text{ ft})^2 \frac{13.23 \text{ lb}}{(1/12 \text{ ft})^2} = 5985 \text{ lb}.$$

(The result in (a), and therefore the numerical result in (b), also follows by noting that the "projected" area of a sphere is just the area of a circle, its cross-section.)

(c) Two teams of horses look more impressive than one, but one could be used if the other hemisphere is attached to one side of a building.

12E

Use Eq.5; the pressure differential is

$$\Delta p = p - p_0 = \rho g h = (900 \text{ kg/m}^3)(9.8 \text{ m/s}^2)(8.2 \text{ m} - 2.1 \text{ m}) = 5.38 \times 10^4 \text{ Pa}.$$

17E

Air pressure inside the submarine annuls the external atmospheric pressure p_0, so that "only" the gauge pressure $\rho g h$ of the water counts. Combining Eqs.2 and 5, and working in SI units, gives

$$\Delta F = p[\Delta A] = [\rho g h][\Delta A] = [(1025)(9.8)(100)][(1.2)(0.6)] = 7.23 \times 10^5 \text{ N}.$$

19P

The net effect is that gravity takes a slab of fluid $(h_2 - h_1)/2$ in thickness and lets it fall a vertical distance equal to this thickness. Thus, the work W done by gravity is

$$W = mg[\tfrac{1}{2}(h_2 - h_1)].$$

But the mass m of the slab equals the density ρ times the volume of the slab, and this volume is the base area A times the thickness; hence,

$$W = [\tfrac{1}{2}(h_2 - h_1)A\rho]g[\tfrac{1}{2}(h_2 - h_1)] = \tfrac{1}{4}A\rho(h_2 - h_1)^2 g.$$

23P

Suppose that the compensation level is at a depth y shown in the sketch on p.134. The distance x is found by equating the depths calculated from point A vertically to the surface, and from point B vertically to the surface; these distances are equal and hence we get

$$y + x + 12 + h = y + 20,$$

$$x = 8 - h.$$

Expressing all densities in g/cm^3 and all distances in km (these mixed units can be used since the conversion factors would all cancel in the next equation), the pressures at A and B are

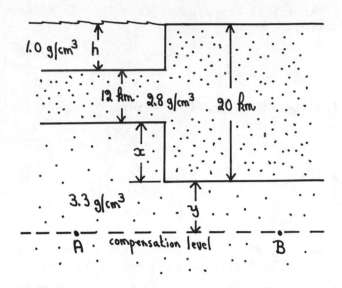

$$p_A = (3.3)g(y + x) + (2.8)g(12) + (1.0)gh,$$

$$p_B = (3.3)gy + (2.8)g(20).$$

Now set $p_A = p_B$ and substitute in the expression for x found above; note that g cancels to give

$$p_A = p_B,$$

$$3.3y + 3.3x + 33.6 + h = 3.3y + 56,$$

$$3.3x + 33.6 + h = 56,$$

$$3.3(8 - h) + 33.6 + h = 56 \rightarrow h = 1.74 \text{ km.}$$

24P

(a) Eq.2 cannot be applied to the entire dam face since the pressure varies with depth. But, the force dF exerted by the water on the very narrow shaded rectangle of the dam face can be computed from Eq.2:

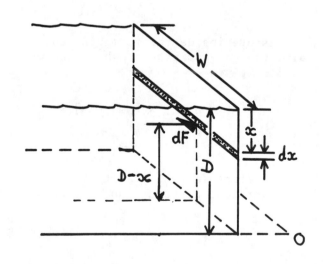

$$dF = pdA = (\rho gx)(Wdx).$$

The force F exerted against the whole dam face is

$$F = \int dF = \rho gW \int_0^D x dx = \frac{1}{2}\rho gWD^2.$$

(b) The torque due to dF is $d\tau = dF(D - x)$, since $D - x$ is the lever arm. Substituting for dF from (a) gives for the total torque

$$\tau = \int_0^D \rho gWx(D - x)dx = \frac{1}{6}\rho gWD^3.$$

(c) The line of action of F from (a) to yield the torque calculated in (b) must be at a distance d above the bottom of the dam, where

$$(\frac{1}{2}\rho gWD^2)d = \frac{1}{6}\rho gWD^3 \rightarrow d = \frac{D}{3}.$$

31E

To find the maximum load m of lead shot, assume that the tin can sinks down in the water so that the water surface is level with the top of the can (can barely floating). The volume of displaced water then equals the volume of the can. Apply Archimedes' principle:

$$F_{buoyant} = W_{total},$$

$$\rho_{water} g V_{can} = (M_{can} + m)g,$$

$$m = \rho_{water} V_{can} - M_{can} = (1\ g/cm^3)(1200\ cm^3) - 130\ g = 1070\ g.$$

34E

(a) The 200 N equals the upward buoyant force F_B, so that

$$F_B = \rho_{water} g V,$$

$$200\ N = (1000\ kg/m^3)(9.8\ m/s^2)V \rightarrow V = 2.04 \times 10^{-2}\ m^3.$$

(b) The weight in air is

$$W = \rho_{iron} g V = (7870\ kg/m^3)(9.8\ m/s^2)(2.04 \times 10^{-2}\ m^3) = 1573\ N.$$

41P

(a) Let R,r be the outer and inner radii of the sphere and ρ, ρ_ℓ the densities of the sphere and of the liquid. The weight W of the sphere equals the buoyant force F_B exerted by the liquid. Since half the sphere is submerged we have, in SI units,

$$F_B = \frac{1}{2}(\frac{4\pi}{3}R^3)\rho_\ell g = (\frac{1}{2})(\frac{4\pi}{3})(0.09)^3(800)(9.8) = 11.97\ N.$$

Hence, the mass of the sphere is

$$M = W/g = F_B/g = (11.97)/(9.8) = 1.22\ kg.$$

(b) The density of the material making up the sphere (not the average density of the sphere) is $\rho = M/V$, where V is the volume of the sphere actually occupied by matter. Therefore,

$$\rho = M/\frac{4\pi}{3}(R^3 - r^3) = 1.22/\frac{4\pi}{3}(0.09^3 - 0.08^3) = 1342\ kg/m^3.$$

43P

Let V be the enclosed volume of the casting, including the volume V_i occupied by iron, and the volume V_c of the empty cavities, so that $V = V_i + V_c$; see the sketch, p.136. The cavities do not contribute to the weight of the casting so that, in SI units (noting that $1\ g/cm^3 = 1000\ kg/m^3$), we have

$$W = m_i g = (\rho_i V_i)g,$$

$$6000 = (7870)V_i(9.8),$$

$$V_i = 0.0778 \text{ m}^3.$$

The buoyant force is 6000 − 4000 = 2000 N. When the casting is submerged, water cannot enter the cavities. Hence, by Archimedes' principle,

$$F_B = (\rho_w V)g,$$

$$2000 = (1000)V(9.8) \rightarrow V = 0.2041 \text{ m}^3.$$

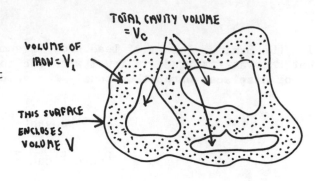

The volume of the cavities must be $V_c = V - V_i = 0.2041 - 0.0778 = 0.126 \text{ m}^3.$

44P

(a) For the minimum area A of ice, let the ice sink the full 0.3 m so that the top surface of the ice is at the water line. The buoyant force must support the weight of the car plus the weight of the ice. The ice weighs $(917 \text{ kg/m}^3)gA(0.3 \text{ m})$, A in m^2; see Table 1 for the needed densities. The car weighs $(1100 \text{ kg})g$. The buoyant force equals the weight of water displaced and so equals $(1024 \text{ kg/m}^3)gA(0.3 \text{ m})$. For equilibrium,

$$(1024)gA(0.3) = (917)gA(0.3) + (1100)g \rightarrow A = 34.3 \text{ m}^2.$$

50P

(a) The forces acting on the log are its weight W and the buoyant force F, which is a function of the displacement x of the log, x equalling the vertical displacement from equilibrium. If a length ℓ of the log is submerged when in equilibrium, then, by Archimedes' principle,

$$W = \rho g A \ell,$$

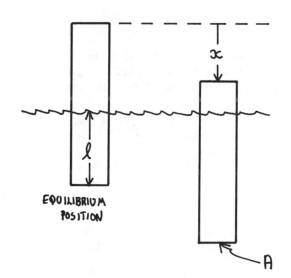

ρ the density of water and A the cross-sectional area of the log. When the log is displaced a distance x, with x positive downwards, Newton's second law gives

$$W - F = ma = (\frac{W}{g})a,$$

$$\rho g A \ell - \rho g A(\ell + x) = (\frac{\rho g A \ell}{g})a,$$

$$a = -(\frac{g}{\ell})x = -\omega^2 x.$$

The last equation describes simple harmonic motion; see Eq.7 in Chapter 14.

(b) The period of oscillation is

$$T = 2\pi/\omega = 2\pi\sqrt{(\frac{2.5}{9.8})} = 3.17 \text{ s},$$

since $\omega = \sqrt{g/\ell}$.

54E

The area A of the hose is

$$A = \pi D^2/4 = \pi(0.75)^2/4 = 0.4418 \text{ in}^2.$$

The total area of the holes is

$$24a = 24[\pi d^2/4] = 24[\pi(0.05)^2/4] = 0.0471 \text{ in}^2.$$

Now apply the equation of continuity, Eq.13:

$$AV = (24a)v,$$

$$(0.4418 \text{ in}^2)(3 \text{ ft/s}) = (0.0471 \text{ in}^2)v \rightarrow v = 28.1 \text{ ft/s}.$$

55P

The work W done by the pump on a mass m of water is

$$W = mgh + \frac{1}{2}mv^2,$$

where v is the ejection speed of the water through the window at a height h above the water line. The power P supplied by the pump is

$$P = \frac{dW}{dt} = \frac{dm}{dt}(gh + \frac{1}{2}v^2).$$

But the mass flow rate, by Eq.9, is

$$\frac{dm}{dt} = (Av)\rho,$$

so that, in SI units,

$$P = Av\rho[gh + \frac{1}{2}v^2] = [\pi(0.01)^2](5)(1000)[(9.8)(3) + \frac{1}{2}(5)^2] = 65.8 \text{ W}.$$

57E

(a) Use the continuity equation, Eq.13:

$$A_1 v_1 = A_2 v_2,$$

$$(4.0 \text{ cm}^2)(5.0 \text{ m/s}) = (8.0 \text{ cm}^2)v_2 \rightarrow v_2 = 2.5 \text{ m/s}.$$

(b) Now apply Bernoulli's equation, Eq.17; put y = 0 at the lower level (so that y is kept positive); use the result from (a) for v_2; substituting all data in SI units (mixed units cannot be used in Bernoulli's equation),

$$P_1 + \frac{1}{2}\rho v_1^2 + \rho g y_1 = P_2 + \frac{1}{2}\rho v_2^2 + \rho g y_2,$$

$$1.5 \times 10^5 + \frac{1}{2}(1000)(5)^2 + (1000)(9.8)(10) = p_2 + \frac{1}{2}(1000)(2.5)^2 + 0,$$

$$p_2 = 2.57 \times 10^5 \text{ Pa.}$$

63E

(a) The speed v at which the water leaves the tank is, by Eq.23,

$$v = \sqrt{[2gD]} = \sqrt{[2(32)(1)]} = 8.0 \text{ ft/s.}$$

The volume flow rate can now be found from Eq.13:

$$R = av = (\frac{1}{144} \text{ ft}^2)(8 \text{ ft/s}) = 0.0555 \text{ ft}^3/s.$$

(b) Let y be vertical distance below the hole; the hole is at $y = 0$. The speed of the water after falling a distance y is the free-fall speed found from Eq.15 in Chapter 2:

$$V^2 = v^2 + 2gy.$$

We must now find V. If A is the cross-sectional area of the stream after falling the distance y, then by the continuity equation, $AV = av$. Setting $A = a/2$ gives $V = 2v$. Putting this result into the free-fall equation yields

$$4v^2 = v^2 + 2gy,$$

$$y = 3v^2/2g = 3(8)^2/2(32) = 3 \text{ ft.}$$

66E

Apply the result of Exercise 65; it is necessary to use a consistent set of units; if we use SI units, then we have

$$\frac{L}{A} = \frac{1}{2}\rho(v_t^2 - v_u^2),$$

$$900 = \frac{1}{2}(1.3)(v_t^2 - 110^2) \rightarrow v_t = 116 \text{ m/s.}$$

72P

The air inside the office is at rest. Let p_{in} be the office air pressure and p_{out} the air pressure outside. Bernoulli's equation requires that

$$p_{in} = p_{out} + \frac{1}{2}\rho v^2;$$

the terms ρgy on each side of the equation cancel since the heights inside and outside the window are equal. The window area is $A = 20 \text{ m}^2$ so that the net force must be

$$F = (p_{in} - p_{out})A = \frac{1}{2}\rho v^2 A = \frac{1}{2}(1.23)(30)^2(20) = 11,070 \text{ N.}$$

75P

The terms ρgh in Bernoulli's equation cancel out since $h_t = h_u$ for all practical purposes so that, with $v_u = 0$ and $v_t = v$ (t = top, u = under),

$$p_t + \frac{1}{2}\rho v^2 = p_u.$$

If A = area of the plate, then

$$F = (p_u - p_t)A = \frac{1}{2}\rho v^2 A,$$

where F is the net upward force exerted by the air on the plate. This must equal the weight W of the plate if it is to be held in position. The plate's weight is W = mg = (0.5)(9.8) = 4.9 N. Hence, with F = W, and working in SI units,

$$\frac{1}{2}\rho v^2 A = W,$$

$$\frac{1}{2}(1.2)v^2(80 \times 10^{-4}) = 4.9 \rightarrow v = 32.0 \text{ m/s}.$$

80P

Apply Bernoulli's equation at points 1 and 2 along the streamline shown:

$$P_1 + \rho g y_1 + \frac{1}{2}\rho v^2 = P_2 + \rho g y_2 + \frac{1}{2}\rho v^2.$$

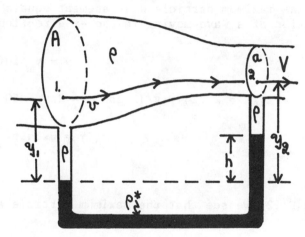

By the continuity equation Av = aV so that, using this to eliminate V gives

$$P_1 + \rho g y_1 + \frac{1}{2}\rho v^2 = P_2 + \rho g y_2 + \frac{1}{2}\rho (A^2/a^2)v^2.$$

Now equate expressions for the pressures at y = 0 in the two sides of the U-tube:

$$P_1 + \rho g y_1 = P_2 + \rho g(y_2 - h) + \rho^* g h,$$

where ρ^* is the density of the liquid in the U-tube. Solve each of the two preceding equations for the quantity $p_1 - p_2 + \rho g(y_1 - y_2)$ and set the results equal; from the resulting expression solve for v to get

$$v = a\sqrt{[\frac{2(\rho^* - \rho)gh}{\rho(A^2 - a^2)}]}.$$

Note that if the central streamline had been selected, then $y_1 = y_2$; in this case, writing $\Delta p = p_1 - p_2$, the difference in pressures at the center of the tube, then the second equation above gives directly

$$v = \sqrt{[\frac{2a^2 \Delta p}{\rho(A^2 - a^2)}]}.$$

CHAPTER 17

3E

(a) The period T is the time required for one oscillation, so that T = (20 s)/(12 osc) = 1.67 s.

(b) The speed is v = (12 m)/(6 s) = 2 m/s.

(c) By Eq.14, λ = vT = (2 m/s)(1.67 s) = 3.34 m.

(d) The amplitude is y_m = 15 cm, the maximum distance the disturbance reaches from the undisturbed position.

15P

(a) The wave speed is given by Eq.14:

$$v = \nu\lambda = (25 \text{ s}^{-1})(24 \text{ cm}) = 600 \text{ cm/s}.$$

(b) The maximum particle displacement equals the amplitude y_m, so that y_m = 0.3 cm. The equation of a wave moving in the −x direction is, from Eq.16,

$$y = y_m \sin(kx + \omega t).$$

But k = $2\pi/\lambda$ = $2\pi/24$ = $\pi/12$ and ω = $2\pi\nu$ = 50π; these lead to

$$y = 0.3\sin(\frac{\pi x}{12} + 50\pi t).$$

16P

From Eq.21 we see that the maximum particle speed is

$$u_{max} = \omega y_m = 2\pi\nu y_m,$$

obtained by putting −cos(kx − ωt) = −(−1) = +1, its maximum value. The wave speed, by Eq.14, is v = $\lambda\nu$. Hence, the desired ratio is

$$u_{max}/v = 2\pi\nu y_m/\lambda\nu = 2\pi y_m/\lambda.$$

Alternatively, we could have used Eq.26 for v, obtaining

$$u_{max}/v = 2\pi\nu y_m \sqrt{\frac{\mu}{\tau}}.$$

For strings of different materials stretched to the same tension τ, this ratio (assuming the same frequency of excitation) is smaller for the lighter materials; i.e., smaller for nylon than for wire.

19E

We do the calculation in SI units. The linear density is $\mu = m/L = (0.06 \text{ kg})/(2 \text{ m}) = 0.03$ kg/m. By Eq.26,

$$v = \sqrt{\left[\frac{\tau}{\mu}\right]} = \sqrt{\left[\frac{500 \text{ N}}{0.03 \text{ kg/m}}\right]} = 129 \text{ m/s}.$$

25E

(a) Comparing the given wave equation with Eq.16, and then using Eq.14 for the wave speed gives

$$v = \frac{\omega}{k} = \left[\frac{30 \text{ s}^{-1}}{2 \text{ m}^{-1}}\right] = 15 \text{ m/s}.$$

(b) By Eq.26,

$$\tau = \mu v^2 = (1.6 \times 10^{-4} \text{ kg/m})(15 \text{ m/s})^2 = 0.036 \text{ N}.$$

31P

The linear density of the wire is $\mu = (0.1 \text{ kg})/(10 \text{ m}) = 0.01$ kg/m. Therefore, by Eq.26,

$$v = \sqrt{(\tau/\mu)} = \sqrt{(250/0.01)} = 158.1 \text{ m/s}.$$

Let the later disturbance be generated at time t = 0. The distance x* travelled by this disturbance in time t is x* = vt. But, in this same time the other disturbance, which was generated earlier, has travelled a distance x = v(t + 0.03). When the disturbances meet, we have x + x* = 10. Hence, at the instant of meeting,

$$v(t + 0.03) + vt = 10,$$

$$2t + 0.03 = \frac{10}{v} = \frac{10}{158.1} \rightarrow t = 0.01662 \text{ s}.$$

The waves, then, meet at a point (158.1 m/s)(0.01662 s) = 2.63 m from that end of the wire from which the later disturbance originated.

33P

(a) If τ is the tension in the rope, we have

$$v(y) = \sqrt{[\tau(y)/\mu]},$$

where $\mu = m/\ell$. Consider a small portion of the rope a distance y from the lower end, the end being at y = 0. The tension in this portion is the weight of that part of the rope hanging beneath it, so that

$$\tau = (\mu y)g.$$

Therefore,

$$v = \sqrt{\left[\frac{\mu g y}{\mu}\right]} = \sqrt{[gy]}.$$

(b) Since v = dy/dt, the result of (a) can be expressed as

$$\frac{dy}{\sqrt{y}} = \sqrt{g}dt.$$

Upon integrating, this becomes

$$2\sqrt{y} = \sqrt{g}t + C,$$

C being the constant of integration. To evaluate C, let the wave start at the bottom (y = 0) at time t = 0, so that

$$2\sqrt{0} = \sqrt{g}(0) + C \rightarrow C = 0.$$

Hence,

$$2\sqrt{y} = t\sqrt{g}.$$

If the wave reaches the top (y = ℓ) in time t*, then the equation above implies that

$$2\sqrt{\ell} = t*\sqrt{g} \rightarrow t* = 2\sqrt{(\frac{\ell}{g})}.$$

36P

(a) In SI units, the given data is: y_m = 0.005 m, ν = 120 Hz, μ = 0.12 kg/m and τ = 90 N. By Eq.21,

$$u_{max} = \omega y_m = 2\pi\nu y_m = 2\pi(120)(0.005) = 3.77 \text{ m/s}.$$

(b) The transverse component of the tension is

$$\tau_{trans} = \tau[\frac{\partial y}{\partial x}] = \tau[ky_m\cos(kx - \omega t)],$$

using Eq.2. Its maximum value is

$$\tau_{trans,max} = \tau k y_m.$$

But k = 2π/λ = 2πν/v = 2πν√(μ/τ), so that

$$\tau_{trans,max} = (\tau)2\pi\nu\sqrt{(\mu/\tau)}y_m = 2\pi\nu y_m\sqrt{[\mu\tau]} = 2\pi(120)(0.005)\sqrt{[(0.12)(90)]} = 12.4 \text{ N}.$$

(c) Both events occur at cos(kx − ωt) = 1, so that y = $y_m$sin(kx − ωt) = 0.

(d) The power (not average power) transferred is, by Eqs.32 and 30,

$$P = \mu v \omega^2 y_m^2 \cos^2(kx - \omega t).$$

For the maximum power, put cos(kx − ωt) = 1. Also, ω = 2πν = 2π(120) = 754.0 rad/s, and v = √(τ/μ) = √(90/0.12) = 27.39 m/s. Hence, in SI units,

$$P_{max} = \mu v \omega^2 y_m^2 = (0.12)(27.39)(754)^2(0.005)^2 = 46.7 \text{ W}.$$

(e) P_{max} occurs when $\cos(kx - \omega t) = 1$, so that, by (c) $y = 0$ here also.

(f) Put $\cos(kx - \omega t) = 0$ in (d) to find $P_{min} = 0$.

(g) If $\cos(kx - \omega t) = 0$, then $\sin(kx - \omega t) = \pm1$. In absolute value, $y = y_m(1) = y_m = 0.005$ m.

37E

By Eq.41, the new amplitude is $[2y_m\cos\frac{1}{2}(90°)] = [2y_m\cos45°] = y_m\sqrt{2}$.

39P

The difference in the lengths of the paths travelled by the direct wave from S and that reflected from height H must equal an integral number of wavelengths for constructive interference:

$$L_H - d = N\lambda.$$

However, waves reflected from the layer when at height H + h are out of phase with the direct wave, so that

$$L_{H+h} - d = N\lambda + \tfrac{1}{2}\lambda;$$

that is, the waves are out of phase by half a wavelength. Subtracting these two equations gives

$$L_{H+h} - L_H = \tfrac{1}{2}\lambda.$$

But the path lengths to the atmospheric layer and back are twice the hypotenuse of right triangles; i.e.,

$$L_H = 2\sqrt{[(\tfrac{1}{2}d)^2 + H^2]} = \sqrt{[d^2 + 4H^2]},$$

$$L_{H+h} = \sqrt{[d^2 + 4(H + h)^2]}.$$

Putting these two results into the previous equation, and then multiplying by two, yields

$$\lambda = 2\sqrt{[d^2 + 4(H + h)^2]} - 2\sqrt{[d^2 + 4H^2]}.$$

41P

Writing $kx + \omega t = \theta$, the equations for the individual waves can be written

$$y_1 = A\sin\theta,$$

$$y_2 = \frac{A}{2}\sin(\theta + \frac{\pi}{2}) = \frac{A}{2}\cos\theta,$$

$$y_3 = \frac{A}{3}\sin(\theta + \pi) = -\frac{A}{3}\sin\theta.$$

The resultant, $y_T = y_1 + y_2 + y_3$ becomes

$$y_T = \frac{2A}{3}\sin\theta + \frac{A}{2}\cos\theta = \frac{2A}{3}(\sin\theta + \frac{3}{4}\cos\theta).$$

If this is expressed in the form

$$y_T = \frac{2A}{3}[(\alpha\sin\beta)\sin\theta + (\alpha\cos\beta)\cos\theta],$$

so that

$$y_T = \frac{2\alpha A}{3}\cos(\theta - \beta),$$

then comparison of the two equations for y_T shows that

$$\alpha\sin\beta = 1,$$
$$\alpha\cos\beta = \frac{3}{4}.$$

Solving these last two equations for α and β (square the equations and add to get α, and divide the equations to get β) gives

$$\alpha = \frac{5}{4}; \quad \beta = \tan^{-1}(\frac{4}{3}) = 0.927 \text{ rad.}$$

Hence,

$$y_T = \frac{5A}{6}\cos(\theta - \beta).$$

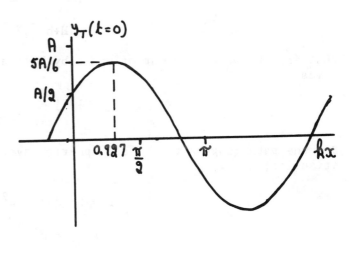

When $t = 0$, $\theta = kx$ so that

$$y_T(t=0) = \frac{5A}{6}\cos(kx - 0.927).$$

At $x = 0$,

$$y_T(x=0,t=0) = \frac{5A}{6}\cos[-\tan^{-1}(\frac{4}{3})] = (\frac{5A}{6})(\frac{3}{5}),$$

$$y_T(x=0,t=0) = 0.5A.$$

A graph of the wave at $t = 0$ is shown on the sketch. As time progresses, the wave moves to the left.

44E

(a) The linear density is $\mu = (0.12 \text{ kg})/(8.4 \text{ m}) = 1.429 \times 10^{-2}$ kg/m. By Eq.26,

$$v = \sqrt{[\tau/\mu]} = \sqrt{[(96 \text{ N})/(1.429 \times 10^{-2} \text{ kg/m})]} = 82.0 \text{ m/s.}$$

(b) For the longest wavelength, select $n = 1$ in Eq.49, so that $\lambda = 2\ell = 2(8.4 \text{ m}) = 16.8$ m.

(c) The corresponding frequency is $\nu = v/\lambda = (82.0 \text{ m/s})/(16.8 \text{ m}) = 4.88$ Hz.

47E

(a) From Eq.50 with $n = 1$, $\nu = v/2\ell = (250 \text{ m/s})/2(0.15 \text{ m}) = 833$ Hz.

(b) The wavelength of the sound wave, not of the wave on the string, is $\lambda = v_s/\nu = $ (348 m/s)/(833 Hz) = 0.418 m.

49E

(a) Calculate the linear density; in SI units μ = (0.002 kg)/(1.25 m) = 0.0016 kg/m. Eq.26 now yields $v = \sqrt{(\tau/\mu)} = \sqrt{(7/0.0016)} = 66.1$ m/s.

(b) Using Eq.50 with n = 1, $\nu = v/2\ell$ = (66.1)/2(1.25) = 26.4 Hz.

53P

(a) If $y_1 = y_m\sin(kx - \omega t)$ and $y_2 = y_m\sin(kx + \omega t)$, then the sum $y = y_1 + y_2$ is given by

$$y = 2y_m\sin(kx)\cos(\omega t) = 0.5\sin(\frac{\pi x}{3})\cos(40\pi t).$$

Therefore the amplitudes are y_m = 0.5/2 = 0.25 cm. The wave speed v is

$$v = \frac{\omega}{k} = \frac{40\pi}{\pi/3} = 120 \text{ cm/s}.$$

(b) The required distance is $\lambda/2$. But

$$\lambda = \frac{2\pi}{k} = \frac{2\pi}{\pi/3} = 6 \text{ cm},$$

and therefore the distance between adjacent nodes is 3 cm.

(c) The particle velocity u is

$$u = \frac{\partial y}{\partial t} = -20\pi\sin(\frac{\pi x}{3})\sin(40\pi t).$$

At the specified place and time,

$$u = -20\pi\sin(\frac{\pi}{2})\sin(45\pi) = -20\pi(+1)(0) = 0.$$

55P

(a) A string stretched between fixed supports has resonant frequencies given by Eq.50:

$$\nu = \frac{vn}{2\ell}.$$

Since n = 1, 2, 3, ..., n changes by 1 from one resonant frequency to the next; that is,

$$\Delta\nu = \frac{v}{2\ell}(\Delta n) = \frac{v}{2\ell}(1),$$

$$420 - 315 = \frac{v}{2\ell} = 105 \text{ Hz}.$$

But, from the first equation above, the frequency of the n = 1 resonance is

$$\nu_1 = \frac{v(1)}{2\ell}.$$

Hence ν_1 = 105 Hz.

(b) With ℓ = 75 cm = 0.75 m,

$$v = 2\ell\nu_1 = 2(0.75 \text{ m})(105 \text{ s}^{-1}) = 157.5 \text{ m/s}.$$

62P

Associated with a length dx is kinetic
energy dK and potential energy dU. The
kinetic energy is

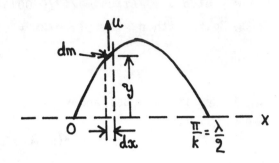

$$dK = \frac{1}{2}(dm)u^2 = \frac{1}{2}(\mu dx)u^2,$$

where u is the transverse speed of the length
dx. From the equation for a standing wave,
Eq.46, this speed is

$$u = \frac{\partial y}{\partial t} = -2y_m\omega(\sin kx)(\sin\omega t),$$

so that

$$dK = 2\mu\omega^2 y_m^2 \sin^2 kx(\sin^2\omega t)dx.$$

The potential energy is that of a simple harmonic oscillator of mass dm; call the force
constant 'a' (we do not use k to avoid confusion with the angular wave number); therefore,

$$dU = \frac{1}{2}ay^2,$$

where, by Eq.11 of Chapter 14,

$$a = (dm)\omega^2 = \mu\omega^2 dx.$$

Hence,

$$dU = \frac{1}{2}\mu\omega^2 y^2 dx = 2\mu\omega^2 y_m^2 \sin^2 kx(\cos^2\omega t)dx.$$

The total energy is dE = dK + dU:

$$dE = 2\mu\omega^2 y_m^2(\sin^2 kx)dx.$$

Finally, integrating over one loop, putting ω = vk = $2\pi\nu$,

$$E = 2\mu\omega^2 y_m^2 \int_0^{\pi/k}(\sin^2 kx)dx = 2\pi^2\mu y_m^2\nu v.$$

63P

(a) From the tension τ in the wires,

$$\tau = mg = (10 \text{ kg})(9.8 \text{ m/s}^2) = 98 \text{ N},$$

the wave speeds in the two wires can be found from

$$v_1 = \sqrt{(\tau/\mu_1)}; \quad v_2 = \sqrt{(\tau/\mu_2)}.$$

In terms of the densities, the linear densities each are given by

$$\mu = \rho A \ell/\ell = \rho A,$$

where A is the cross-sectional area of the wire and ρ is the appropriate density. Numerically,

$$\mu_1 = (2600 \text{ kg/m}^3)(10^{-6} \text{ m}^2) = 2.6 \times 10^{-3} \text{ kg/m},$$

$$\mu_2 = 7.8 \times 10^{-3} \text{ kg/m}.$$

With these, the equations for the wave speeds yield v_1 = 194.1 m/s and v_2 = 112.1 m/s. The distance between adjacent nodes is $\lambda/2$, so that if it is required that the joint be a node, then

$$n_1 \lambda_1/2 = \ell_1; \quad n_2 \lambda_2/2 = \ell_2.$$

It is given that ℓ_1 = 0.6 m and ℓ_2 = 0.866 m and therefore, since $v = \lambda \nu$,

$$n_1 = 0.00618 \nu_1; \quad n_2 = 0.0155 \nu_2.$$

It can be found by trial and error that the smallest integers n_1 and n_2 giving $\nu_1 = \nu_2$ are n_1 = 2 and n_2 = 5. The frequency is 2/0.00618 = 5/0.0155 = 323 Hz.

(b) There are 5 + 2 = 7 loops or 8 nodes in all. If the two at the ends are not counted, the number of nodes is 8 - 2 = 6.

CHAPTER 18

2E

(a) The wavelength is given by Eq.14 in Chapter 17:

$$\lambda = \frac{v}{\nu} = \frac{343 \text{ m/s}}{4.5 \times 10^6 \text{ s}^{-1}} = 7.62 \times 10^{-5} \text{ m} = 0.0762 \text{ mm.}$$

(b) With the different sound speed,

$$\lambda = \frac{v}{\nu} = \frac{1500 \text{ m/s}}{4.5 \times 10^6 \text{ s}^{-1}} = 3.33 \times 10^{-4} \text{ m} = 0.333 \text{ mm.}$$

8E

(a) The distances from the player to the two spectators are $d_1 = (343 \text{ m/s})(0.23 \text{ s}) = 78.9$ m, and $d_2 = (343 \text{ m/s})(0.12 \text{ s}) = 41.2$ m.

(b) Due to the presence of the right angle at the player, the distance bwteen the two spectators must be $\sqrt{[(78.9 \text{ m})^2 + (41.2 \text{ m})^2]} = 89.0$ m.

14P

Let D be the required distance. The travel times for the P and S waves are

$$t_P = D/v_P,$$

$$t_S = D/v_S.$$

The seismograph does not tell us these times separately, but it does tell us their difference. Recording t in seconds and measuring D in km, we have

$$t_S - t_P = D[\frac{1}{v_S} - \frac{1}{v_P}],$$

$$180 = D[\frac{1}{4.5} - \frac{1}{8}] \rightarrow D = 1850 \text{ km.}$$

15P

The time t_1 required for a stone to fall to the bottom of a well of depth d is

$$t_1 = \sqrt{(2d/g)},$$

and the time t_2 needed for sound, traveling at speed v, to cover this same distance d is

$$t_2 = d/v.$$

The total time t that elapses between dropping the stone and hearing the splash is t = $t_1 + t_2$:

$$t = \sqrt{(2d/g)} + d/v.$$

Rearranging and then squaring the equation above gives a quadratic equation for d in terms of t:

$$gd^2 - d[(2v)(gt + v)] + (vt)^2 g = 0.$$

Solve the equation using the quadratic formula (see Appendix G); we get

$$d = v[(\frac{v}{g} + t) \pm \sqrt{\{(\frac{v}{g})(\frac{v}{g} + 2t)\}}].$$

The negative sign is appropriate since t = 0 should indicate d = 0. With g = 9.8 m/s^2, v = 343 m/s, a time t = 3 s gives d = 40.7 m.

22E

Use Eq.24 in Sample Problem 5 (which deals with spherical waves):

$$P = I[4\pi r^2] = (1.91 \times 10^{-4} \text{ W/m}^2)4\pi(2.5 \text{ m})^2 = 1.50 \times 10^{-2} \text{ W} = 15.0 \text{ mW}.$$

23E

Use the result of Exercise 20:

$$I = 2\pi^2 \rho v \nu^2 s_m^2,$$

$$10^{-6} \text{ W/m}^2 = 2\pi^2(1.29 \text{ kg/m}^3)(343 \text{ m/s})(300 \text{ s}^{-1})^2 s_m^2 \rightarrow s_m = 3.57 \times 10^{-8} \text{ m} = 35.7 \text{ nm}.$$

25E

(a) The sound level β in decibels is given by Eq.18:

$$\beta = 10\log[I/I_0].$$

Hence, for two different sound intensities,

$$\beta_1 = 10\log[I_1/I_0], \quad \beta_2 = 10\log[I_2/I_0].$$

Therefore,

$$\beta_2 - \beta_1 = 10(\log[I_2/I_0] - \log[I_1/I_0]) = 10\log[I_2/I_1].$$

For an increase of 30 dB,

$$30 = 10\log[I_2/I_1] \rightarrow I_2/I_1 = 1000.$$

(b) By Eq.17,

$$s_m = \sqrt{[2I/\rho v \omega^2]}.$$

Therefore,

$$s_{m2}/s_{m1} = \sqrt{(I_2/I_1)} = \sqrt{1000} = 31.6.$$

29E

(a) Combine Eqs.11 and 17 by eliminating s_m to get

$$\Delta p_m^2 = 2I \rho v.$$

Now apply this equation to air (a) and water (w); take the ratio, assuming equal values of the intensity I, to find

$$\frac{\Delta p_{mw}}{\Delta p_{ma}} = \sqrt{[\frac{\rho_w v_w}{\rho_a v_a}]}.$$

Substituting numerical values from Table 1 gives

$$\frac{\Delta p_{mw}}{\Delta p_{ma}} = \sqrt{[\frac{(1000)(1482)}{(1.29)(343)}]} = 57.9.$$

(b) Return to the first equation in (a), again taking a ratio but this time with equal pressure amplitudes and different intensities; we find

$$\frac{I_w}{I_a} = \frac{\rho_a v_a}{\rho_w v_w} = \frac{(1.29)(343)}{(1000)(1482)} = 2.99 \times 10^{-4}.$$

30P

(a) The path difference between the two waves going via SBD in the two positions must be half a wavelength:

$$\frac{1}{2}\lambda = 2(1.65 \text{ cm}) \rightarrow \lambda = 0.066 \text{ m}.$$

The frequency ν must be

$$\nu = \frac{v}{\lambda} = \frac{343 \text{ m/s}}{0.066 \text{ m}} = 5200 \text{ Hz}.$$

(b) Let A be the amplitude of the wave when at D that went via route SAD, and B the amplitude at D of the wave that traversed SBD in either of the two positions. Since the intensity is proportional to the square of the resultant combined amplitude, we have

$$(A + B)^2 = 900,$$

$$(A - B)^2 = 100.$$

Thus,

$$\frac{A + B}{A - B} = 3,$$

$$A + B = 3A - 3B \rightarrow B - 0.5A.$$

(c) The waves going via SAD and SBD travel different distances and therefore lose different amounts of energy by, for example, gas friction with the walls of the tubes.

32P

(a) We assume a plane-parallel wave. This is a good approximation far enough away from a transmitter of spherical or cylindrical waves. In a time t, the wave travels forward a distance vt, and therefore the energy contained in a cylinder of base area A and length vt crosses area A in the same time t. The energy in the cylinder is the energy density u multiplied by the volume of the cylinder A(vt):

$$energy = u(Avt).$$

By definition (see just above Eq.17), the energy crossing per unit area per unit time is the intensity, so that

$$\frac{energy}{tA} = I = uv.$$

(b) For a spherical wavefront a distance r from a source of power P, we have by Eq.24,

$$I = P/4\pi r^2,$$

so that, in SI units,

$$u = \frac{P}{4\pi r^2 v} = \frac{50,000}{4\pi(4.8 \times 10^5)^2(3 \times 10^8)} = 5.76 \times 10^{-17} \; J/m^3.$$

34P

(a) Assuming no absorption, the power crossing a sphere centered on the source does not depend on the radius of the sphere. By the definition of intensity, this means that for two spheres,

$$4\pi r_1^2 I_1 = 4\pi r_2^2 I_2 = constant.$$

But the intensity I is proportional to the square of the amplitude A of the wave (see Eq.17); thus also,

$$4\pi r_1^2 A_1^2 = 4\pi r_2^2 A_2^2 = constant,$$

where the constants in the two equations are different. In general, then, for any point,

$$A = \frac{constant}{r} = \frac{Y}{r},$$

Y being the name of the constant. Let the source be at r = 0, as implied above. The oscillatory part of the particle displacement is

$$\sin[kr - \omega t] = \sin[k(r - \frac{\omega}{k} t)] = \sin[k(r - vt)],$$

giving

$$y = \frac{Y}{r} \sin[k(r - vt)],$$

when combined with the previous result for the amplitude.

(b) The sine factor in the above is dimensionless. Since both y and r have the dimensions of length, Y must have dimensions of the square of length.

41P

The waves arrive at a point between the sources from opposite directions; this is the equivalent of a 180° phase difference in the arriving longitudinal waves. Added to the intrinsic 180° phase difference in the sources of the waves and we have a 360° phase difference, equivalent in effect to a 0° phase difference. Hence, only the difference in the distances traveled by the waves from the two sources has an effect. Consider a point a distance y meters from one source; this point is 5 - y meters from the other source. Therefore, the path difference is

$$\Delta \text{path} = y - (5 - y) = 2y - 5.$$

For maxima, this path difference should equal an integral number of wavelengths. Since $\lambda = v/\nu$, this condition becomes

$$2y - 5 = n(\frac{343}{300}),$$

in SI units; n = 0, 1, 2, ... Putting n = 0, 1, 2, 3, 4 gives y = 2.5, 3.07, 3.64, 4.22, 4.79 meters from each source.

43P

Given that the angle of incidence equals the angle of reflection for the sound waves reflected off the wall, we have, by examining two of the right triangles, that

$$\tan\theta = \frac{90}{50 - s} = \frac{10}{s} \rightarrow s = 5.00 \text{ feet.}$$

The path lengths of the two interfering waves are

$$x_1 = \sqrt{(80^2 + 50^2)} = 94.34 \text{ ft,}$$
$$x_2 = \sqrt{(10^2 + 5^2)} + \sqrt{(90^2 + 45^2)} = 111.80 \text{ ft.}$$

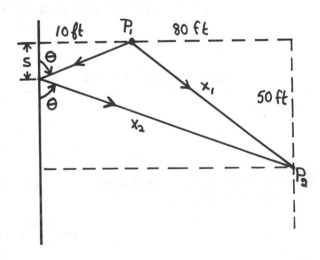

Thus, the path difference is 17.46 ft. For constructive interference this must equal an integral number of wavelengths, there being no phase change on reflection; that is,

$$n\lambda = 17.46 \text{ ft.}$$

(i) $n = 1$: $\lambda = 17.46$ ft; $\nu = v/\lambda = (1125$ ft/s$)/(17.46$ ft$) = 64.4$ Hz. (ii) $n = 2$: $\lambda = 8.73$ ft; $\nu = 129$ Hz.

45P

Let d be the limiting distance for the whispered conversation to be audible with the reflector. If S is the power output of the conversation, then the intensity at the reflector is $I_s = S/4\pi d^2$. Sound energy enters the reflector at the rate $I_s(\pi R^2)$, where R is the radius of the reflector. The intensity at the tube opening is $I_s\pi R^2/\pi r^2$, r the radius of the tube. But only 12% of this actually passes down the tube, so that the actual intensity I available to the ear is $0.12 I_s R^2/r^2$, or

$$I = 0.12(S/4\pi d^2)(R^2/r^2) = 1200S/4\pi d^2.$$

The last result follows after substituting $R = 50$ cm and $r = 0.5$ cm. Now consider a whisper at a distance of 1 m. Its intensity I_w at the reflector is $S_w/4\pi(1)^2 = S_w/4\pi$, where S_w is the power of the whisper. The sound level of the whisper is

$$\beta_w = 10\log(S_w/4\pi I_0) = 20 \text{ dB}.$$

But the conversation is also whispered, so that $S = S_w$. Hence, the sound level of the conversation at the earpiece of the reflector is

$$\beta = 10\log(1200S_w/4\pi d^2 I_0) = 10\log(S_w/4\pi I_0) + 10\log(1200/d^2),$$

$$\beta = \beta_w + 10\log 1200 - 20\log d,$$

$$0 = 20 + 10\log(1200) - 20\log d,$$

$$\log d = 2.5396 \rightarrow d = 346 \text{ m}.$$

49E

(a) For the fundamental, put $n = 1$ into Eq.50 in Chapter 17, so that

$$v = 2\ell\nu_1 = 2(0.22 \text{ m})(920 \text{ s}^{-1}) = 405 \text{ m/s}.$$

(b) The wave speed and the tension are related by $v = \sqrt{(\tau/\mu)}$. The linear density of the string is

$$\mu = \frac{m}{\ell} = \frac{0.8 \times 10^{-3} \text{ kg}}{0.22 \text{ m}} = 3.636 \times 10^{-3} \text{ kg/m}.$$

Hence,

$$\tau = \mu v^2 = (3.636 \times 10^{-3} \text{ kg/m})(405 \text{ m/s})^2 = 596 \text{ N}.$$

54P

(a) In the fundamental mode,

$$L = \lambda_0/2 = v/2\nu_0.$$

If ℓ is the length by which the string is shortened, then

$$L - \ell = v(2 \cdot r \nu_0).$$

Eliminating ν_0 between these equations gives

$$\ell = L(1 - \frac{1}{r}).$$

(b) For L = 80 cm, successive substitution of r = 6/5, 5/4, 4/3, 3/2 gives ℓ = 13.3, 16, 20, 26.7 cm.

55P

(a) Displacement antinodes occur at each open end of the pipe. Since the distance between adjacent antinodes is half a wavelength, we have

$$L = n(\lambda/2), \quad n = 1, 2, 3 \ldots \qquad \text{(b)}$$

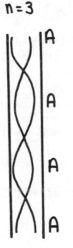 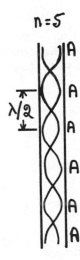

In terms of the frequency $\nu = v/\lambda$,

$$\nu = \frac{nv}{2L} = n \frac{1130}{2(1.5)} = n \frac{1130}{3} \text{ Hz.}$$

With the frequency limited to 1000 Hz to 2000 Hz, acceptable values of n and the associated frequencies are

$$n = 3, \nu = 1130 \text{ Hz,}$$

$$n = 4, \nu = 1507 \text{ Hz,}$$

$$n = 5, \nu = 1883 \text{ Hz.}$$

56P

A displacement node exists at the bottom of the well and an antinode is present at the top. Since the frequency is the smallest allowable, the wavelength must be the longest permitted for resonance. The distance between a node and an adjacent antinode is one-quarter of a wavelength. Thus, if v is the speed of sound and D is the depth of the well,

$$D = \lambda/4 = \frac{v}{4\nu}.$$

The speed of sound is given by Eq.3:

$$v = \sqrt{[\frac{B}{\rho}]} = \sqrt{[\frac{1.33 \times 10^5 \text{ N/m}^2}{1.1 \text{ kg/m}^3}]} = 347.7 \text{ m/s.}$$

Hence, by the first equation,

$$D = \frac{347.7 \text{ m/s}}{4(7 \text{ s}^{-1})} = 12.4 \text{ m.}$$

59P

(a) For spherically symmetric pulsations, the center of the star remains at rest and is a displacement node.

(b) In the fundamental mode of oscillation, antinodes exist at the surface and nowhere else. Since the distance between adjacent antinodes is $\lambda/2$ and $\lambda = v/\nu$, with v the speed of sound, we have

$$v/2\nu = 2R,$$

$$\frac{1}{\nu} = T = 4R/v.$$

(c) By Eq.3, the speed of sound is

$$v = \sqrt{[B/\rho]} = \sqrt{[(1.33 \times 10^{22}\ Pa)/(10^{10}\ kg/m^3)]} = 1.1533 \times 10^6\ m/s.$$

The radius of the sun is found in Appendix C. The radius R of the star is $(0.009)R_{sun} = (0.009)(6.96 \times 10^8\ m) = 6.264 \times 10^6\ m$. Thus,

$$T = \frac{(4)(6.264 \times 10^6\ m)}{1.1533 \times 10^6\ m/s} = 21.7\ s.$$

60P

Possible frequencies are given by Eq.50 in Chapter 17:

$$\nu = \frac{v}{2L},\ \frac{v}{L},\ \frac{3v}{2L}\ etc.,$$

among which must be 880 Hz and 1320 Hz as successive frequencies. But $880/1320 = 2/3$ and therefore

$$880 = \frac{v}{L};\quad 1320 = \frac{3}{2}\frac{v}{L}.$$

Either of these gives

$$v = 880L = (880)(0.30) = 264\ m/s.$$

But $\mu = 0.00065$ kg/m so that

$$\tau = \mu v^2 = (6.5 \times 10^{-4}\ kg/m)(264\ m/s)^2 = 45.3\ N.$$

61E

The beat frequency is given by Eq.31:

$$\nu_{beat} = \nu_1 - \nu_2.$$

In the formula, $\nu_1 > \nu_2$. If $\nu_{beat} = 3$ and either ν_1 or $\nu_2 = 384$ Hz, the frequency of the other fork is $384 \pm 3 = 387$ Hz or 381 Hz. Putting a small piece of wax on the fork will decrease its frequency. Hence, if the fork had a frequency of 387 Hz, the wax brings its frequency closer to 384 Hz. This would decrease the beat frequency as noted. Thus, the fork frequency was 387 Hz.

64P

For the fundamental mode, put n = 1 in Eq.50 in Chapter 17; using $v = \sqrt{(\tau/\mu)}$ we get

$$\nu = \frac{\sqrt{\tau}}{2L\sqrt{\mu}}.$$

We assume that μ does not change. The first frequency is 600 Hz, so we have

$$600 = \frac{\sqrt{\tau_1}}{2L\sqrt{\mu}} \rightarrow 2L\sqrt{\mu} = \sqrt{\tau_1}/600.$$

By Eq.31, the new frequency is 600 + 6 = 606 Hz, so that for this frequency,

$$606 = \frac{\sqrt{\tau_2}}{2L\sqrt{\mu}} = (600)\frac{\sqrt{\tau_2}}{\sqrt{\tau_1}} \rightarrow \frac{\tau_2}{\tau_1} = (\frac{606}{600})^2 = 1.0201.$$

Therefore, the fractional change in tension is

$$\frac{\Delta\tau}{\tau_1} = \frac{\tau_2 - \tau_1}{\tau_1} = \frac{1.0201 - 1}{1} = 0.0201,$$

or 2.01%.

69E

Apply Eq.36. We have $\nu = 16,000$ Hz, $v = 343$ m/s, $V_D = 250$ m/s, and $V_S = 200$ m/s. The source is moving away from the detector, so use the lower sign on V_S; the detector is moving toward the source, so use the upper sign on V_D. Thus,

$$\nu' = \nu(\frac{v + V_D}{v + V_S}) = (16,000 \text{ Hz})(\frac{343 + 250}{343 + 200}) = 17,473 \text{ Hz}.$$

74E

Use Eq.38 for the Mach cone. We get

$$V_S = \frac{v}{\sin\theta} = \frac{0.75(3 \times 10^8 \text{ m/s})}{\sin 60°} = 2.60 \times 10^8 \text{ m/s}.$$

78P

(a) By Eq.38,

$\sin\theta = v/V_S = v/1.5v = 2/3 \rightarrow \theta = 41.8°.$

(b) The shock wave travels with the plane at a speed of Mach 1.5 = (1.5)(331 m/s) = 496.5

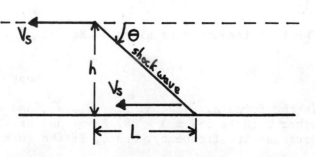

m/s. It must cover a horizontal distance L = hcotθ = (5000 m)cot41.8° = 5592 m to reach
the ground observer. This will take a time t = (5592 m)/(496.5 m/s) = 11.3 s.

91P

(a) We have $V_D = 0$, so that the frequency the uncle hears from the source which is moving
away is

$$\nu' = \nu(\frac{v}{v + V_S}) = (500)(\frac{343}{343 + 10}) = 485.8 \text{ Hz}.$$

(b) Since the relative velocity between the train and the girl is zero, she hears the
rest frequency of the whistle; i.e., 500 Hz (there is no Doppler effect).

(c) The velocities relative to the medium enter into the Doppler shift equations. The
train moves at 20 m/s and the uncle at 10 m/s relative to the air, both to the east.
The train (source) is therefore moving away from the uncle (detector), and the uncle
(detector) is moving toward the train (source). Eq.36 gives

$$\nu' = (500)\frac{343 + 10}{343 + 20} = 486.2 \text{ Hz}.$$

(d) The girl is still at rest relative to the source and thus hears 500 Hz, as before.

93E

Eq.40 applies to light waves; assuming that V is much less than c, we have

$$V = (\frac{\Delta\lambda}{\lambda})c = (\frac{0.004\lambda}{\lambda})(3 \times 10^8 \text{ m/s}) = 1.2 \times 10^6 \text{ m/s}.$$

2E

The rate of change of the output voltage V (in millivolts) with temperature (in °C) is

$$\frac{\Delta V}{\Delta T} = \frac{(28 - 0)mV}{(510 - 0)°C} = 0.0549 \text{ mV/°C.}$$

Hence, in general,

$$V = (0.0549)T,$$

so that a voltage of 10.2 mV indicates a temperature of

$$T = \frac{V}{0.0549} = \frac{10.2}{0.0549} = 186°C.$$

10P

(a) The cooling of an object can occur by radiation, conduction, convection, evaporation, etc., and these processes will be affected by the nature of the surface material, the object's surface area and temperature, the orientation of the object, the temperature of the surroundings, the presence or absence of air currents, etc. The units in the equation

$$\frac{d\Delta T}{dt} = -A(\Delta T),$$

are

$$\text{kelvins/second} = A(\text{kelvins});$$

therefore, the units of A must be 1/second, or dimensions of 1/time.

(b) Rearrange Newton's law of cooling to read

$$\frac{d\Delta T}{\Delta T} = -Adt,$$

and integrate to get

$$\Delta T = Ce^{-At},$$

where C is a constant (i.e., independent of T and of t). To evaluate C, set $\Delta T(t = 0) = \Delta T_0$. The equation then gives for C, $C = \Delta T_0$. Hence,

$$\Delta T = \Delta T_0 e^{-At}.$$

17E

(a) By Eq.8,

$$T_F = 32 + (\frac{9}{5})T_C,$$

so that setting $T_F = T_C = T$ gives

$$T = 32 + (\frac{9}{5})T \rightarrow T = -40°.$$

(b) The relation between the Kelvin and Fahrenheit scales is, by Eqs.7 and 8,

$$T_C = T_K - 273.15 = \frac{5}{9}(T_F - 32).$$

If $T_K = T_F = T$, then

$$T - 273.15 = (\frac{5}{9})T - 17.78 \rightarrow T = 575°.$$

(c) The Kelvin and Celsius scales are identical except for the choice of a zero point. Therefore, they never can give the same reading, for their readings will always differ by the difference in their zero points.

20E

The change in mirror diameter is given by Eq.9, with α extracted from Table 3:

$$\Delta L = \alpha L \Delta T,$$

$$\Delta L = (3.2 \times 10^{-6} \ °C^{-1})(200 \ in)[50°C - (-10°C)] = 0.0384 \ in.$$

27E

At 10°C the area of the window is $A = LW = (30 \ cm)(20 \ cm) = 600 \ cm^2$. The length and width at 40°C can be calculated from Eq.9:

$$\Delta L = \alpha L \Delta T = (9 \times 10^{-6} \ °C^{-1})(30 \ cm)(30°C) = 0.0081 \ cm,$$

$$\Delta W = \alpha W \Delta T = (9 \times 10^{-6} \ °C^{-1})(20 \ cm)(30°C) = 0.0054 \ cm.$$

Thus the new length is 30.0081 cm and the new width is 20.0054 cm. The increase in area is therefore

$$\Delta A = (30.0081 \ cm)(20.0054 \ cm) - 600 \ cm^2 = 0.324 \ cm^2.$$

29E

The change in radius is

$$\Delta R = \alpha R \Delta T = (23 \times 10^{-6} \ °C^{-1})(10 \ cm)(100°C) = 0.023 \ cm.$$

The new radius is $R + \Delta R = 10.023$ cm, so that the volume $V*$ at the higher temperature is

$$V* = \frac{4\pi}{3}(R + \Delta R)^3 = \frac{4\pi}{3}(10.023 \ cm)^3 = 4217.76 \ cm^3.$$

The volume at 0°C is

$$V = \frac{4\pi}{3}R^3 = \frac{4\pi}{3}(10 \text{ cm})^3 = 4188.79 \text{ cm}^3.$$

The change in volume is just $\Delta V = V* - V = 29.0 \text{ cm}^3$.

32E

The capacity V of the cup increases by $\Delta V = 3\alpha V\Delta T$, α the linear expansion coefficient for aluminum; see Eq.12. The volume of the glycerine, which at 22°C equals V (the capacity of the cup), increases by $\Delta V = \beta V\Delta T$. Thus, glycerine will spill if $\beta V\Delta T > 3\alpha V\Delta T$. Since, in fact, $\beta > 3\alpha$, some of the glycerine will spill, the amount being

$$\text{spill} = (\beta - 3\alpha)V\Delta T,$$

$$\text{spill} = (5.1 \times 10^{-4} - 6.9 \times 10^{-5})(100)(6) = 0.265 \text{ cm}^3.$$

34P

Let Δd be the changes in diameter upon heating. It is desired that

$$d_s + \Delta d_s = d_b + \Delta d_b,$$

when the brass (b) ring just slides onto the steel (s) rod; d_s and d_b are the diameters of the steel rod and brass ring at 25°C. Since for each object

$$\Delta d = \alpha d\Delta T,$$

the first equation becomes

$$d_s(1 + \alpha_s \Delta T) = d_b(1 + \alpha_b \Delta T).$$

Numerically, this is

$$3(1 + 11 \times 10^{-6}\Delta T) = 2.992(1 + 19 \times 10^{-6}\Delta T) \rightarrow \Delta T = 335°C.$$

Hence, the common temperature must be 25 + 335 = 360°C.

39P

Let A be the cross-sectional area of the glass tube. Then,

$$\Delta V = \beta V\Delta T = \beta(Ah)\Delta T,$$

h the original height. But

$$V = Ah,$$

$$\Delta V = A\Delta h,$$

provided that $\Delta A = 0$ (compare 3α for glass with β for mercury, the usual liquid in a barometer; for Hg, $\beta = 180 \times 10^{-6}/°C$). It follows that

$$A\Delta h = \beta Ah\Delta T \rightarrow \Delta h = \beta h\Delta T.$$

41P

The period P of a pendulum is given by Eq.25 in Chapter 14; $P = 2\pi\sqrt{(L/g)}$. Hence, the period at the higher temperature can be written as $P + \Delta P$ where, from calculus,

$$\Delta P \simeq \frac{\partial P}{\partial L}(\Delta L) = \frac{\pi}{\sqrt{(L/g)}}\frac{\Delta L}{g}.$$

But $\sqrt{(L/g)} = P/2\pi$, and $g = 4\pi^2 L/P^2$. Also $\Delta L = \alpha L\Delta T$. Putting all this together gives

$$\Delta P = \frac{1}{2}\alpha P\Delta T.$$

Numerically,

$$\Delta P = \frac{1}{2}(7 \times 10^{-7}/°C)(0.5 \text{ s})(10°C) = 1.75 \times 10^{-6} \text{ s}.$$

Since there are 86,400 s in one day, the clock makes 5,184,000 oscillations in 30 days. This leads to a total correction of $(5.184 \times 10^6)(1.75 \times 10^{-6}) = 9.07$ s, the clock running slow, since the pendulum is longer at the higher temperature.

45P

The increase in length of each half of the bar is

$$\Delta L = \alpha(L/2)\Delta T = (25 \times 10^{-6}/°C)(\frac{3.77 \text{ m}}{2})(32°C) = 0.1508 \times 10^{-2} \text{ m}.$$

Thus the new half-length is $1.885 + 0.001508 = 1.8865$ m. Taking note of the right-triangle configuration of the buckled bar, the buckling distance x is

$$x = \sqrt{(1.8865^2 - 1.885^2)} = 0.0752 \text{ m} = 7.52 \text{ cm}.$$

47P

Note that this problem is an extension of Problem 39, only in this problem we do not ignore the expansion of the glass. The volume of the liquid at 20°C is $V_0 = A_0 h_0$, where A_0 is the cross-sectional area of the liquid column and h_0 is the height of the liquid column ($h_0 = 0.64$ m). At 30°C the corresponding volume is $V = Ah$. We know that

$$V - V_0 = \beta V_0 \Delta T,$$

by Eq.11. The change in cross-sectional area of the liquid column is governed by the expansion of the glass. We use the result of Problem 35 to write

$$A - A_0 = 2\alpha A_0 \Delta T \rightarrow A = A_0 + 2\alpha A_0 \Delta T.$$

Using this last result, the first equation can be written

$$Ah - A_0 h_0 = \beta A_0 h_0 \Delta T,$$

$$(A_0 + 2\alpha A_0 \Delta T)h - A_0 h_0 = \beta A_0 h_0 \Delta T,$$

$$h = h_0 \left[\frac{1 + \beta \Delta T}{1 + 2\alpha \Delta T} \right].$$

Hence, the change in height is

$$\Delta h = h - h_0 = \left[\frac{\beta - 2\alpha}{1 + 2\alpha \Delta T} \right] h_0 \Delta T = \frac{4 \times 10^{-5} - 2(1 \times 10^{-5})}{1 + 2(1 \times 10^{-5})(10)}(64 \text{ cm})(10) = 0.0128 \text{ cm}.$$

51P

For each rod

$$L = L_0(1 + \alpha \Delta T),$$

so that, deleting the α^2 term,

$$L^2 = L_0^2(1 + 2\alpha \Delta T).$$

The new lengths are related by the law of cosines (see Appendix G):

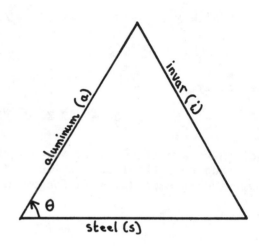

$$L_i^2 = L_a^2 + L_s^2 - 2L_a L_s \cos\theta.$$

In terms of the old lengths (see the second equation), this equation becomes

$$L_{io}^2(1 + 2\alpha_i \Delta T) = L_{ao}^2(1 + 2\alpha_a \Delta T) + L_{so}^2(1 + 2\alpha_s \Delta T) - 2L_{ao} L_{so}(1 + \alpha_a \Delta T + \alpha_s \Delta T)\cos\theta.$$

Originally the triangle was equilateral so that

$$L_{io} = L_{ao} = L_{so}.$$

Therefore, the lengths in the long equation above cancel, leaving

$$2\alpha_i \Delta T = 1 + 2\alpha_a \Delta T + 2\alpha_s \Delta T - 2(1 + \alpha_a \Delta T + \alpha_s \Delta T)\cos\theta.$$

This can be solved for ΔT:

$$\Delta T = \frac{1}{2} \frac{1 - 2\cos\theta}{\alpha_i - (\alpha_a + \alpha_s)(1 - \cos\theta)},$$

$$\Delta T = \frac{1}{2}(10^6) \frac{1 - 2\cos 59.95°}{0.7 - (23 + 11)(1 - \cos 59.95°)} = 46.4°.$$

Thus, the desired temperature is $20°C + 46.4°C = 66.4°C$.

53P

Let x be the length of each edge of the cube, and let the cube be submerged to depth y.

The weight of the cube is $x^3\rho_a g$. Before heating, the buoyant force is $x^2 y \rho_m g$ and thus

$$x^2 y \rho_m g = x^3 \rho_a g,$$

$$y \rho_m = x \rho_a.$$

All quantities in the last equation change as the temperature rises, so that

$$y \Delta \rho_m + \rho_m \Delta y = x \Delta \rho_a + \rho_a \Delta x.$$

But the mass of the cube is $M = x^3 \rho_a$, and this remains constant, so that

$$\Delta M = 0 = 3x^2 \rho_a \Delta x + x^3 \Delta \rho_a,$$

$$x \Delta \rho_a = -3 \rho_a \Delta x.$$

Therefore,

$$y \Delta \rho_m + \rho_m \Delta y = -2 \rho_a \Delta x.$$

If β is the coefficient of volume expansion of the mercury, then, by Problem 37, we have $\Delta \rho_m = -\beta \rho_m \Delta T$. Substitute this into the equation above and solve for Δy:

$$\Delta y = [\beta y - 2(\frac{\rho_a}{\rho_m})\frac{\Delta x}{\Delta T}]\Delta T.$$

But $\Delta x = \alpha x \Delta T$, so that

$$\Delta y = [\beta y - 2(\frac{\rho_a}{\rho_m})\alpha x]\Delta T.$$

Since, as shown above, $y = x\rho_a/\rho_m$, this becomes

$$\Delta y = [\beta - 2\alpha]x(\frac{\rho_a}{\rho_m})\Delta T.$$

Putting in the numbers ($\rho_a = 2.7$ g/cm^3, $\rho_m = 13.6$ g/cm^3, etc.) gives $\Delta y = 0.266$ mm.

54P

(a) Let $2y$ be the horizontal span and x the sag of the cable at the midpoint. For a parabola, the length of the cable is (and this you are apparently expected to look up)

$$L = \sqrt{(4x^2 + y^2)} + \frac{y^2}{2x} \ln[\frac{2x + \sqrt{(4x^2 + y^2)}}{y}].$$

The half-span $y = 4200/2 = 2100$ ft and is assumed to remain unchanged. At 50°F, $x = 470$ ft, so that the length L of the cable at this temperature is given by the equation above with $y = 2100$ ft, $x = 470$ ft. This turns out to be 4336 ft. Hence,

$$\Delta L = \alpha L \Delta T = (6.5 \times 10^{-6}/°F)(4336 \text{ ft})(80°F) = 2.25 \text{ ft.}$$

(b) The change in the sag x cannot be found from $\Delta x = \alpha x \Delta T$ because of the constraint on y; i.e., Δx is determined by the geometry of the situation; a given length of cable hung in a specified curve between supports a specified distance apart determines the sag. From calculus, write

$$\Delta L \simeq \frac{dL}{dx}(\Delta x) \quad \rightarrow \quad \Delta x = \frac{\Delta L}{dL/dx}.$$

Use the first equation to compute dL/dx; the result is

$$\frac{dL}{dx} = \frac{4x}{\sqrt{(4x^2 + y^2)}} - \frac{y^2}{2x^2} \ln\left[\frac{2x + \sqrt{(4x^2 + y^2)}}{y}\right] + \frac{y^2}{2x} \frac{2 + 4x/\sqrt{(4x^2 + y^2)}}{2x + \sqrt{(4x^2 + y^2)}}.$$

Now verify that, numerically, dL/dx = 0.56436, so that

$$\Delta x = (2.25 \text{ ft})/0.56436 = 3.99 \text{ ft.}$$

CHAPTER 20

2E

(a) Use Eq.3:

$$Q = cm(T_f - T_i),$$

$$75 \text{ cal} = c(30 \text{ g})(45°C - 25°C) \rightarrow c = 0.125 \text{ cal/g·K}.$$

(Note that a temperature change of 1°C equals a temperature change of 1 K.)

(b) The number of moles is

$$n = \frac{\text{bulk mass}}{\text{molecular mass}} = \frac{30 \text{ g}}{50 \text{ g/mol}} = 0.600 \text{ mole.}$$

(c) In Eq.3, replace the mass m with the number of moles n; then c stands for the molar specific heat capacity. We get

$$Q = cn(T_f - T_i),$$

$$75 \text{ cal} = c(0.6 \text{ mol})(45°C - 25°C) \rightarrow c = 6.25 \text{ cal/mol·K}.$$

By Eq.1, this can also be expressed as c = (6.25)(4.19) = 26.2 J/mol·K.

7E

To completely melt the silver, we first must heat it to the melting temperature; the needed heat is found from Eq.3. Then, we must actually melt the silver; for this, Eq.5 gives us the required heat. The specific heat is found in Table 1. The melting temperature and latent heat are found in Table 2. Putting all this together, we obtain

$$Q = cm(T_f - T_i) + Lm,$$

$$Q = (236 \text{ J/kg·K})(0.130 \text{ kg})(1235 \text{ K} - 288 \text{ K}) + (1.05 \times 10^5 \text{ J/kg})(0.130 \text{ kg}),$$

$$Q = 29,054 \text{ J} + 13,650 \text{ J} = 42,704 \text{ J.}$$

12E

(a) The heat generated is (see Appendix G for the unit conversions)

$$Q = Pt = (0.40 \text{ hp})(550 \text{ ft·lb/s·hp})(120 \text{ s}) = 26,400 \text{ ft·lb,}$$

$$Q = \frac{26,400 \text{ ft·lb}}{777.9 \text{ ft·lb/Btu}} = 33.9 \text{ Btu.}$$

(b) The specific heat of copper is 0.0923 cal/g·°C = 0.0923 Btu/lb·°F. Noting that only 75% of the heat warms the copper, Eq.3, written in British units, gives

$$\Delta T = \frac{Q}{wc} = \frac{(0.75)(33.9 \text{ Btu})}{(1.6 \text{ lb})(0.0923 \text{ Btu/lb} \cdot \text{°F})} = 172\text{°F}.$$

15E

The boiling temperature is 100°C, so that the heat required to reach this temperature is

$$Q = cm\Delta T = (4190 \text{ J/kg} \cdot \text{K})(0.1 \text{ kg})(77 \text{ K}) = 32,263 \text{ J}.$$

The time to reach the boiling point is therefore

$$t = \frac{Q}{P} = \frac{32263 \text{ J}}{200 \text{ J/s}} = 161 \text{ s}.$$

23P

In 500 cm^3 of water there are 500 g = 0.5 kg of water, since the density of water is 1.0 g/cm^3. To begin boiling, this water must be brought to 100°C starting from 59°F = 15°C; that is, the water must be heated by 85°C. The heat needed is

$$Q = mc\Delta T = (0.5 \text{ kg})(4190 \text{ J/kg} \cdot \text{K})(85 \text{ K}) = 1.781 \text{ X } 10^5 \text{ J}.$$

If any additional energy is supplied, boiling commences. Each shake produces

$$E = mg\Delta h = (0.5 \text{ kg})(9.8 \text{ m/s}^2)(0.3048 \text{ m}) = 1.494 \text{ J},$$

of energy. Therefore 1.781 X 10^5/1.494 = 1.192 X 10^5 shakes are needed. At the rate of 30 shakes per minute = 0.5 shakes/s, this will take 2(1.192 X 10^5) = 2.384 X 10^5 s, or 238,400/86,400 = 2.76 days.

30P

(a) As instructed, the glass is ignored throughout. Here, we choose to use grams and cal as units of mass and heat energy. The heat that is needed to raise the temperature of the ice from –15°C to 0°C is

$$Q_1 = mc\Delta T = (100)(0.53)(15) = 795 \text{ cal}.$$

Melting the ice at 0°C requires

$$Q_2 = mL = (100)(79.5) = 7950 \text{ cal},$$

of heat. Thus, 8745 cal of heat are required to convert 100 g of ice initially at –15°C to liquid water at 0°C. However, if the 200 g of water originally in the glass at 25°C is cooled to 0°C, it would release

$$Q_3 = mc\Delta T = (200)(1)(25) = 5000 \text{ cal}.$$

Now this is greater than Q_1 but less than $Q_1 + Q_2$. Hence, the equilibrium temperature is 0°C, but not all of the ice melts. In fact, the mass M of ice that does melt is

$$M = \frac{(5000 - 795)cal}{79.5 \ cal/g} = 52.9 \ g.$$

(b) With one ice cube, m = 50 g; this gives Q_1 = 397.5 cal and Q_2 = 3975 cal. Q_3 is not changed so that in this case we have $Q_3 > Q_1 + Q_2$, which indicates that all of the ice melts, with the final temperature of the system being between 0°C and 25°C. Let T be this final temperature. The water formed from the melted ice cube must be warmed from 0°C to T. With this in mind, to find T set

$$Q_{lost} = Q_{gained},$$

$$(200)(1)(25 - T) = (50)(0.53)(15) + (50)(79.5) + (50)(1)(T - 0),$$

$$T = 2.51°C.$$

The first term on the right-hand side of the main equation above represents the heat needed to bring the ice to the melting temperature, the second term is the heat needed to melt the ice, and the third term is the heat needed to warm the resulting liquid water from 0°C to T.

31P

The sphere, mass M, will pass through the ring, mass m, when the sphere's diameter D is equal to the inner diameter d of the ring. Let T be the equilibrium temperature at which this condition is met. We must have

$$Q_{gained} = Q_{lost},$$

$$mc_c(T - 0) = Mc_a(100 - T),$$

$$M = \frac{c_c T}{c_a(100 - T)} m = \frac{0.0923T}{21.5 - 0.215T} m,$$

after substituting numerical values (using cal for heat units) for the specific heat c_c of copper and c_a for aluminum. The diameters of the ring and of the sphere at the temperature T will be (with d,D the initial diameters),

$$d + \Delta d = d + \alpha_c d(T) = d(1 + \alpha_c T),$$

$$d + \Delta d = (1.0)(1 + 17 \times 10^{-6}T);$$

$$D + \Delta D = D[1 - \alpha_a(100 - T)],$$

$$D + \Delta D = (1.002)[1 - 23 \times 10^{-6}(100 - T)].$$

(See Table 3 in Chapter 19.) Setting $d + \Delta d = D + \Delta D$, so that the sphere just passes through the ring, gives

$$1 + (17 \times 10^{-6})T = 1.002 - 23.046 \times 10^{-4} + (23.046 \times 10^{-6})T \rightarrow T = 50.38°C.$$

Finally, put this result for T into the third equation above to find M = 0.436m = 8.72 g since m = 20 g.

35E

In each case, the work done equals the area (down to p = 0) under the particular curve, or path. The work is positive since the gas expands. By path A, $W_A = (40 \text{ N/m}^2)(3 \text{ m}^3) = 120$ J. By path B, $W_B = \frac{1}{2}(30 \text{ N/m}^2)(3 \text{ m}^3) + (10 \text{ N/m}^2)(3 \text{ m}^3) = 75$ J. Finally, path C gives $W_C = (10 \text{ N/m}^2)(3 \text{ m}^3) = 30$ J.

39E

Suppose we use cal for the energy unit. Then $W_{BCA} = 15$ J $= 15(0.239) = 3.585$ cal, by Eqs. 1. Over a complete cycle $\Delta U = 0$ so that the first law, Eq.12, reduces to

$$Q = W,$$

$$Q_{CA} + Q_{AB} + Q_{BC} = W_{BCA} + W_{AB},$$

$$Q_{CA} + 4.77 + 0 = 3.585 + 0,$$

$$Q_{CA} = -1.19 \text{ cal.}$$

Note that $W_{AB} = 0$ since AB is a constant volume process.

40E

Over a complete cycle the change in internal energy is $\Delta U = 0$, so that the first law now reads $Q = W$. W is the work done by the gas and this equals the area inside the curve of pressure p versus volume V. This curve is a triangle and therefore

$$W = -\frac{1}{2}(\Delta p)(\Delta V) = -\frac{1}{2}(20 \text{ N/m}^2)(3 \text{ m}^3) = -30 \text{ N·m} = -30 \text{ J.}$$

The negative sign is necessary since the triangle is traversed in the counterclockwise direction. It follows that the heat added is $Q = W = -30$ J $= -30(0.239) = -7.17$ cal, with heat flowing out of the gas.

42P

(a) The change in internal energy $\Delta U_{fi} = U_f - U_i$ between points i and f is the same regardless of the path over which it is computed. Along path iaf, $Q = 50$ cal and $W = 20$ cal; hence $\Delta U_{fi} = Q - W = 50 - 20 = 30$ cal. Along path ibf,

$$\Delta U_{fi} = Q - W,$$

$$30 = 36 - W \rightarrow W = 6 \text{ cal.}$$

(b) From (a), $\Delta U_{if} = U_i - U_f = -30$ cal. Hence, $\Delta U_{if} = -30 = Q - (-13)$, or $Q = -43$ cal.

(c) If $U_i = 10$ cal, then since from (a), $U_f - U_i = 30$, we have $U_f = 40$ cal.

(d) If $U_b = 22$ cal, $\Delta U_{bi} = U_b - U_i = 22 - 10 = 12$ cal. For the process ib then, $12 = Q - W$, by the first law. Now, $W = 6$ cal for ibf, but $W = 0$ along bf, since bf occurs at constant volume; hence $W = 6$ cal for ib. This gives $12 = Q - 6$, or $Q = 18$ cal. Along bf, $U_b = 22$ cal and $U_f = 40$ cal, so that $U_f - U_b = 18$ cal. But $W = 0$ along bf, as noted, so that $Q = 18$ cal here also.

43P

(a) Let m, ρ be the mass and density of the steam and V* its volume; V is the volume of the chamber and M, A, v are the mass, area and speed of the piston. Although, due to the presence of the water, V* ≠ V, we have dV*/dt = dV/dt if the rate at which the water level rises, due to steam condensation, is negligible with the speed at which the piston falls. Under this assumption, and disregarding the minus sign that indicates descreasing volumes, we have

$$\frac{dV*}{dt} = \frac{dV}{dt} = Av = \frac{d}{dt}\left(\frac{m}{\rho}\right) = \frac{1}{\rho}\frac{dm}{dt},$$

$$\frac{dm}{dt} = \rho Av = (6 \times 10^{-4} \text{ g/cm}^3)(2 \text{ cm}^2)(0.3 \text{ cm/s}) = 3.6 \times 10^{-4} \text{ g/s}.$$

(b) Since, by Eq.6, 539 cal are liberated for each gram of steam that condenses, Eq.5 gives for the rate at which heat leaves the chamber

$$\frac{dQ}{dt} = -L\left(\frac{dm}{dt}\right) = -(539 \text{ cal/g})(3.6 \times 10^{-4} \text{ g/s}) = -0.1940 \text{ cal/s} = -0.8130 \text{ J/s}.$$

(c) Differentiate the first law with respect to time to obtain

$$\frac{dQ}{dt} = \frac{dU}{dt} + \frac{dW}{dt}.$$

We apply this to the system water+steam. To evaluate dW/dt, note that

$$\frac{dW}{dt} = p\frac{dV}{dt} = p(-Av).$$

(In this part, the signs must be accounted for.) The pressure p is given by

$$p = p_{atm} + \frac{Mg}{A},$$

so that

$$\frac{dW}{dt} = -(p_{atm}A + Mg)v.$$

Numerically, $p_{atm}A = (1.01 \times 10^5 \text{ N/m}^2)(2 \times 10^{-4} \text{ m}^2) = 20.2 \text{ N}$, and Mg = 19.6 N. Since v = 0.003 m/s, we get

$$\frac{dW}{dt} = -(20.2 + 19.6)(0.003) = -0.1194 \text{ J/s}.$$

Using the result from (b) for dQ/dt in the first law as written above gives

$$\frac{dU}{dt} = \frac{dQ}{dt} - \frac{dW}{dt} = -0.8130 - (-0.1194) = -0.694 \text{ J/s}.$$

47E

Apply Eq.18. The thermal conductivity of copper is found in Table 4. Using SI units,

$$\frac{Q}{t} = kA\left(\frac{T_H - T_C}{L}\right) = (401 \text{ W/m·K})(90 \times 10^{-4} \text{ m}^2)\frac{125°C - 10°C}{0.25 \text{ m}} = 1660 \text{ W}.$$

56P

(a) Use Eq.18. The temperature difference is $\Delta T = T_H - T_C = 72°F - (-20°F) = 92°F = \frac{5}{9}(92) = 51.11°C$. Hence,

$$\frac{H}{A} = k(\frac{\Delta T}{L}) = (1.0 \text{ W/m·K})(\frac{51.11°C}{0.003 \text{ m}}) = 1.70 \times 10^4 \text{ W/m}^2.$$

(b) The two glass panes are equivalent to a single pane 6 mm thick. Thus, by Eq.19,

$$R_{glass} = \frac{L}{k} = \frac{0.006 \text{ m}}{1 \text{ W/m·K}} = 0.006 \text{ m}^2\text{·K/W}.$$

For the trapped air,

$$R_{air} = \frac{L}{k} = \frac{0.075 \text{ m}}{0.026 \text{ W/m·K}} = 2.885 \text{ m}^2\text{·K/W}.$$

Eq.24 now yields

$$\frac{H}{A} = \frac{\Delta T}{R_{glass} + R_{air}} = \frac{51.11 \text{ K}}{(0.006 + 2.885)\text{m}^2\text{·K/W}} = 17.7 \text{ W/m}^2.$$

57P

Consider a thickness of ice with a surface area of 1 cm^2. Since its thickness is 5 cm, the rate $Q/t = H$ at which heat flows through it to the outside is

$$H = kA(\frac{\Delta T}{L}) = (0.004)(1)\frac{0 - (-10)}{5} = 0.008 \text{ cal/s}.$$

Therefore, it takes (79.5 cal/g)/(0.008 cal/s) = 9938 s/g, or 9938 s to freeze 1 g of water. In 9938 s, then, the layer of ice will grow in thickness by x cm, where x is found from

$$\rho V = \rho(Ax) = m,$$

$$(0.92 \text{ g/cm}^3)(1 \text{ cm}^2)(x) = 1 \text{ g} \rightarrow x = 1.087 \text{ cm}.$$

Hence, the hourly growth rate of the ice layer is

$$\frac{dx}{dt} = \frac{3600 \text{ s/h}}{9938 \text{ s}}(1.087 \text{ cm}) = 0.394 \text{ cm/h}.$$

60P

(a) The R-value is defined by R = L/k. For the foam, if R = 5.9 for L = 1 inch in thickness, then R = (5.9)(3.75) = 22.125 for the actual 3.75 inch thickness. Similarly, the R-value of the wood studs is (1.3)(3.75) = 4.875 for each stud. The rate of heat flow H = Q/t through the multilayered section is given by Eq.24:

$$H = \frac{A(T_H - T_C)}{\Sigma R}.$$

For flow through the studs, the area is

$$A = (16)(\frac{1.75}{12} \text{ ft})(12 \text{ ft}) = 28 \text{ ft}^2.$$

Hence, the rate H_s of heat flow through the studded portion is

$$H_s = \frac{(28)(30)}{0.47 + 4.875 + 0.30 + 0.98} = 126.79 \text{ Btu/h},$$

the units of R being $\text{ft}^2 \cdot {}^\circ F \cdot h/\text{Btu}$. The foam sections each have a width of $16 - 1.75 = 14.25$ inch $= 1.1875$ ft between adjacent studs, and there are $16 - 1 = 15$ such sections. They occupy a total area

$$A = (15)(1.1875 \text{ ft})(12 \text{ ft}) = 213.75 \text{ ft}^2.$$

(Note that $213.75 + 28 = 241.75 \simeq 240 \text{ ft}^2$; evidently, the various dimensions have been rounded-off a bit.) The heat flow rate H_f through the total foam section is

$$H_f = \frac{(213.75)(30)}{0.47 + 22.125 + 0.3 + 0.98} = 268.59 \text{ Btu/h}.$$

The total heat flow rate is

$$H = H_s + H_f = 126.79 + 268.59 = 395.4 \text{ Btu/h}.$$

(b) The R-value of the whole wall is defined by

$$H = \frac{A_{tot}\Delta T}{R},$$

$$395.4 = \frac{(240)(30)}{R} \rightarrow R = 18.2.$$

(c) From (a), the ratio is $(28 \text{ ft}^2)/(240 \text{ ft}^2) = 0.117$.

(d) Also from (a), $H_s/H = (126.79)/(395.4) = 0.32$.

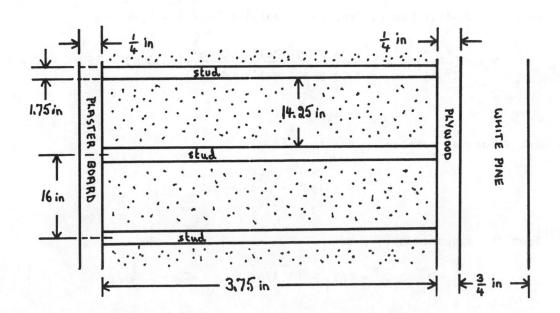

CHAPTER 21

2E

By Eq.2, the number of moles in the sample is

$$n = \frac{7.5 \times 10^{24}}{6.02 \times 10^{23}} = 1.246 \text{ mol.}$$

Eq.3 now gives for the mass M of the sample,

$$M = n(\text{molecular mass}) = (1.246 \text{ mol})(74.9 \text{ g/mol}) = 93.3 \text{ g.}$$

7E

(a) Eq.4 yields

$$n = \frac{pV}{RT} = \frac{(100 \text{ N/m}^2)(10^{-6} \text{ m}^3)}{(8.31 \text{ J/mol·K})(220 \text{ K})} = 5.47 \times 10^{-8} \text{ mol.}$$

(b) By Eq.2, the number of molecules is

$$N = nN_A = (5.47 \times 10^{-8} \text{ mol})(6.02 \times 10^{23} \text{ mol}^{-1}) = 3.29 \times 10^{16}.$$

10E

(a) Solve Eq.4 for n:

$$n = \frac{pV}{RT} = \frac{(1.01 \times 10^5 \text{ Pa})(10^{-3} \text{ m}^3)}{(8.31 \text{ J/mol·K})(313 \text{ K})} = 0.0388 \text{ mol.}$$

(b) Use the method of Sample Problem 1; solving Eq.7 for T_f gives

$$T_f = [\frac{P_f V_f}{P_i V_i}]T_i = [\frac{(1.06 \times 10^5 \text{ Pa})(1500 \text{ cm}^3)}{(1.01 \times 10^5 \text{ Pa})(1000 \text{ cm}^3)}](313 \text{ K}) = 493 \text{ K} = 220°C.$$

14P

The work W done at constant pressure p is

$$W = \int_{V_1}^{V_2} p dV = p\int_{V_1}^{V_2} dV = p(V_2 - V_1).$$

But, from the given equation,

$$V_1 = (AT_1 - BT_1^2)/p; \quad V_2 = (AT_2 - BT_2^2)/p.$$

We have set $p_1 = p_2 = p$. Substitute these into the expression for W; p cancels, leaving

$$W = (AT_2 - BT_2^2) - (AT_1 - BT_1^2) = A(T_2 - T_1) - B(T_2^2 - T_1^2).$$

15P

We work with the British units. The initial absolute pressure is $p_i = 14.7 + 15 = 29.7$ lb/in^2. Since the expansion is isothermal (T constant), Eq.7 with $T_i = T_f$ gives

$$p_i V_i = p_f V_f,$$

$$(29.7 \text{ lb/in}^2)(5 \text{ ft}^3) = (14.7 \text{ lb/in}^2)V_f \rightarrow V_f = 10.10 \text{ ft}^3.$$

By Eq.6, the work W_e done during this expansion is

$$W_e = (nRT)\ln(V_f/V_i) = p_i V_i \ln(V_f/V_i),$$

$$W_e = (29.7 \text{ lb/in}^2)(5 \text{ ft}^3)\ln(\frac{10.1}{5}) = 104.44 \text{ ft}^3 \cdot \text{lb/in}^2.$$

For the compression at constant pressure,

$$W_c = p\Delta V = p_f(V_i - V_f) = (14.7 \text{ lb/in}^2)(5 \text{ ft}^3 - 10.10 \text{ ft}^3) = -74.97 \text{ ft}^3 \cdot \text{lb/in}^2.$$

Therefore, the total work is $W = W_e + W_c$, and is

$$W = 29.47 \text{ ft}^3 \cdot \text{lb}/(\frac{1}{144} \text{ ft}^2) = 4240 \text{ ft} \cdot \text{lb}.$$

19P

Apply the method of Sample Problem 1. We must use absolute pressures. In its initial position at the bottom of the lake, the pressure inside the bubble, which must equal the water pressure at that depth if the bubble's volume is stable, is

$$p_i = p_{atm} + \rho gh = 1.01 \times 10^5 \text{ Pa} + (10^3 \text{ kg/m}^3)(9.8 \text{ m/s}^2)(40 \text{ m}) = 4.93 \times 10^5 \text{ Pa}.$$

At the surface (h = 0) the pressure is $p_f = p_{atm}$. The temperatures must be in kelvin. We have, by Eq.7,

$$V_f = (\frac{p_i}{p_f})(\frac{T_f}{T_i})V_i = (\frac{4.93 \times 10^5 \text{ Pa}}{1.01 \times 10^5 \text{ Pa}})(\frac{293 \text{ K}}{277 \text{ K}})(20 \text{ cm}^3) = 103 \text{ cm}^3.$$

25E

The rms speed is given by Eq.12, with the molecular mass M found in Table 1. We find

$$v_{rms} = \sqrt{[\frac{3RT}{M}]} = \sqrt{[\frac{3(8.31 \text{ J/mol} \cdot \text{K})(2.7 \text{ K})}{0.00202 \text{ kg/mol}}]} = 183 \text{ m/s}.$$

31P

The momentum imparted to the wall on each collision is

$$\Delta p = m \Delta v_p = 2mv_p = 2mv\cos 55°,$$

where v_p is the component of the velocity perpendicular to the wall. If N is the number of collisions per unit time (so that 1/N is the time between collisions), then the pressure P exerted on the wall is

$$P = \frac{F}{A} = \frac{N\Delta p}{A} = \frac{2Nmv\cos 55°}{A},$$

$$P = [2(10^{23} \text{ s}^{-1})(3.3 \text{ X } 10^{-24} \text{ g})(10^5 \text{ cm/s})\cos 55°]/(2 \text{ cm}^2) = 18930 \text{ dyne/cm}^2 = 1893 \text{ Pa.}$$

32E

The average kinetic energy is given by Eq.14. Choose the appropriate value of k from Eq. 15. Thus we have,

$$\overline{K} = \frac{3}{2}kT = \frac{3}{2}(1.38 \text{ X } 10^{-23} \text{ J/K})(1600 \text{ K}) = 3.31 \text{ X } 10^{-20} \text{ J};$$

$$\overline{K} = \frac{3}{2}kT = \frac{3}{2}(8.62 \text{ X } 10^{-5} \text{ eV/K})(1600 \text{ K}) = 0.207 \text{ eV.}$$

37P

(a) The molecular mass of water, H_2O, is M = 2(1) + 16 = 18 g/mol. This problem calls the number of molecules n; in the chapter text, however, n = the number of moles instead. To avoid confusion, we will use the chapter notation here. The number N of molecules of water in 1 gram of water is, by Eq.1,

$$N = nN_A = (\frac{1}{18} \text{ mol})(6.02 \text{ X } 10^{23} \text{ mol}^{-1}) = 3.344 \text{ X } 10^{22}.$$

Therefore,

$$\epsilon = \frac{539}{N} = \frac{(539 \text{ cal/g})(4.19 \text{ J/cal})}{3.344 \text{ X } 10^{22} \text{ g}^{-1}} = 6.75 \text{ X } 10^{-20} \text{ J.}$$

(b) The average kinetic energy per molecule is found from Eq.14:

$$\overline{K} = \frac{3}{2}kT = \frac{3}{2}(1.38 \text{ X } 10^{-23} \text{ J/K})(305 \text{ K}) = 6.31 \text{ X } 10^{-21} \text{ J.}$$

Comparing this to (a) we see that $\epsilon/\overline{K} = 10.7$.

38P

Combine Eqs.2 and 4 to write the ideal gas law in terms of N, the number of molecules, rather than n, the number of moles; we get

$$pV = nRT = (\frac{N}{N_A})RT = NkT \rightarrow N = \frac{pV}{kT},$$

since $k = R/N_A$. Hence, for two different gases 1 and 2, if $p_1 = p_2$, $V_1 = V_2$ and $T_1 = T_2$, we have $N_1 = N_2$. Clearly, then, the ideal gas law "contains" Avogadro's law. But the Avogadro law does not necessarily imply the ideal gas law, since Avogradro's law would be satisfied by

$$N = (\frac{pV}{kT})^2,$$

for example. Thus, Avogadro's law is not equivalent to the ideal gas law.

40E

(a) For the typical molecular diameter of $d = 2 \times 10^{-8}$ cm, Eq.16 yields

$$\lambda = \frac{1}{\sqrt{2}\pi N d^2} = \frac{1}{\sqrt{2}\pi(1 \text{ cm}^{-3})(2 \times 10^{-8} \text{ cm})^2} = 5.63 \times 10^{14} \text{ cm} = 5.63 \times 10^9 \text{ km.}$$

(b) The answer to (a) has little significance because, at this altitude, nearly all molecules follow collisionless ballistic paths in the earth's gravitational field, and many would escape from the earth's atmosphere.

41E

The number N of jelly beans per unit volume is $N = 15/1000 = 0.015$ cm^{-3}, and the diameter of each bean is $d = 1$ cm. Hence, assuming a Maxwellian speed distribution, the mean free path λ will be given by Eq.16:

$$\lambda = [\sqrt{2}\pi N d^2]^{-1} = [\sqrt{2}\pi(0.015 \text{ cm}^{-3})(1 \text{ cm})^2]^{-1} = 15.0 \text{ cm.}$$

43P

(a) In terms of the number N of gas molecules, the ideal gas law can be written

$$pV = nRT = (\frac{N}{N_A})RT = NkT.$$

Set $V = 1$ cm$^3 = 10^{-6}$ m^3. Since 1 atm = 76 cm Hg (see Appendix F), we have

$$p = \frac{10^{-6} \text{ mm Hg}}{760 \text{ mm Hg}}(1.01 \times 10^5 \text{ Pa}) = 1.329 \times 10^{-4} \text{ Pa.}$$

Therefore,

$$N = \frac{pV}{kT} = \frac{(1.329 \times 10^{-4} \text{ Pa})(10^{-6} \text{ m}^3)}{(1.38 \times 10^{-23} \text{ J/K})(295 \text{ K})} = 3.26 \times 10^{10}.$$

(b) By Eq.16,

$$\lambda = [\sqrt{2}\pi N d^2]^{-1} = [\sqrt{2}\pi(3.26 \times 10^{10} \text{ cm}^{-3})(2 \times 10^{-8} \text{ cm})^2]^{-1} = 1.73 \times 10^4 \text{ cm} = 173 \text{ m.}$$

48E

(a) The speed distribution is not Maxwellian, and therefore the basic definition of the average speed must be used, rather than the formula developed in Sample Problem 9. This definition gives

$$\bar{v} = \frac{1}{10}[2 + 3 + 4 + 5 + 6 + 7 + 8 + 9 + 10 + 11] = 6.5 \text{ km/s}.$$

(b) The root-mean-square speed is the square root of the average of the squared speeds; that is,

$$\overline{v^2} = \frac{1}{10}[2^2 + 3^2 + 4^2 + 5^2 + 6^2 + 7^2 + 8^2 + 9^2 + 10^2 + 11^2] = 50.5,$$

$$v_{rms} = \sqrt{(\overline{v^2})} = 7.11 \text{ km/s}.$$

53P

(a) The escape speed and the root-mean-square speeds are given by the equations

$$v_{esc} = \sqrt{[\frac{2GM_e}{R_e}]} = \sqrt{[2gR_e]}; \quad v_{rms} = \sqrt{[\frac{3RT}{M}]};$$

R_e is the radius of the earth and M_e is its mass. If $v_{esc} = v_{rms}$, then

$$2gR_e = \frac{3RT}{M} \rightarrow T = 2gR_e M/3R.$$

For the earth, $R_e = 6.4 \times 10^6$ m and $g = 9.8$ m/s^2. H_2 has a molecular mass of M = 0.00202 kg/mol (see Table 1); the last equation above gives T = 10,200 K. Oxygen O_2 has a molecular mass of 0.032 kg/mol, which yields T = 161,000 K.

(b) For the moon, put g = 0.16(9.8) = 1.568 m/s^2, and R_e = radius of the moon = 1.74 $\times$ 10^6 m (see Appendix C). The same equation now gives, for hydrogen T = 442 K, and for oxygen, T = 7000 K.

(c) Although 1000 K is much less than 10,000 K, there are enough molecules in the "tail" of the Maxwell speed distribution to ensure depletion of molecular hydrogen over the three or four billion years since the atmosphere was formed. Most of the O_2 has been retained, fortunately. The reduced value of g, associated with the height of "high in the atmosphere" has little effect on these conclusions.

54P

(a) Use Eq. 12 with the molecular mass M taken from Table 1 to obtain

$$v_{rms} = \sqrt{[\frac{3RT}{M}]} = \sqrt{[\frac{3(8.31 \text{ J/mol} \cdot \text{K})(4000 \text{ K})}{0.00202 \text{ kg/mol}}]} = 7.03 \times 10^3 \text{ m/s}.$$

(b) When the hydrogen molecule and the argon atom touch, their center to center distance is

$$d = r_H + r_A = \frac{1}{2}(1 + 3) \times 10^{-8} = 2 \times 10^{-8} \text{ cm}.$$

This assumes that they can be pictured as rigid spheres.

(c) By the definition of mean free path, the collision frequency ν is given by

$$\nu = \frac{v_{rms}}{\lambda} = \sqrt{2}\pi N d^2 v_{rms} = \sqrt{2}\pi(4 \times 10^{19} \text{ cm}^{-3})(2 \times 10^{-8} \text{ cm})^2(7.03 \times 10^5 \text{ cm/s}),$$

$$\nu = 5.00 \times 10^{10} \ s^{-1}.$$

57P

(a) The total area under the distribution curve must equal unity (i.e., one). This area is a rectangle plus a triangle, so we have

$$av_0 + av_0/2 = 1 \ \rightarrow \ a = 2/3v_0.$$

(b) The area under the curve between $v = 1.5v_0$ and $v = 2v_0$ equals the fraction of gas particles with speeds between these limits. Hence, this fraction f is

$$f = a(v_0/2) = (2/3v_0)(v_0/2) = 1/3.$$

Thus, the number of particles in this speed range is $fN = N/3$.

(c) Apply Eq.20. Note that $P(v) = 0$ for $v > 2v_0$. Also, the distribution function has different functional forms over the two parts of its range where it is non-zero; these forms (obtained by applying the equation $y = mx + b$ of a straight line to each segment) are

$$P(v) = av/v_0, \ \ 0 \le v \le v_0,$$

$$P(v) = a, \ \ v_0 \le v \le 2v_0.$$

Thus, the integral must be broken into these two parts, and the appropriate form of $P(v)$ used in each. With this in mind, Eq.20 becomes

$$\overline{v} = \int vP(v)dv = \int_0^{v_0} v(av/v_0)dv + \int_{v_0}^{2v_0} v(a)dv = (av_0^2/3) + (3av_0^2/2) = (11av_0^2/6).$$

Substituting $a = 2/3v_0$ from (a) gives

$$\overline{v} = \frac{11}{9} v_0 = 1.222v_0.$$

(d) The rms speed is defined from Eq.22:

$$v^2_{rms} = \int v^2 P(v)dv.$$

Breaking up the integral as in (c),

$$v^2_{rms} = \int_0^{v_0} (av/v_0)v^2 dv + \int_{v_0}^{2v_0} (a)v^2 dv = a[\frac{1}{4}(v_0^3) + \frac{7}{3}(v_0^3)] = \frac{31}{12}(v_0^3)(\frac{2}{3v_0}) = \frac{31}{18}(v_0^2),$$

$$v_{rms} = v_0 \sqrt{(\frac{31}{18})} = 1.312v_0.$$

59E

For an isothermal process $\Delta U = 0$, so that the first law becomes $Q = W$. But W is given by Eq.6 and therefore so is Q; for $n = 1$ mol, then,

$$Q = RT\ln(V_f/V_i).$$

67P

(a) For a diatomic gas $f = 5$ (see Table 4). By Eq.32, $C_V = fR/2 = 5R/2$. Eq.31 then gives $C_p = R + C_V = R + 5R/2 = 7R/2 = 7(8.31)/2 = 29.085$ J/mol·K. Therefore,

$$Q = nC_p\Delta T = (4 \text{ mol})(29.085 \text{ J/mol·K})(60 \text{ K}) = 6980 \text{ J}.$$

(b) By Eq.26,

$$\Delta U = nC_V\Delta T = (4 \text{ mol})[(2.5)(8.31 \text{ J/mol·K})](60 \text{ K}) = 4986 \text{ J}.$$

(c) By the first law, Eq.12 of Chapter 20,

$$W = Q - \Delta U = 6980 - (4986) = 1994 \text{ J}.$$

(d) The internal translational kinetic energy of any gas is given by Eq.24. (For a gas that is monatomic, this also equals the total internal energy.) Thus we have

$$\Delta K_{int} = \frac{3}{2}nR\Delta T = \frac{3}{2}(4 \text{ mol})(8.31 \text{ J/mol·K})(60 \text{ K}) = 2992 \text{ J}.$$

69P

Since the pressure does not change, the ideal gas law yields

$$p_0 = \frac{n_1 R_1 T_1}{V} = \frac{n_2 R_2 T_2}{V} \rightarrow n_1 T_1 = n_2 T_2.$$

The internal energy of the air in the room before heating is, by Eq.26,

$$U_1 = n_1 C_V T_1.$$

After heating, it is

$$U_2 = n_2 C_V T_2 = C_V(n_2 T_2) = C_V(n_1 T_1) = U_1,$$

as asserted. Although the internal energy is unchanged, the temperature of the remaining air has increased so that, provided too much air did not escape, the room feels warmer.

70E

(a) For an adiabatic process, use Eq.33:

$$p_1 V_1^\gamma = p_2 V_2^\gamma,$$

$$(1.2 \text{ atm})(4.3 \text{ L})^{1.4} = p_2 (0.76 \text{ L})^{1.4},$$

$$p_2 = (\frac{4.3}{0.76})^{1.4} (1.2 \text{ atm}) = 13.6 \text{ atm}.$$

(b) We could use Eq.34, but with p_2 now known, we could use instead Eq.7, which does not involve exponents. Choosing this latter alternative, we have

$$T_2 = (\frac{p_2 V_2}{p_1 V_1})T_1 = \frac{(13.6 \text{ atm})(0.76 \text{ L})}{(1.2 \text{ atm})(4.3 \text{ L})}(310 \text{ K}) = 621 \text{ K}.$$

72E

For an adiabatic process $Q = 0$. The first law therefore requires that $W = -\Delta U$. By Eq.26, $\Delta U = nC_V \Delta T$. Thus, for $n = 1$ mol, $W = -C_V \Delta T = C_V(T_1 - T_2)$.

83P

(a) <u>Process 1→2</u>: since the volume does not change, $W = \int p dV = 0$. Also, by Eq.26 and Eq.27

$$\Delta U = \frac{3}{2}nR\Delta T = \frac{3}{2}(1)(8.314)(600 - 300) = 3741.3 \text{ J}.$$

From the first law, $Q = \Delta U + W = 3741.3 + 0 = 3741.3$ J.
<u>Process 2→3</u>: this is adiabatic, meaning that $Q = 0$. Also, $\Delta U = (3/2)R\Delta T = (3/2)(8.314)$ $(455 - 600) = -1808.295$ J. The first law gives $W = -\Delta U = +1808.295$ J.
<u>Process 3→1</u>: the pressure is constant so that $W = p\Delta V = nR\Delta T = (1)(8.314)(300 - 455) = -1288.67$ J. In addition $\Delta U = (3/2)(8.314)(300 - 455) = -1933.005$ J. The first law now gives $Q = \Delta U + W = -3221.675$ J.
<u>Whole cycle</u>: adding the results above yields $\Delta U = 3741.3 - 1808.295 - 1933.005 = 0$, as expected; $Q = 3741.3 + 0 - 3221.675 = 519.625$ J; $W = 0 + 1808.295 - 1288.67 = 519.625$ J. Thus $W = Q$, as anticipated from the first law applied to a complete cycle.

(b) The volume at point 1 is given from the ideal gas law; using SI units,

$$pV = nRT,$$

$$(1.013 \times 10^5)(V_1) = (1)(8.314)(300) \rightarrow V_1 = 0.02462 \text{ m}^3.$$

At point 2, $V_2 = V_1 = 0.02462$ m^3. Also,

$$\frac{p_1 V_1}{p_2 V_2} = \frac{T_1}{T_2},$$

$$\frac{(1)}{p_2} = \frac{300}{600} \rightarrow p_2 = 2 \text{ atm}.$$

At point 3, $p_3 = p_1 = 1$ atm, so that

$$\frac{p_1 V_1}{p_3 V_3} = \frac{T_1}{T_3},$$

$$\frac{0.02462}{V_3} = \frac{300}{455} \rightarrow V_3 = 0.03734 \text{ m}^3.$$

As a check, since 2→3 is an adiabat, the values found above should satisfy Eq.33:

$$p_2 V_2^\gamma = p_3 V_3^\gamma,$$

$$(2)(0.02462)^{5/3} = (1)(0.03734)^{5/3},$$

$$(2)(0.00208) = 0.00417,$$

equal, that is, within round-off error.

CHAPTER 22

<u>4P</u>

Along the path ab,

$$Q_{ab} = nC_v\Delta T = n(3R/2)\Delta T,$$

where $\Delta T = T_b - T_a$. But

$$pV = nRT,$$

$$(\Delta p)V_0 = nR\Delta T,$$

since along ab the volume is constant and equals $V_0 = 1\ m^3$. Hence,

$$Q_{ab} = \frac{3}{2}(\Delta p)V_0 = \frac{3}{2}(p_b - p_a)V_0.$$

Now $p_a = p_c$ and

$$p_b V_b^\gamma = p_c V_c^\gamma,$$

$$(10\ atm)(V_0)^{5/3} = p_c(8V_0)^{5/3} \rightarrow p_c = p_a = \frac{5}{16}\ atm.$$

Therefore,

$$Q_{ab} = \frac{3}{2}(10 - \frac{5}{16})(1)\ atm\cdot m^3 = \frac{465}{32}(1.01 \times 10^5)\ J = 1.468 \times 10^6\ J.$$

Since the path bc is an adiabat, $Q_{bc} = 0$. Along ca,

$$Q_{ca} = nC_p\Delta T = n(5R/2)(T_a - T_c).$$

But, for this constant pressure process,

$$p\Delta V = nR\Delta T,$$

so that

$$Q_{ca} = \frac{5}{2}(\Delta V)p = \frac{5}{2}(V_a - V_c)p_c,$$

$$Q_{ca} = \frac{5}{2}(1 - 8)V_0 p_c = -\frac{35}{2}(1)(\frac{5}{16})\ atm\cdot m^3 = -5.523 \times 10^5\ J.$$

From the above results, the following conclusions can be drawn:

(a) The heat added $= Q_{ab} = 1.468 \times 10^6\ J.$

(b) The heat removed $= Q_{ca} = -5.523 \times 10^5\ J.$

(c) Over a complete cycle $\Delta U = 0$. The first law then predicts that $W = Q = Q_{ab} + Q_{bc} + Q_{ca} = 1.468 \times 10^6 + 0 - 5.523 \times 10^5 = 9.16 \times 10^5\ J.$

(d) Although, unlike the Carnot cycle, heat in this cycle is not added or removed at constant temperatures, an efficiency can still be calculated from Eq.2, as follows:

$$e = 1 - |Q_C|/|Q_H| = 1 - (5.523 \times 10^5)/(1.468 \times 10^6) = 0.624.$$

8E

(a) We are told that $T_C = 115 + 273 = 388$ K and $T_H = 235 + 273 = 508$ K. From Eq.4,

$$e = 1 - T_C/T_H = 1 - 388/508 = 0.236.$$

(b) By Eq.2,

$$W = e|Q_H| = (0.236)(6.3 \times 10^4 \text{ cal})(4.19 \text{ J/cal}) = 62.3 \text{ kJ}.$$

12E

Use Eq.4. Note that $\Delta T = 75°C = 75$ K, so we have

$$e = 1 - \frac{T_C}{T_H} = 1 - \frac{T_C}{T_C + \Delta T},$$

$$0.22 = 1 - \frac{T_C}{T_C + 75} \rightarrow T_C = 266 \text{ K} = -7°C.$$

Therefore, $T_H = -7°C + 75°C = 68°C$.

14E

The air conditioner is a refrigerator. We have $T_C = 70°F = 21.11°C = 294.1$ K, and $T_H = 96°F = 35.56°C = 308.6$ K. If we assume an ideal refrigerator, then by Eq.5,

$$K = \frac{T_C}{T_H - T_C} = \frac{294.1 \text{ K}}{308.6 \text{ K} - 294.1 \text{ K}} = 20.28.$$

(Do not confuse the K for coefficient of performance and the K for kelvins.) By Eq.3,

$$|Q_C| = K|W| = (20.28)(1 \text{ J}) = 20.3 \text{ J}.$$

22P

In 10 minutes, the work performed by the motor is

$$W = Pt = (200 \text{ W})(600 \text{ s}) = 1.2 \times 10^5 \text{ J}.$$

The coefficient of performance is, by Eq.5,

$$K = \frac{T_C}{T_H - T_C} = \frac{270 \text{ K}}{300 \text{ K} - 270 \text{ K}} = 9,$$

so that, by Eq.3,

$$|Q_C| = K|W| = 9(1.2 \times 10^5 \text{ J}) = 1.08 \text{ MJ}.$$

27P

We dispense with the absolute value signs. The first law of thermodynamics, applied to the engine and to the refrigerator, requires that over one cycle,

$$Q_1 = W + Q_2,$$

$$W + Q_4 = Q_3.$$

Eliminating W between these equations yields

$$Q_1 - Q_2 = Q_3 - Q_4.$$

This equation can be successively rearranged as follows:

$$\frac{Q_1 - Q_2}{Q_3 - Q_4} = 1,$$

$$\frac{Q_1 - Q_2}{Q_3 - Q_4}\frac{Q_3}{Q_1} = \frac{Q_3}{Q_1},$$

$$(\frac{Q_1 - Q_2}{Q_1})/(\frac{Q_3 - Q_4}{Q_3}) = (\frac{Q_3}{Q_1}),$$

$$[1 - (Q_2/Q_1)]/[1 - (Q_4/Q_3)] = \frac{Q_3}{Q_1}.$$

But, for Carnot engines and Carnot refrigerators, these heat ratios are equal to the corresponding absolute temperature ratios (e.g., see Eq.8), so that

$$Q_3/Q_1 = [1 - (T_2/T_1)]/[1 - (T_4/T_3)].$$

32P

(a) Apply the ideal gas law to points 1 and 2. Since $p_2 = 3p_1$ and $V_1 = V_2$, we get

$$p_1 V_1/p_2 V_2 = T_1/T_2,$$

$$p_1 V_1/(3p_1)V_1 = T_1/T_2 \rightarrow T_2 = 3T_1.$$

Points 2 and 3 are connected by the adiabatic relation

$$p_2 V_2^\gamma = p_3 V_3^\gamma,$$

$$(3p_1)(V_1)^\gamma = p_3(4V_1)^\gamma \rightarrow p_3 = (3/4^\gamma)p_1.$$

Also,

$$p_1V_1/p_3V_3 = T_1/T_3,$$

$$p_1V_1/(3/4^\gamma)p_1(4V_1) = T_1/T_3 \rightarrow T_3 = (3/4^{\gamma-1})T_1.$$

For point 4,

$$p_1V_1^\gamma = p_4V_4^\gamma,$$

$$p_1V_1^\gamma = p_4(4V_1)^\gamma \rightarrow p_4 = p_1/4^\gamma.$$

Finally,

$$p_1V_1/p_4V_4 = T_1/T_4,$$

$$p_1V_1/(4^{-\gamma}p_1)(4V_1) = T_1/T_4 \rightarrow T_4 = T_1/4^{\gamma-1}.$$

(b) From (a), $T_1 < T_2$ and $T_4 < T_1$. Hence, heat Q_{12} enters on path 1-2 and heat $Q_{34} > 0$ leaves along path 3-4. No heat enters or leaves along adiabats. Although the paths along which heat is transferred are not isotherms, an efficiency e can still be defined by

$$e = 1 - Q_{34}/Q_{12}.$$

The paths 1-2 and 3-4 are constant volume processes; hence,

$$Q_{12} = nC_V(T_2 - T_1).$$

But $C_p - C_V = R$ and $C_p/C_V = \gamma$ so that

$$\gamma C_V - C_V = R \rightarrow C_V = R/(\gamma - 1).$$

With $T_2 = 3T_1$ found above, this gives

$$Q_{12} = 2nRT_1/(\gamma - 1).$$

Similarly,

$$Q_{34} = 2nRT_1(4^{1-\gamma})/(\gamma - 1).$$

Thus,

$$Q_{34}/Q_{12} = 4^{1-\gamma},$$

so the efficiency is

$$e = 1 - 4^{1-\gamma}.$$

If the gas is diatomic, then $\gamma = 1.4$, giving e = 0.426.

<u>34E</u>

(a) Go back to Eq.3 of Chapter 20:

$$Q = cm(T_f - T_i) = (0.092 \text{ cal/g} \cdot {}^{\circ}C)(10^3 \text{ g})(75{}^{\circ}C) = 6900 \text{ cal.}$$

(b) Apply Eq.16 to obtain

$$\Delta S = \int_{T_i}^{T_f} \frac{dQ}{T} = \int_{T_i}^{T_f} \frac{cmdT}{T} = cm\int_{T_i}^{T_f} dT/T = cm\ln(T_f/T_i).$$

The temperatures in the final expression must be expressed in kelvins; we get

$$\Delta S = (0.092 \text{ cal/g} \cdot {}^{\circ}C)(10^3 \text{ g})\ln(\frac{373}{298}) = 20.7 \text{ cal/K.}$$

<u>39E</u>

Eq.20 is valid for isothermal processes. We have

$$Q = T\Delta S = (405 \text{ K})(46 \text{ J/K}) = 18630 \text{ J} = 4446 \text{ cal.}$$

<u>42E</u>

(a) By the definition of a heat reservoir, its temperature does not change even though heat is added or removed. The temperatures of the two reservoirs are T_H = 403 K and T_C = 297 K. The heat leaves the hot reservoir and enters the cold reservoir; this determines the signs of Q for each. Thus, the entropy changes of the reservoirs are, by Eq.20,

$$\Delta S_H = \frac{Q}{T} = \frac{-1200}{403} = -2.978 \text{ cal/K,}$$

$$\Delta S_C = \frac{Q}{T} = \frac{+1200}{297} = +4.040 \text{ cal/K.}$$

The entropy change of the system, then, is 4.040 - 2.978 = +1.06 cal/K.

(b) Since the rod simply transmits the heat, absorbing none itself, its entropy change is zero, as assumed above in calculating the entropy change of the system.

<u>43E</u>

(a) Set the heat lost by the aluminum equal to the heat gained by the water; each is given by $Q = mc\Delta T$, so we have

$$(200)(0.215)(100 - T) = (50)(1)(T - 20) \rightarrow T = 57.0{}^{\circ}C = 330 \text{ K.}$$

(b) The aluminum was the hot object, so that Eq.23 applies. The aluminum originally was at $T + \Delta T$ = 100°C = 373 K. Therefore,

$$\Delta S_H = (200 \text{ g})(0.215 \text{ cal/g} \cdot {}^{\circ}C)\ln\frac{330}{373} = -5.267 \text{ cal/K.}$$

(c) The water was the originally cold object, at $T - \Delta T$ = 20°C = 293 K. Thus, by Eq.24,

$$\Delta S_C = (50 \text{ g})(1 \text{ cal/g} \cdot {}^\circ C) \ln \frac{330}{293} = +5.946 \text{ cal/K}.$$

(d) The system consists of the aluminum plus the water, so for the system $\Delta S = +5.946 - 5.267 = 0.679 \text{ cal/K}.$

44P

The equilibrium temperature can be safely taken as the temperature of the lake, $15^\circ C = 288$ K. By Eq.23,

$$\Delta S = (mc) \ln(T_f/T_i),$$

so that

$$\Delta S_{ice} = (10)(0.5) \ln \frac{273}{263} = 0.1866 \text{ cal/K},$$

$$\Delta S_{water} = (10)(1) \ln \frac{288}{273} = 0.5349 \text{ cal/K},$$

the water being the liquid formed from the melted ice. During the melting itself, $Q = mL$ so that

$$\Delta S_{melting} = \frac{Q}{T} = \frac{(10)(80)}{273} = 2.9304 \text{ cal/K}.$$

(The melting takes place at the constant temperature $0^\circ C = 273$ K.) Turning now to the lake, although its temperature does not change, it loses heat ΔQ required to warm the ice to $0^\circ C$, melt the ice at $0^\circ C$, and then bring the water formed from $0^\circ C$ to the lake's temperature of $15^\circ C$. This total heat is

$$\Delta Q = (10)(0.5)(10) + (10)(80) + (10)(1)(15) = 1000 \text{ cal}.$$

Hence,

$$\Delta S_{lake} = \frac{-\Delta Q}{T} = -\frac{1000}{288} = -3.4722 \text{ cal/K}.$$

The total entropy change is the sum of the four entropy changes given above, and this is $\Delta S = 0.18 \text{ cal/K}.$

46P

(a) The work is given by Eq.21:

$$W = nRT \ln(V_f/V_i) = (4 \text{ mol})(8.31 \text{ J/mol} \cdot K)(400 \text{ K}) \ln(2V_1/V_1) = 9216 \text{ J}.$$

(b) For an isothermal process $W = Q$, so that

$$\Delta S = \frac{Q}{T} = \frac{W}{T} = \frac{9216 \text{ J}}{400 \text{ K}} = 23.0 \text{ J/K}.$$

(c) For such an adiabatic process $Q = 0$, and therefore $\Delta S = \Sigma Q/T = 0$ also.

51P

(a) In equilibrium, $(1773 + 227)/2 = 1000$ g of ice are present, a gain of $1000 - 227 = 773$ g over the amount present initially. This indicates that $(773$ g$)(79.5$ cal/g$) = 61453.5$ cal of heat is liberated to the environment. The entropy change of the ice+water is $Q/T = (-61453.5$ cal$)/(273$ K$) = -225.10$ cal/K.

(b) The entropy change can be calculated by substituting the process used in (a), i.e., gradual, infinitesimal increments in temperature, but in reverse from (a). In this case heat flows into the system, the entropy change therefore being $+225.10$ cal/K.

(c) In (a), the entropy change of the environment will be $+225.10$ cal/K for, to keep the process reversible, the temperature of the environment can differ from the temperature of the system only by a very small degree. In (b), the entropy change of the environment will be greater than -225.10 cal/K, so that the total entropy change of the system plus environment will exceed zero, as required by the second law.

53P

(a) The work $W_{abc} = W_{ab} + W_{bc} = W_{ab}$; $W_{bc} = 0$ since bc is a constant volume process. The path ab takes place at constant pressure and therefore

$$W_{abc} = W_{ab} = p\Delta V = p_0(4V_0 - V_0) = 3p_0V_0.$$

(b) The change in internal energy, with $n = 1$, is, by Eqs.26 and 27 in Chapter 21,

$$\Delta U_{bc} = \frac{3}{2}R(T_c - T_b).$$

To find the temperatures, use the ideal gas law:

$$p_bV_b/p_aV_a = T_b/T_a,$$

$$p_0(4V_0)/p_0V_0 = T_b/T_0 \rightarrow T_b = 4T_0;$$

$$p_cV_c/p_aV_a = T_c/T_a,$$

$$(2p_0)(4V_0)/p_0V_0 = T_c/T_0 \rightarrow T_c = 8T_0.$$

With these, the internal energy change becomes

$$\Delta U_{bc} = (3R/2)(8T_0 - 4T_0) = 6RT_0.$$

But $p_0V_0 = RT_0$, so that

$$\Delta U_{bc} = 6p_0V_0.$$

The entropy change over bc is

$$\Delta S_{bc} = \int_b^c dQ/T = \int_b^c nC_V dT/T = (3R/2)\int_{4T_0}^{8T_0} dT/T = (3R/2)\ln 2 = \frac{3}{2}(p_0V_0/T_0)\ln 2.$$

(c) Over a complete cycle, both ΔS and $\Delta U = 0$.

CHAPTER 23

2E

By Newton's third law, the magnitudes of the forces exerted by each charge on the other are equal. Each is given by Eq.4; as in Sample Problem 1, we note that only the absolute values of the charges are needed to find the magnitude of the force.

$$F = \frac{1}{4\pi\epsilon_0} \frac{q_1 q_2}{r^2} = (8.99 \times 10^9 \ N\cdot m^2/C^2)\frac{(3 \times 10^{-6} \ C)(1.5 \times 10^{-6} \ C)}{(0.12 \ m)^2} = 2.81 \ N.$$

3E

Rearrange Eq.4 to find

$$r = \sqrt{[\frac{1}{4\pi\epsilon_0} \frac{q_1 q_2}{F}]} = \sqrt{[(8.99 \times 10^9 \ N\cdot m^2/C^2)\frac{(26 \times 10^{-6} \ C)(47 \times 10^{-6} \ C)}{(5.7 \ N)}]} = 1.39 \ m.$$

12P

Assume that the spheres are small enough so that the charges are uniformly distributed over their surfaces, allowing them to be treated as point charges (radii << 0.5 m). Let the initial charges be Q_1, $-Q_2$, with Q_1 and Q_2 both positive numbers. Before being connected by the wire,

$$F = \frac{1}{4\pi\epsilon_0} \frac{Q_1 Q_2}{r^2},$$

where r = 0.5 m. Thus, in SI units,

$$Q_1 Q_2 = (0.108)(0.5)^2/(8.99 \times 10^9) = 3 \times 10^{-12} \ C^2.$$

When the spheres are connected with the wire, they and the wire form a single conductor carrying a charge $(Q_1 - Q_2)$. When equilibrium is reached, the wire will carry virtually no charge since its surface area is very small compared with the surface areas of the spheres. These spheres, being identical, will each carry half the net charge, which is $(Q_1 - Q_2)/2$. Coulomb's law, Eq.4, then gives, in SI units,

$$F = (8.99 \times 10^9) \frac{[(Q_1 - Q_2)/2]^2}{(0.5)^2} = 0.036 \ \rightarrow \ Q_1 - Q_2 = \pm 2 \times 10^{-6} \ C.$$

But it has been found above that

$$Q_1 = 3 \times 10^{-12}/Q_2.$$

Combining these last two equations yields

$$Q_2^2 \pm (2 \times 10^{-6})Q_2 - 3 \times 10^{-12} = 0,$$

$$Q_2 = 1 \times 10^{-6} \text{ C}, \quad Q_1 = 3 \times 10^{-6} \text{ C}, \text{ upper sign;}$$

$$Q_2 = 3 \times 10^{-6} \text{ C}, \quad Q_1 = 1 \times 10^{-6} \text{ C}, \text{ lower sign.}$$

Hence, the initial charges on the spheres are 3 µC and 1 µC in magnitude, but of opposite sign.

15P

(a) The third charge must lie between the two charges +q, +4q (i.e., on the line joining them), for if it did not it would feel a force of repulsion if it was positive and a force of attraction if negative. In fact, the charge must be negative: each of the given charges +q, +4q, could not be in equilibrium themselves otherwise, since each would feel repulsive forces in the same direction from the other two charges. So let the third charge be −Q, with Q > 0, located a distance x from the +q charge. Then for this charge to be in equilibrium,

$$\frac{1}{4\pi\epsilon_0} \frac{qQ}{x^2} = \frac{1}{4\pi\epsilon_0} \frac{Q(4q)}{(L - x)^2} \rightarrow x = \frac{L}{3}.$$

To evaluate Q, examine the requirement that one of the given charges, the +q charge say, must also be in equilibrium:

$$\frac{1}{4\pi\epsilon_0} \frac{qQ}{x^2} = \frac{1}{4\pi\epsilon_0} \frac{q(4q)}{L^2} \rightarrow Q = 4q\left(\frac{x}{L}\right)^2 = \frac{4}{9} q,$$

using the first result obtained above. The same result is derived if the equilibrium of the +4q charge is examined.

(b) The equilibrium is unstable, because a displacement of the −Q charge along the line joining +q, +4q, results in a net force directed away from the equilibrium position.

16P

(a) It is desired that

$$GM_e M_m/r^2 = (1/4\pi\epsilon_0)q^2/r^2 \rightarrow q = \sqrt{[G4\pi\epsilon_0 M_e M_m]}.$$

Taking data from Appendix C, in SI units,

$$q = \sqrt{\left[\left(\frac{6.67 \times 10^{-11}}{8.99 \times 10^9}\right)(5.98 \times 10^{24})(7.36 \times 10^{22})\right]} = 5.71 \times 10^{13} \text{ C.}$$

(b) Each hydrogen atom contributes 1.6×10^{-19} C of positive charge. Thus, the number of hydrogen atoms needed is $(5.71 \times 10^{13})/(1.6 \times 10^{-19}) = 3.569 \times 10^{32}$. In mass, this is $(3.569 \times 10^{32} \text{ H-atoms})(1.67 \times 10^{-27} \text{ kg/H-atom}) = 5.96 \times 10^5 \text{ kg} = 596$ metric tons.

18P

The force between the two pieces of charge is

$$F = \frac{1}{4\pi\epsilon_0} \frac{q(Q - q)}{r^2}.$$

To find the maximum force, set $dF/dq = 0$:

$$\frac{dF}{dq} = (\frac{1}{4\pi\epsilon_0 r^2})(Q - 2q) = 0 \quad \rightarrow \quad q = Q/2.$$

This must correspond to a maximum force, since the minimum force ($F = 0$) is achieved with $q = 0$.

19P

(a) Draw a free-body diagram of one of the balls; the force F is the Coulomb force of repulsion. Assuming equilibrium, we have

$$T\sin\theta = F,$$

$$T\cos\theta = mg.$$

Dividing these equations yields

$$\tan\theta = F/mg.$$

Now put

$$\tan\theta \simeq \sin\theta = \frac{x/2}{L},$$

$$F = (1/4\pi\epsilon_0)q^2/x^2,$$

to obtain

$$\frac{x/2}{L} = \frac{1}{4\pi\epsilon_0}(\frac{q^2}{x^2})/(mg) \quad \rightarrow \quad x = [\frac{1}{2\pi\epsilon_0} \frac{q^2 L}{mg}]^{1/3}.$$

(b) Setting $x = 5$ cm $= 0.05$ m, $L = 120$ cm $= 1.2$ m, $m = 10$ g $= 0.01$ kg and $g = 9.8$ m/s^2, and $1/4\pi\epsilon_0 = 8.99 \times 10^9$ N·m^2/C^2 gives $q = 2.4 \times 10^{-8}$ C $= 24$ nC. Since repulsive Coulomb forces are involved, the charges on the balls must be either both positive or both negative.

21P

(a) The electrostatic forces are

$$F_1 = \frac{1}{4\pi\epsilon_0} \frac{qQ}{h^2},$$

$$F_2 = \frac{1}{4\pi\epsilon_0} \frac{2qQ}{h^2} = 2F_1,$$

on the left and right end. A normal force N acts at the pivot. The two conditions for equilibrium (see Eqs.3 and 5 in Chapter 13) require that

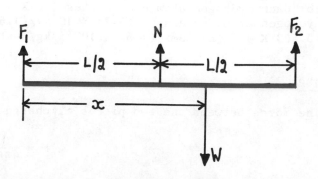

$$\Sigma F = 0 = F_1 + F_2 + N - W,$$

$$\Sigma \tau = 0 = F_2(\frac{L}{2}) - W(x - \frac{L}{2}) - F_1(\frac{L}{2}),$$

where the torques have been taken about the pivot. The torque equation does not involve N (moment arm = 0), so use that equation to find x:

$$x = \frac{L}{2}[\frac{W + F_2 - F_1}{W}] = \frac{L}{2}[1 + \frac{F_1}{W}] = \frac{L}{2}[1 + \frac{1}{4\pi\epsilon_0}\frac{qQ}{Wh^2}].$$

(b) If N = 0, the force equation reduces to

$$W = F_1 + F_2 = 3F_1,$$

$$W = 3\frac{1}{4\pi\epsilon_0}\frac{qQ}{h^2} \rightarrow h = \sqrt{[\frac{3}{4\pi\epsilon_0}\frac{qQ}{W}]}.$$

24E

The mass and charge of an electron are m = 9.11 X 10^{-31} kg and q = -e = -1.60 X 10^{-19} C. Therefore,

$$Q = (\text{# of electrons})(-e) = \frac{75 \text{ kg}}{9.11 \text{ X } 10^{-31} \text{ kg}}(-1.6 \text{ X } 10^{-19} \text{ C}) = -1.32 \text{ X } 10^{13} \text{ C}.$$

26E

(a) By Eq.4, in SI units,

$$q = r\sqrt{[(4\pi\epsilon_0)F]} = (5 \text{ X } 10^{-10})\sqrt{[(8.99 \text{ X } 10^9)^{-1}(3.7 \text{ X } 10^{-9})]} = \pm3.21 \text{ X } 10^{-19} \text{ C}.$$

(b) If electrons are missing, then the remaining charge must be positive. The number n of missing electrons is, by Eq.8,

$$n = \frac{-3.21 \text{ X } 10^{-19} \text{ C}}{-1.60 \text{ X } 10^{-19} \text{ C/electron}} = 2 \text{ electrons}.$$

27E

(a) By Eq.4, in SI units,

$$F = \frac{1}{4\pi\epsilon_0}\frac{q^2}{r^2} = (8.99 \text{ X } 10^9)\frac{(10^{-16})^2}{(0.01)^2} = 8.99 \text{ X } 10^{-19} \text{ N}.$$

(b) Assuming that q < 0, Eq.8 gives

$$n = \frac{-1.0 \text{ X } 10^{-16} \text{ C}}{-1.60 \text{ X } 10^{-19} \text{ C/electron}} = 625 \text{ electrons}.$$

31P

One mole of electrons carries a total charge dq that, in magnitude, is

$$dq = (6.02 \text{ X } 10^{23})(1.6 \text{ X } 10^{-19} \text{ C}) = 9.632 \text{ X } 10^4 \text{ C}.$$

Eq.3 now gives for the required time

$$dt = \frac{dq}{i} = \frac{9.632 \text{ X } 10^4 \text{ C}}{0.83 \text{ C/s}} = 1.160 \text{ X } 10^5 \text{ s} = 1.34 \text{ days},$$

since 1 day = 86,400 s and 1 A = 1 C/s. (Note that we could just as well have written q for charge and t for the time.)

1E

Combine Eq.2 with Newton's second law:

$$E = F/q = ma/q,$$

$$a = \frac{qE}{m} = \frac{(1.6 \times 10^{-19} \text{ C})(2 \times 10^4 \text{ N/C})}{9.11 \times 10^{-31} \text{ kg}} = 3.51 \times 10^{15} \text{ m/s}^2.$$

5E

(a) By Eq.2,

$$E = \frac{F}{q} = \frac{3 \times 10^{-6} \text{ N}}{2 \times 10^{-9} \text{ C}} = 1500 \text{ N/C}.$$

(b) The force on the proton, in magnitude, is

$$F = qE = (1.6 \times 10^{-19} \text{ C})(1500 \text{ N/C}) = 2.40 \times 10^{-16} \text{ N}.$$

Since the charge of a proton is positive, the force acts in the same direction as the electric field. The electric field points up, since the particle in (a), which has a negative charge, feels a downward force. Hence, the force on the proton is directed up.

(c) The gravitational force is $W = mg = (1.67 \times 10^{-27} \text{ kg})(9.8 \text{ m/s}^2) = 1.64 \times 10^{-26} \text{ N}.$

(d) From (b) and (c), $F/W = (2.40 \times 10^{-16} \text{ N})/(1.64 \times 10^{-26} \text{ N}) = 1.46 \times 10^{10}.$ Evidently, in these kinds of situations, gravity can be ignored.

11E

Very close to the surface of the disk, at distances much less than the radius R, the lines of force must resemble those from an infinite plane of charge; that is, they leave the disk at a right-angle to the surface. Since the net charge on the disk is not zero, very far away the lines will appear to emanate from a point charge, equal to the total charge on the disk, and located at the center of the charge distribution. In the sketch, we assume the charge is positive, and only lines in the upper half plane are drawn.

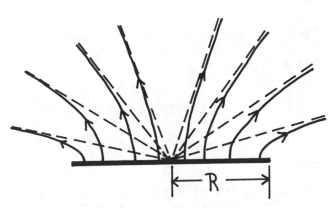

16E

Eq.6 gives directly (in SI units)

$$q = (4\pi\varepsilon_0)Er^2 = (8.99 \times 10^9)^{-1}(2)(0.5)^2 = 5.56 \times 10^{-11} \text{ C} = 55.6 \text{ pC}.$$

23P

At points between the charges, their electric
fields point in the same direction, and
therefore the total field cannot be zero at
any of these points. At points to the left of
q_1 and points to the right of q_2, the fields
are oppositely directed. Since in magnitude
$q_2 > q_1$, the desired point must be to the

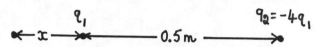

left of q_1, where its greater distance to q_2 compensates for the larger charge on q_2.
Hence, at P a distance x to the left of q_1, set

$$E_1 = E_2,$$

$$\frac{1}{4\pi\epsilon_0}\frac{q_1}{x^2} = \frac{1}{4\pi\epsilon_0}\frac{(4q_1)}{(x+0.5)^2},$$

$$\frac{1}{x} = \frac{2}{x+0.5} \;\rightarrow\; x = 0.5 \text{ m}.$$

27P

Since the point P is at the same distance
$(= a\sqrt{2}/2)$ from each charge, the strengths of
the electric fields at P due to each charge
will be proportional to the charges. The
directions will be toward the negative
charges and away from the positive charges.
Thus, if we write

$$E_1 = \frac{1}{4\pi\epsilon_0}\frac{q}{(a\sqrt{2}/2)^2},$$

then we have

$$E_2 = 2E_1; \quad E_3 = 2E_1; \quad E_4 = E_1.$$

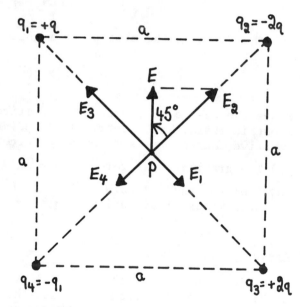

From the resulting diagram and the rules of vector addition, the vector sum E of the four
fields will point "vertically upward" (i.e., the "horizontal" components cancel), and
has a magnitude

$$E = 2E_1\cos 45° = 2\frac{(8.99 \times 10^9)(10^{-8})}{[\sqrt{2}(0.05)/2]^2}(\sqrt{2}/2) = 1.02 \times 10^5 \text{ N/C}.$$

28P

The magnitudes of the charges on the proton and electron are each equal to 1.6×10^{-19} C
$= e$. Therefore (see just below Eq.10),

$$p = qd = ed = (1.6 \times 10^{-19} \text{ C})(4.3 \times 10^{-9} \text{ m}) = 6.88 \times 10^{-28} \text{ C·m}.$$

32P

The distances of P to the positive charges are $(z - d)$ and $(z + d)$, and the distance to the two negative charges is z. Taking account of the directions of the three fields to the net field at P, and carrying out the approximation suitable for large distances used in Section 5 out to the third term in the series yields (see Appendix G, Binomial Theorem)

$$E = \frac{1}{4\pi\epsilon_0}[\frac{q}{(z - d)^2} + \frac{q}{(z + d)^2} - \frac{2q}{z^2}],$$

$$E = \frac{q}{4\pi\epsilon_0 z^2}[\frac{1}{(1 - d/z)^2} + \frac{1}{(1 + d/z)^2} - 2],$$

$$E \simeq \frac{q}{4\pi\epsilon_0 z^2}[(1 + 2d/z + 3d^2/z^2) + (1 - 2d/z + 3d^2/z^2) - 2],$$

$$E = \frac{q}{4\pi\epsilon_0 z^2}[6d^2/z^2] = \frac{3Q}{4\pi\epsilon_0 z^4},$$

where $Q = 2qd^2$.

35P

The equilibrium position of the electron is at the center of the ring, since the net electrical force on it when there is zero. The force on the electron when it is at a distance z away from this point along the axis of the ring is F = eE, with E given by Eq.17, so that

$$F = eE = \frac{1}{4\pi\epsilon_0} \frac{eqz}{(R^2 + z^2)^{3/2}}.$$

This force is attractive, being directed back toward the center of the ring. If z << R, then the z^2 in the denominator is very small compared to R^2 and can be deleted, leaving

$$F = \frac{1}{4\pi\epsilon_0} \frac{eq}{R^3} z.$$

By Newton's second law, and noting that F is directed to x = 0,

$$\frac{1}{4\pi\epsilon_0} \frac{eq}{R^3} z = -ma \rightarrow a = -(\frac{1}{4\pi\epsilon_0} \frac{eq}{mR^3})z.$$

This has the same form as the equation describing simple harmonic motion; see Eq.7 in Chapter 14; we have, then,

$$a = -\omega^2 z \rightarrow \omega = \sqrt{[\frac{eq}{4\pi\epsilon_0 mR^3}]}.$$

39P

Due to the element of charge λdz located at S as shown, the contribution to the electric field at P is

$$dE = (1/4\pi\epsilon_0)\lambda dz/r^2,$$

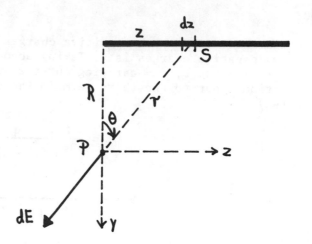

directed along SP. The components of this field are

$$dE_y = (dE)\cos\theta, \quad dE_z = -(dE)\sin\theta.$$

Also note that

$$\cos\theta = R/r, \quad \sin\theta = z/r, \quad r^2 = R^2 + z^2.$$

Therefore,

$$E_y = \int_0^\infty \frac{\lambda}{4\pi\epsilon_0} \frac{1}{R^2 + z^2} \frac{R}{\sqrt{(R^2 + z^2)}} \, dz = \frac{\lambda}{4\pi\epsilon_0 R};$$

$$E_z = - \int_0^\infty \frac{\lambda}{4\pi\epsilon_0} \frac{1}{R^2 + z^2} \frac{z}{\sqrt{(R^2 + z^2)}} \, dz = - \frac{\lambda}{4\pi\epsilon_0 R}.$$

In magnitude then, $E_y = E_z$ and therefore $\tilde{\theta} = 45°$, where $\tilde{\theta}$ is the angle made by E with the rod.

42P

Eq.21 gives the field on the axis of a charged disk. The surface field is found by substituting $z = 0$ to get

$$E_{surf} = \frac{\sigma}{2\epsilon_0}.$$

We want to find z so that $E = E_{surf}/2$; that is,

$$\frac{\sigma}{4\epsilon_0} = \frac{\sigma}{2\epsilon_0}[1 - \frac{z}{\sqrt{(z^2 + R^2)}}],$$

$$\frac{z}{\sqrt{(z^2 + R^2)}} = \frac{1}{2},$$

$$\frac{z^2}{z^2 + R^2} = (\frac{1}{2})^2 = \frac{1}{4} \rightarrow z = R/\sqrt{3}.$$

<u>44E</u>

(a) By Eq.2,

$$a = \frac{F}{m} = \frac{eE}{m} = \frac{(1.6 \times 10^{-19} \text{ C})(2 \times 10^4 \text{ N/C})}{1.67 \times 10^{-27} \text{ kg}} = 1.92 \times 10^{12} \text{ m/s}^2.$$

(b) By Eq.15 in Chapter 2, with $v_0 = 0$,

$$v = \sqrt{[2a(x - x_0)]} = \sqrt{[2(1.92 \times 10^{12} \text{ m/s}^2)(10^{-2} \text{ m})]} = 196 \text{ km/s}.$$

<u>45E</u>

(a) We use SI units throughout. The acceleration of the electron is

$$a = \frac{F}{m} = \frac{eE}{m} = \frac{(1.6 \times 10^{-19})(10^3)}{9.11 \times 10^{-31}} = 1.756 \times 10^{14} \text{ m/s}^2.$$

Noting that the acceleration is directed to stop the electron, i.e., is in the direction opposite to that of the initial velocity, Eq.15 of Chapter 2 gives

$$v^2 = v_0^2 + 2ax,$$

$$0^2 = (5 \times 10^6)^2 + 2(-1.756 \times 10^{14})x \rightarrow x = 7.12 \times 10^{-2} \text{ m} = 7.12 \text{ cm}.$$

(b) The time required to stop is given from Eq.10 of Chapter 2:

$$v = v_0 + at,$$

$$0 = 5 \times 10^6 + (-1.756 \times 10^{14})t \rightarrow t = 2.85 \times 10^{-8} \text{ s} = 28.5 \text{ ns}.$$

(c) The initial kinetic energy is

$$K_0 = \tfrac{1}{2}mv_0^2 = \tfrac{1}{2}(9.11 \times 10^{-31})(5 \times 10^6)^2 = 1.139 \times 10^{-17} \text{ J}.$$

The force on the electron is $F = eE$. Use the work-energy theorem:

$$W = \Delta K = -eEx.$$

But

$$eEx = (1.6 \times 10^{-19})(10^3)(0.008) = 0.128 \times 10^{-17} \text{ J},$$

so that

$$\frac{\Delta K}{K_0} = \frac{|W|}{K_0} = \frac{0.128 \times 10^{-17}}{1.139 \times 10^{-17}} = 0.112 = 11.2\%.$$

47E

See Sample Problem 7, to find for the charge, in SI units,

$$q = \frac{1}{E}[\frac{4}{3}\pi R^3 \rho g] = \frac{1}{1.92 \text{ X } 10^5}[\frac{4}{3}\pi(1.64 \text{ X } 10^{-6})^3(851)(9.8)] = 8.026 \text{ X } 10^{-19} \text{ C}.$$

By Eq.23,

$$n = \frac{q}{e} = \frac{8.026 \text{ X } 10^{-19} \text{ C}}{1.6 \text{ X } 10^{-19} \text{ C}} = 5,$$

so that q = 5e.

54P

Let the x axis be horizontal and to the right in Fig.36, the y axis vertically up, with the origin at the projection point of the electron. Since the charge on the electron is negative, the force on it will be directed downward, opposite to the electric field. Gravity can be ignored, as the magnitude of the electrical acceleration will indicate. This acceleration will be (and we work in SI units, as usual, throughout)

$$a = \frac{F}{m} = \frac{eE}{m} = \frac{(1.6 \text{ X } 10^{-19})(2000)}{9.11 \text{ X } 10^{-31}} = 3.513 \text{ X } 10^{14} \text{ m/s}^2.$$

To determine if the electron strikes the top plate, solve for the time t required from projection for the electron to reach a height y = 2 cm = 0.02 m. Since the acceleration is constant, we have

$$y = \frac{1}{2}at^2 + (v_0 \sin\theta)t.$$

Now,

$$v_0 \sin\theta = (6 \text{ X } 10^6)\sin 45° = 4.243 \text{ X } 10^6 \text{ m/s}.$$

Use this, together with y = 0.02 m to obtain

$$0.02 = \frac{1}{2}(-3.513 \text{ X } 10^{14})t^2 + (4.243 \text{ X } 10^6)t \rightarrow t = 6.420 \text{ X } 10^{-9} \text{ s}; 1.774 \text{ X } 10^{-8} \text{ s}.$$

The smaller of the two solutions of the quadratic equation is chosen (the other refers to the descending part of the trajectory if the motion is not interrupted by the upper plate). The horizontal distance traveled in this time is

$$x = (v_0 \cos\theta)t = (4.243 \text{ X } 10^6)(6.42 \text{ X } 10^{-9}) = 0.0272 \text{ m} = 2.72 \text{ cm}.$$

Since 2.72 cm < 10 cm, (a) the electron will strike the upper plate (b) at a distance of 2.72 cm from the left edge.

58P

The torque on a dipole is $\tau = pE\sin\theta$, where θ is the angular displacement from the equilibrium position. But $\tau = I\alpha = -pE\sin\theta$, the minus sign added since the torque tends to return the dipole to its equilibrium position (i.e., toward smaller values of θ).

The equilibrium position is $\theta = 0$. For small displacements $\sin\theta \simeq \theta$ and therefore

$$d^2\theta/dt^2 = -(pE/I)\theta = -\omega^2\theta,$$

so that

$$\nu = \omega/2\pi = \frac{1}{2\pi}\sqrt{(pE/I)},$$

the preceding equation having the general form of the equation describing simple harmonic motion (see Eq.7 in Chapter 14) written for angular variables.

CHAPTER 25

2E

At all points on the surface the magnitude of the electric field is E = 1800 N/C and the angle between the electric field and the outward pointing normal is $\theta = (180° - 35°) = 145°$. Although the square is not a closed surface, we can still apply the definition of flux contained in Eq.6:

$$\Phi_E = \int \vec{E} \cdot d\vec{A} = \int E\cos\theta dA = E\cos\theta \int dA = EA\cos\theta = (1800 \text{ N/C})(0.0032 \text{ m})^2 \cos 145°,$$

$$\Phi_E = -0.0151 \text{ N·m}^2/\text{C}.$$

7E

By Gauss' law, Eq.9, the flux is

$$\Phi_E = \oint \vec{E} \cdot d\vec{A} = q/\epsilon_0 = (1.8 \times 10^{-6} \text{ C})/(8.85 \times 10^{-12} \text{ C}^2/\text{N·m}^2) = 2.03 \times 10^5 \text{ N·m}^2/\text{C}.$$

11P

Along the vertical sides of the cube E is perpendicular to dA so that $\vec{E} \cdot d\vec{A} = EdA\cos 90°$ = 0. Along the upper face of the cube, the electric field E and dA point in opposite directions. Since the field has the same magnitude at all points on the upper surface

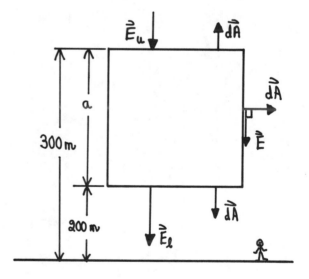

$$\Phi_u = E_u A\cos 180° = -E_u A = -E_u a^2.$$

Along the bottom surface the electric field E and dA point in the same direction, so that

$$\Phi_\ell = E_\ell A\cos 0° = E_\ell a^2.$$

Hence, by Gauss' law, in SI units,

$$\Phi = E_\ell a^2 - E_u a^2 = (E_\ell - E_u)a^2 = q/\epsilon_0,$$

$$q = \epsilon_0(E_\ell - E_u)a^2 = (8.85 \times 10^{-12})(100 - 60)(100)^2 = 3.54 \times 10^{-6} \text{ C} = 3.54 \text{ μC}.$$

13P

Refer to the sketch on p.201. The flux through those three faces (such as number 2) that meet at the charge is zero, since E and dA are at a 90° angle on these surfaces. The flux through the other surfaces (represented by surface 1) are identical and can be computed from the definition of flux (without recourse to Gauss' law) as follows. The electric field strength at a point on the surfaces is

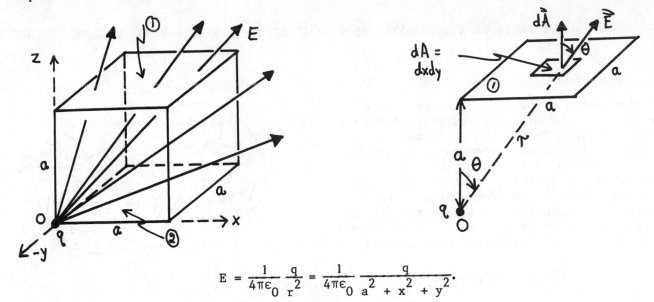

$$E = \frac{1}{4\pi\epsilon_0} \frac{q}{r^2} = \frac{1}{4\pi\epsilon_0} \frac{q}{a^2 + x^2 + y^2}.$$

Note that

$$\cos\theta = a/r; \quad dA = dxdy.$$

Therefore the flux is

$$\Phi_E = \int \vec{E} \cdot d\vec{A} = \frac{aq}{4\pi\epsilon_0} \int_0^a\int_0^a \frac{dxdy}{(a^2 + x^2 + y^2)^{3/2}},$$

since $\vec{E} \cdot d\vec{A} = EdA\cos\theta$. Integrating over x first gives

$$\Phi_E = \frac{aq}{4\pi\epsilon_0} \int_0^a \frac{xdy}{(a^2 + y^2)\sqrt{(a^2 + x^2 + y^2)}}\Big|_0^a,$$

$$\Phi_E = \frac{a^2 q}{4\pi\epsilon_0} \int_0^a \frac{dy}{(a^2 + y^2)\sqrt{(2a^2 + y^2)}},$$

$$\Phi_E = \frac{q}{4\pi\epsilon_0}\tan^{-1}\Big[\frac{y}{\sqrt{(2a^2 + y^2)}}\Big]_0^a = \frac{q}{24\epsilon_0}.$$

Hence, the total flux through the three faces not touching the charge is $q/8\epsilon_0$, and since the flux through each of the other faces is zero, the net flux through the cube is $q/8\epsilon_0$. This agrees with Gauss' law since the cube subtends at q a solid angle equal to 1/8th that subtended if the charge lay inside (one quadrant versus eight quadrants). Thus the flux must be $(1/8)(q/\epsilon_0)$, as obtained by direct calculation above.

<u>15E</u>

(a) The charge on the sphere is

$$q = \sigma A = (8.1 \times 10^{-6} \text{ C/m}^2)[4\pi(0.6 \text{ m})^2] = 3.66 \times 10^{-5} \text{ C} = 36.6 \text{ } \mu\text{C}.$$

(b) By Gauss's law, the flux is

$$\Phi_E = q/\varepsilon_0 = (3.66 \text{ X } 10^{-5} \text{ C})/(8.85 \text{ X } 10^{-12} \text{ C}^2/\text{N} \cdot \text{m}^2) = 4.14 \text{ X } 10^6 \text{ N} \cdot \text{m}^2/\text{C}.$$

18P

(a) The sketch shows the conductor with a cavity. Also shown is the Gaussian surface S which lies just inside the conductor and "parallel" to the cavity surface. Q_1 and Q_2 are the charges on the inner and outer surfaces and q is the enclosed charge. (We know that there is no free charge in the body of the conductor.) Apply Gauss' law to the surface S; be aware that E = 0 everywhere on S since S lies inside the conductor. Thus,

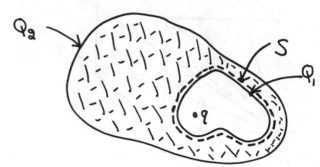

$$\Phi_E = \oint_S \vec{E} \cdot d\vec{A} = (q + Q_1)/\varepsilon_0,$$

$$0 = (q + Q_1)/\varepsilon_0 \rightarrow Q_1 = -q = -3.0 \text{ } \mu\text{C}.$$

(b) We have been told that the net charge on the conductor is 10 μC. Therefore,

$$Q_1 + Q_2 = 10 \text{ } \mu\text{C},$$

$$Q_2 = 10 - Q_1 = 10 - (-3) = +13 \text{ } \mu\text{C}.$$

20E

Eq.14 applies to a line of charge. Solve for λ and substitute the data in SI units to obtain

$$\lambda = (2\pi\varepsilon_0)rE = \frac{1}{2}(4\pi\varepsilon_0)rE = \frac{1}{2}(8.99 \text{ X } 10^9)^{-1}(2)(4.5 \text{ X } 10^4) = 5.01 \text{ } \mu\text{C/m}.$$

29P

(a) By symmetry, it is expected that the field will be azimuthally symmetric around the cylinder axis, the lines being radially inward if the charge is negative and outward if the charge is positive. Positive charge is assumed here. Construct a coaxial, cylindrical Gaussian surface (ends are included) of radius r and length L. By Gauss' law,

$$\varepsilon_0 \oint \vec{E} \cdot d\vec{A} = q_{enc},$$

where q_{enc} is the charge inside the Gaussian surface over which the integral is taken. For convenience, the integral is divided into two parts, one over the curved portion of the cylindrical surface, and the second part over the two end caps:

$$\varepsilon_0 \int_{\substack{\text{curved} \\ \text{surface}}} \vec{E} \cdot d\vec{A} + \varepsilon_0 \int_{\text{ends}} \vec{E} \cdot d\vec{A} = q_{enc}.$$

Now, $\vec{E} \cdot d\vec{A} = EdA\cos\angle(\vec{E}, d\vec{A})$. On the flat ends the angle between E and dA is 90° and cos90°

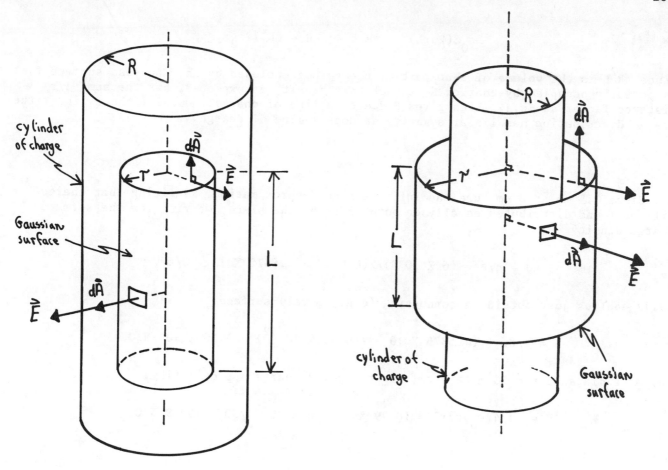

= 0. Hence, the integral over the ends (indeed, over each end) is zero. On the curved surface, however, the angle is zero (or 180° if the charge is negative), the cosine of which is +1 (-1). Also, since all points on the curved surface are at the same distance from the axis of the rod, E is independent of location on the curved surface, that is, the strength (magnitude) of E is constant. Apply these considerations to the curved surface to get

$$\int \vec{E} \cdot d\vec{A} = \int E dA = E \int dA = E(2\pi rL).$$

Since the entire Gaussian surface is inside the rod, the charge enclosed is, with ρ = the charge density,

$$q_{enc} = (\rho)(\pi r^2 L),$$

so that Gauss' law gives

$$\varepsilon_0 E(2\pi rL) = \rho \pi r^2 L \quad \rightarrow \quad E = \rho r/2\varepsilon_0.$$

(b) For R < r, draw the Gaussian surface outside the rod; now only part of the volume that is enclosed by the Gaussian surface is occupied by charge. Thus for this case, by Gauss' law,

$$\varepsilon_0 E(2\pi rL) = \rho\pi R^2 L \;\rightarrow\; E = \rho R^2/2\varepsilon_0 r,$$

since $\pi R^2 L$ is the volume of that part of the volume enclosed by the Gaussian surface that is actually occupied by charge. Note that this result, as expected, has the same $1/r$ distance factor as Eq.14 which gives E due to a line of charge; outside the cylinder the charge distribution has linear symmetry as does a line of charge.

31E

(a) Since 0.5 mm << 8 cm, use the infinite plane approximation. Realizing that charge will be found distributed equally on both sides of the plate, we find for the surface charge density

$$\sigma = q/A = (6 \times 10^{-6})/2(0.08)^2 = 4.6875 \times 10^{-4} \text{ C/m}^2.$$

Eq.12 applies just outside a conducting (e.g., metal) surface:

$$E = \sigma/\varepsilon_0 = (4.6875 \times 10^{-4})/(8.85 \times 10^{-12}) = 5.30 \times 10^7 \text{ N/C.}$$

(b) At a distance r = 30 m >> 8 cm, use the point charge approximation:

$$E = (1/4\pi\varepsilon_0)q/r^2 = (8.99 \times 10^9)(6 \times 10^{-6})/30^2 = 59.9 \text{ N/C.}$$

35E

The restriction on points to be considered is tantamount to regarding the plates as infinite in two dimensions (the thickness is unaffected). The field lines, by symmetry, must pass perpendicular to the plates from the positive to the negative charges. Under electrostatic conditions no lines can penetrate into the conducting plates, and since there is no fringing of the lines around the edges, there being no edges in effect, the field must be zero at all points not between the plates; to wit, (a) E = 0, and (c) E = 0. (b) To calculate E at points between the plates, construct a right-rectangular cylindrical "pillbox" Gaussian surface (any shape cross-section of area A), with end caps parallel to the plates, as shown. The flux, $\int \vec{E} \cdot d\vec{A}$, over that part of the Gaussian surface lying inside the plate is zero since E = 0 there. The flux is also zero over the curved part of the Gaussian surface found outside the plate: this is because here E and dA are perpendicular. Over the left end cap, however, E and dA are parallel, and since all points on the left end cap are at the same distance from the plate, Gauss' law yields

$$\varepsilon_0 \oint \vec{E} \cdot d\vec{A} = \varepsilon_0 \int_{\substack{\text{left} \\ \text{cap}}} \vec{E} \cdot d\vec{A} = \varepsilon_0 \int E dA = \varepsilon_0 E \int dA = \varepsilon_0 E A = q_{enc}.$$

The charge enclosed by the Gaussian surface resides on that part of the plate lying inside the surface; i.e., $q_{enc} = \sigma A$. Therefore,

$$\varepsilon_0 EA = \sigma A \rightarrow E = \sigma/\varepsilon_0.$$

This is directed to the left (toward the negative plate). It may appear as though the presence of the negative plate has been ignored, but this is not so for, if that plate was absent, the positive charge on the remaining plate (now isolated) would then be found distributed on both sides of the plate, making $E = \sigma_{isol}/\varepsilon_0 = (\sigma/2)/\varepsilon_0$ (by Eq.12) at all points not inside the plate.

38P

(a) Consider the slab to be of very large area and the charge positive. By symmetry, the field inside the slab will be directed at 90° to the slab surface but will vary in strength with the distance x. On the median plane (x = 0) we expect E = 0. Construct a cylindrical Gaussian surface oriented as shown. The flux over the curved part of the surface is zero since E and dA are here at a 90° angle. Since each end cap is at the same distance from the median plane E is the same on each, so that the flux is

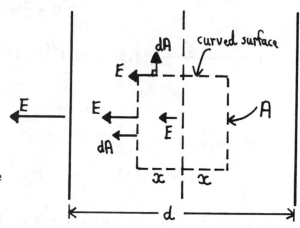

$$\Phi_E = \oint \vec{E} \cdot d\vec{A} = 2EA,$$

where E is the electric field at a distance x from the median plane and A is the area of each end cap. The charge enclosed by the Gaussian surface is the volume charge density of the slab multiplied by the volume that is enclosed by the Gaussian surface, since this entire volume contains charge; hence, $q_{enc} = \rho(2xA)$, so that Gauss' law gives

$$\varepsilon_0(2EA) = \rho(2xA) \rightarrow E = \rho x/\varepsilon_0.$$

(b) Since the field outside the slab is at 90° to the slab surface, the electric field outside a single nonconducting sheet, given in Eq.15, is valid here, so that $E = \sigma/2\varepsilon_0$. But we have

$$q_{enc} = \sigma A = \rho V = \rho A d \rightarrow \sigma = \rho d.$$

Therefore $E = \rho d/2\varepsilon_0$.

39E

Since the field is directed inward, the charge must be negative. The magnitude q of the charge is given by Eq.16 since we are outside the sphere. In SI units,

$$q = (4\pi\varepsilon_0)Er^2 = (8.99 \times 10^9)^{-1}(3000)(0.15)^2 = 7.51 \text{ nC}.$$

50P

For uniform circular motion $a = v^2/r$ (Eq.18 in Chapter 6). Outside a charged sphere the electric field due to the sphere is given by Eq.16. Let Q be the magnitude of the charge on the sphere. (An attractive force is required, so the actual charge on the sphere is $-Q$.) By Newton's second law

$$eE = ma,$$

$$\frac{1}{4\pi\varepsilon_0} \frac{Qe}{r^2} = m\frac{v^2}{r},$$

$$Q = \frac{1}{e}(4\pi\varepsilon_0)mv^2 r,$$

$$Q = \frac{1}{1.6 \times 10^{-19}}(8.99 \times 10^9)^{-1}(1.67 \times 10^{-27})(3 \times 10^5)^2(10^{-2}) = 1.04 \text{ nC}.$$

52P

The field due to q is

$$E_q = \frac{1}{4\pi\varepsilon_0}\frac{q}{r^2}.$$

The field due to the sphere, for $a < r < b$, is found from Gauss' law by constructing a spherical Gaussian surface centered at $r = 0$, the center of the sphere. There is no charge between $r = 0$ and $r = a$. Hence,

$$\varepsilon_0 E \cdot 4\pi r^2 = q_{enc} = \int \rho dV = \int_a^r (\frac{A}{r})4\pi r^2 dr = 2\pi A(r^2 - a^2) \rightarrow E = \frac{A}{2\varepsilon_0}(1 - \frac{a^2}{r^2}),$$

radially in or out. In order that $\vec{E}_q + \vec{E}$ be independent of r the following equation, representing the sum of the r-dependent terms, must be satisfied:

$$-Aa^2/2\varepsilon_0 + q/4\pi\varepsilon_0 = 0 \rightarrow A = q/2\pi a^2.$$

54P

Assume that no charge is located in the immediate vicinity of P. Imagine a spherical Gaussian surface of very small radius with its center at P. Let the test charge +q be displaced from P to any point on the Gaussian surface. For stable equilibrium the charge must experience a force that tends to push it back toward P; i.e., the electrostatic field E must have an inward directed normal component everywhere on the Gaussian surface. But, by Gauss' law, this means that net negative charge must reside inside the Gaussian surface. This is contrary to assumption. A similar argument can be made for negative test charges.

CHAPTER 26

2E

(a) Since 1 A = 1 C/s, we have

$$q = 84 \text{ A} \cdot \text{h} = 84\left(\frac{C}{s}\right)(3600 \text{ s}) = 3.02 \text{ X } 10^5 \text{ C}.$$

(b) By Eq.4, the energy (work) is

$$W = qV = (3.02 \text{ X } 10^5 \text{ C})(12 \text{ V}) = 3.62 \text{ MJ}.$$

3P

(a) The energy E released in the flash is, by Eq.4,

$$E = qV = (30 \text{ C})(10^9 \text{ V}) = 3 \text{ X } 10^{10} \text{ J}.$$

(b) Set the energy in (a) equal to the final kinetic energy of the car:

$$E = K = \frac{1}{2}mv^2,$$

$$3 \text{ X } 10^{10} \text{ J} = \frac{1}{2}(1000 \text{ kg})v^2 \rightarrow v = 7.75 \text{ X } 10^3 \text{ m/s}.$$

(c) Using the energy E found in (a), the mass m of ice that can be melted is, by Eq.5 in Chapter 20,

$$m = \frac{Q}{L} = \frac{3 \text{ X } 10^{10} \text{ J}}{3.3 \text{ X } 10^5 \text{ J/kg}} = 9.10 \text{ X } 10^4 \text{ kg}.$$

9E

As noted in Table 1 for the infinite sheet,

$$V = V_0 - (\sigma/2\epsilon_0)z.$$

Hence, for two equipotentials a distance Δz apart,

$$\Delta V = -(\sigma/2\epsilon_0)\Delta z.$$

In absolute value (and working in SI units),

$$\Delta z = 2\epsilon_0(\Delta V)/\sigma = 2(8.85 \text{ X } 10^{-12})(50)/(0.1 \text{ X } 10^{-6}) = 8.85 \text{ X } 10^{-3} \text{ m} = 8.85 \text{ mm}.$$

12P

The radius of the cylinder is b = 1.0 cm = 0.01 m and the radius of the wire is 0.65 X 10^{-4} cm = 6.5 X 10^{-7} m. By Problem 24 in Chapter 25, the electric field between the wire and the cylinder is

$$E = \lambda/2\pi\epsilon_0 r.$$

Hence, the potential difference, in absolute value, between the wire and the cylinder is, by Eq.9,

$$V = \int E dr = \int_a^b \lambda dr/2\pi\epsilon_0 r = (\frac{\lambda}{2\pi\epsilon_0}) \ln(\frac{b}{a}).$$

Numerically,

$$850 = (\frac{\lambda}{2\pi\epsilon_0}) \ln(\frac{10^{-2}}{6.5 \times 10^{-7}}) \rightarrow \lambda/2\pi\epsilon_0 = 88.164 \text{ V.}$$

Therefore,

$$E = \lambda/2\pi\epsilon_0 r = 88.164/r.$$

(a) At the surface of the wire, $E = 88.164/(6.5 \times 10^{-7}) = 136$ MV/m.

(b) At the cylinder, $E = 88.164/0.01 = 8.82$ kV/m.

13P

(a) Let the point P at which the potential is desired be at a distance $r = a$ from the center of the sphere of charge, with $a < R$. Then,

$$V_P - V_\infty = - \int_\infty^P \vec{E} \cdot d\vec{s};$$

choosing $V_\infty = 0$ reduces this to

$$V_P = - \int_\infty^P \vec{E} \cdot d\vec{s}.$$

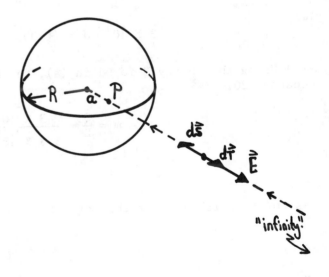

For convenience, integrate along a straight line from infinity to P. Now ds points in the direction of integration, opposite to dr which points outward from the center of the sphere. Also, $r = 0$ at the center of the sphere, but $s = 0$ at infinity. Therefore,

$$\vec{E} \cdot d\vec{s} = (E)(ds)\cos 180° = -(E)(ds) = -E(-dr) = (E)(dr),$$

assuming that the charge is positive, so that E points outward. Substituting the last result into the integral for V_P yields

$$V_P = - \int_\infty^a E dr.$$

From Chapter 25 (Sample Problem 7) it is realized that E has different analytical forms for $r < R$ and for $r > R$:

$$E(r) = \frac{1}{4\pi\epsilon_0} \frac{q}{r^2}, \quad R < r; \quad E(r) = \frac{1}{4\pi\epsilon_0} \frac{qr}{R^3}, \quad r < R.$$

For this reason, the integral must be divided into two parts, one for points exterior, and one for points interior, to the sphere of charge, and the appropriate equation for E used in each:

$$V_P = -\int_\infty^R (q/4\pi\epsilon_0)dr/r^2 - \int_R^a (q/4\pi\epsilon_0 R^3)rdr,$$

$$V_P = -\frac{q}{4\pi\epsilon_0}[(-\frac{1}{R} + 0) + \frac{1}{R^3}(\frac{1}{2}a^2 - \frac{1}{2}R^2)] = \frac{q(3R^2 - a^2)}{8\pi\epsilon_0 R^3}.$$

Substitution of r for a gives the desired form of the result.

(b) In this problem V = 0 at infinity; in Problem 11, V = 0 at the center of the sphere.

18E

The potential due to a sphere is, by Sample Problem 5,

$$V = q/4\pi\epsilon_0 R.$$

But q = ne (see Eq.8 of Chapter 23), where n = the number of excess electrons. Hence,

$$n = (4\pi\epsilon_0)VR/e = (8.99 \times 10^9)^{-1}(400)(10^{-6})/(1.6 \times 10^{-19}) = 278,000.$$

23P

From Sample Problem 5 we learn that Eq.12 applies at points external to a sphere of charge, so that $V = (1/4\pi\epsilon_0)q/r$. But for a sphere the electric field at external points is, by Eq.16 of Chapter 25, $E = (1/4\pi\epsilon_0)q/r^2$. Hence,

$$V = Er = (100 \text{ V/m})(6.37 \times 10^6 \text{ m}) = 6.37 \times 10^8 \text{ V} = 637 \text{ MV}.$$

25P

(a) For a spherical charge distribution the potential at the surface, by Eq.12, is

$$V = q/4\pi\epsilon_0 R = (8.99 \times 10^9)(30 \times 10^{-12})/R = 500 \rightarrow R = 5.39 \times 10^{-4} \text{ m} = 0.539 \text{ mm}.$$

(b) The radius r of the combined spherical drop is found from

$$4\pi r^3/3 = 2(4\pi R^3/3) \rightarrow r = 2^{1/3}R.$$

The total charge on the combined drop is $2q = 6 \times 10^{-11}$ C. If the charge is spread uniformly throughout the drop, the surface potential becomes

$$V^* = 2q/4\pi\epsilon_0 2^{1/3}R = (6 \times 10^{-11})(8.99 \times 10^9)/2^{1/3}(5.39 \times 10^{-4}) = 794 \text{ V}.$$

29P

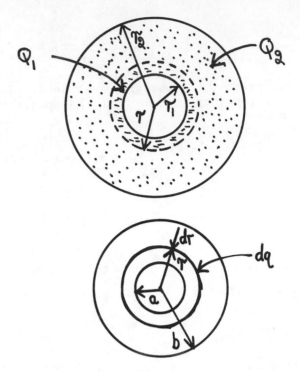

(a) Let Q be the total charge on the shell. For $r > r_2$, the potential is identical with that of a point charge at the center of the actual distribution: $V = Q/4\pi\epsilon_0 r$.

(b) In this region, the charge distribution can be considered as made up of two parts: a charge Q_1 interior to r and a charge Q_2 exterior to r. The potential V at r is the sum $V_1 + V_2$ of the potentials due to each of these two parts. The point considered is at the surface of Q_1 so $V_1 = Q_1/4\pi\epsilon_0 r$. To find V_2 it is necessary to calculate the potential V^* in the cavity of a thick shell of charge. Consider such a shell carrying charge q and of inner radius a and outer radius b. The shell can be divided into many very thin shells, such as the one shown of radius r and thickness dr. The charge on the thin shell is dq, so the potential dV^* due to the thin shell is $dV^* = dq/4\pi\epsilon_0 r$. But

$$dq = [q/(\tfrac{4}{3}\pi b^3 - \tfrac{4}{3}\pi a^3)](4\pi r^2 dr),$$

the term in square brackets being the charge per unit volume on the thick shell; $4\pi r^2 dr$ is the volume of the thin shell of radius r. Hence,

$$dV^* = \frac{qrdr}{\tfrac{4}{3}\pi\epsilon_0(b^3 - a^3)},$$

and

$$V^* = \int_a^b dV^* = \frac{3q}{8\pi\epsilon_0}\frac{b^2 - a^2}{b^3 - a^3}.$$

Applying this result to V_2, put $b = r_2$, $a = r$, $q = Q_2$ to get

$$V_2 = \frac{3Q_2}{8\pi\epsilon_0}\frac{r_2^2 - r^2}{r_2^3 - r^3}.$$

It remains to calculate Q_1 and Q_2 in terms of Q. The charge per unit volume ρ on the original thick shell is

$$\rho = Q/[\tfrac{4}{3}\pi r_2^3 - \tfrac{4}{3}\pi r_1^3],$$

so that

$$Q_1 = \rho(\tfrac{4}{3}\pi r^3 - \tfrac{4}{3}\pi r_1^3) = Q(r^3 - r_1^3)/(r_2^3 - r_1^3),$$

$$Q_2 = Q - Q_1 = Q(r_2^3 - r^3)/(r_2^3 - r_1^3).$$

Substituting these into the expressions for V_1 and V_2 and adding gives

$$V = \frac{Q}{4\pi\epsilon_0(r_2^3 - r_1^3)}(\frac{3}{2}r_2^2 - \frac{1}{2}r^2 - \frac{r_1^3}{r}).$$

(c) For $r < r_1$, i.e., inside the cavity, use the result from (b) but now set $r = r_1$ (the inner boundary of the original shell), so that

$$V = \frac{3Q}{8\pi\epsilon_0(r_2^3 - r_1^3)}(r_2^2 - r_1^2).$$

(d) The appropriate potentials agree at the boundaries of the regions in which they are valid; i.e., $V_a = V_b$ at $r = r_2$ and $V_b = V_c$ at $r = r_1$.

31P

Put the origin at the middle charge. Then

$$V = \frac{1}{4\pi\epsilon_0}\frac{q}{r - d} + \frac{1}{4\pi\epsilon_0}\frac{q}{r} + \frac{1}{4\pi\epsilon_0}\frac{-q}{r + d} = \frac{q}{4\pi\epsilon_0}\frac{r^2 + 2rd - d^2}{(r^2 - d^2)r}.$$

If $r \gg d$ (distant point), the d^2 terms can be ignored, so the above becomes

$$V = \frac{q}{4\pi\epsilon_0}(\frac{r + 2d}{r^2}) = \frac{1}{4\pi\epsilon_0}(\frac{q}{r} + \frac{2qd}{r^2}).$$

The first term in the last expression is the potential due to a point charge (the middle charge), and the second term is the potential due to a dipole (the outer charges) at large distances (compared to the dipole charge separation) along the dipole axis.

32P

(a) Let $d\ell$ be an element of length on the ring. The linear charge density on the ring is $q/2\pi R$. Hence, due to the small element the potential at P is

$$dV = \frac{1}{4\pi\epsilon_0}\frac{dq}{r} = \frac{1}{4\pi\epsilon_0}\frac{(q/2\pi R)d\ell}{\sqrt{(R^2 + z^2)}}.$$

The total potential at P due to the entire ring is

$$V = \int dV = \frac{1}{4\pi\epsilon_0}\frac{(q/2\pi R)}{\sqrt{(R^2 + z^2)}}\int d\ell,$$

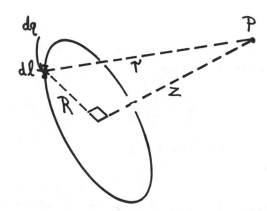

noting that R and z each are independent of the element's position on the ring. Now, $\int d\ell = 2\pi R$, the circumference of the ring, and therefore

$$V = \frac{1}{4\pi\epsilon_0}\frac{q}{\sqrt{(R^2 + z^2)}}.$$

(b) By symmetry, the component of E at the axis is zero in any direction perpendicular to the axis. Thus

$$E = E_z = -\frac{dV}{dz} = \frac{q}{4\pi\epsilon_0} \frac{z}{(R^2 + z^2)^{3/2}}.$$

39P

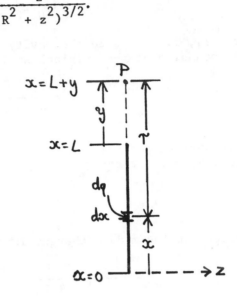

(a) With $dq = \lambda dx$ = charge on an element of length dx, the potential at P is found from

$$4\pi\epsilon_0 V = \int\frac{\lambda dx}{r} = \int_0^L \frac{\lambda dx}{(L + y) - x},$$

$$4\pi\epsilon_0 V = -\lambda \ln(L + y - x)\Big|_0^L,$$

$$4\pi\epsilon_0 V = -\lambda[\ln y - \ln(L + y)],$$

$$V = (\frac{\lambda}{4\pi\epsilon_0})\ln(\frac{L + y}{y}).$$

(b) Now call the axis the y axis and move the origin from the lower end to the upper end of the charged segment, as shown on the sketch. Then

$$E_y = -\frac{dV}{dy} = \frac{\lambda}{4\pi\epsilon_0} \frac{L}{y(L + y)}.$$

(c) The direction of the electric field at P due to any element of the segment of charge is directed along the y (or x) axis; hence $E_z = 0$. This result cannot be derived from the potential obtained in (a) for that expression is valid only at $z = 0$ and therefore dV/dz, which requires knowledge of V in a neighborhood in z about P, cannot be computed from it.

42E

(a) The distance of C from each of the original two charges is $r = \sqrt{[(d/2)^2 + (d/2)^2]} = d/\sqrt{2}$. Therefore, in SI units,

$$V_C = 2[\frac{1}{4\pi\epsilon_0}\frac{q}{r}] = 2(8.99 \times 10^9)\frac{2 \times 10^{-6}}{(0.02)/\sqrt{2}} = 2.54 \text{ MV}.$$

(b) See Sample Problem 10. Calculate the potential energy of the final assembly of the 3 charges:

$$U_f = \frac{q^2}{4\pi\epsilon_0}[\frac{1}{d} + \frac{1}{d/\sqrt{2}} + \frac{1}{d/\sqrt{2}}] = \frac{q^2}{4\pi\epsilon_0 d}[1 + 2\sqrt{2}] = \frac{(2 \times 10^{-6})^2}{0.02}(8.99 \times 10^9)(1 + 2\sqrt{2}),$$

$$U_f = 6.884 \text{ J}.$$

The potential energy of the system before the third charge is brought up is

$$U_i = \frac{1}{4\pi\epsilon_0} \frac{q^2}{d} = (8.99 \times 10^9)\frac{(2 \times 10^{-6})^2}{0.02} = 1.798 \text{ J.}$$

(The third charge is at infinity and thereby contributes nothing to U_i.) Hence, the work that you (external agent) must do to bring up the third charge is $W = U_f - U_i = 6.884 - 1.798 = 5.09$ J.

(c) From (b), $U_f = 6.88$ J.

50P

Let W_e = work to be done against electrostatic forces; this is

$$W_e = (-q)(V_2 - V_1) = -q(\frac{1}{4\pi\epsilon_0} \frac{Q}{r_2} - \frac{1}{4\pi\epsilon_0} \frac{Q}{r_1}) = \frac{Qq}{4\pi\epsilon_0}(\frac{1}{r_1} - \frac{1}{r_2}).$$

The initial kinetic energy of the electron can supply energy to do some of this work. For uniform circular motion,

$$F = \frac{1}{4\pi\epsilon_0} \frac{Qq}{r^2} = \frac{mv^2}{r} \rightarrow K = \frac{1}{2}mv^2 = \frac{1}{2}\frac{1}{4\pi\epsilon_0} \frac{Qq}{r}.$$

At the larger radius the electron still has kinetic energy of revolution. Thus the kinetic energy available to do work against the Coulomb force is

$$K_1 - K_2 = \frac{1}{2}\frac{Qq}{4\pi\epsilon_0}(\frac{1}{r_1} - \frac{1}{r_2}).$$

This implies that the work W that the external agent must do is

$$W = W_e - (K_1 - K_2) = \frac{Qq}{8\pi\epsilon_0}(\frac{1}{r_1} - \frac{1}{r_2}).$$

53P

Since the second particle is released from rest at P, thereby having zero initial kinetic energy, the conservation of energy (kinetic + potential) gives

$$\frac{1}{4\pi\epsilon_0} \frac{q^2}{r_1} + 0 = \frac{1}{4\pi\epsilon_0} \frac{q^2}{r_2} + \frac{1}{2}mv^2 \rightarrow v^2 = \frac{2}{m} \frac{q^2}{4\pi\epsilon_0}[\frac{1}{r_1} - \frac{1}{r_2}],$$

v being the speed at r_2. Putting in the numbers (in SI units, of course) yields

$$v^2 = (\frac{2}{2 \times 10^{-5}})(8.99 \times 10^9)(3.1 \times 10^{-6})^2[\frac{1}{9} - \frac{1}{25}]10^4 \rightarrow v = 2.48 \times 10^3 \text{ m/s.}$$

54P

(a) The initial potential energy is

$$U_i = \frac{1}{4\pi\epsilon_0} \frac{q^2}{d} = (8.99 \times 10^9)\frac{(5 \times 10^{-6})^2}{1} = 0.225 \text{ J.}$$

(b) The force F on each sphere after the string is cut is just the electrostatic Coulomb force:

$$F = \frac{1}{4\pi\epsilon_0} \frac{q^2}{d^2} = (8.99 \times 10^9)\frac{(5 \times 10^{-6})^2}{1^2} = 0.22475 \text{ N.}$$

By Newton's third law, this is the magnitude of the force on each sphere. Therefore, the accelerations are

$$a_1 = F/m_1 = (0.22475)/(0.005) = 45.0 \text{ m/s}^2,$$

$$a_2 = F/m_2 = (0.22475)/(0.010) = 22.5 \text{ m/s}^2.$$

(c) A long time after the string is cut the spheres are far enough apart so that the electrostatic potential energy is close to zero. Conservation of energy then implies that

$$U_i = \frac{1}{2}m_1 v_1^2 + \frac{1}{2}m_2 v_2^2,$$

since the initial speeds, and therefore the initial kinetic energies, are zero. The conservation of momentum provides a second equation between the speeds; since the spheres move off in opposite directions and initially were at rest,

$$0 = m_1 v_1 - m_2 v_2.$$

But $m_1 = m_2/2$, so that this last relation gives $v_1 = 2v_2$. Putting this into the energy equation gives

$$0.225 = \frac{1}{2}(m_2/2)(2v_2)^2 + \frac{1}{2}m_2 v_2^2 = \frac{3}{2}m_2 v_2^2 = \frac{3}{2}(0.01)v_2^2 \rightarrow v_2 = 3.873 \text{ m/s,}$$

$$v_1 = 2v_2 = 7.746 \text{ m/s.}$$

59P

Let v be the sought-after speed and D = 0.02 m the distance between the fixed spheres. Conservation of kinetic + potential energy requires that

$$\frac{1}{2}mv^2 + 0 = 0 + 2[\frac{1}{4\pi\epsilon_0}\frac{(-e)^2}{D/2}],$$

$$v^2 = 8(\frac{1}{4\pi\epsilon_0})\frac{e^2}{mD} = 8(8.99 \times 10^9)\frac{(1.6 \times 10^{-19})^2}{(9.11 \times 10^{-31})(0.02)} \rightarrow v = 318 \text{ m/s.}$$

64E

Since r << a, the disturbing effect on the assumed uniform charge distribution on each sphere due to the electric field from the other sphere can be disregarded; that is, V = ±1500 = $Q/4\pi\epsilon_0 r$, so that with r = 0.15 m, Q = ±25.0 nC.

68P

(a) Let r be the radius of each sphere are d their distance apart. Then, at the point midway (d/2) between them,

$$V = \frac{q_1}{4\pi\epsilon_0 (d/2)} + \frac{q_2}{4\pi\epsilon_0 (d/2)} = -180 \text{ V},$$

since $q_1 = 10^{-8}$ C, $q_2 = -3 \times 10^{-8}$ C and d = 2 m.

(b) In evaluating the potential of each sphere, the potential due to the other sphere (d >> r) should be ignored; if it is not, then the charge distributions on the spheres will not be uniform, and the potential for spherically symmetric charge distributions cannot be used. With this in mind, then,

$$V_1 = q_1/4\pi\epsilon_0 r = (10^{-8})(8.99 \times 10^9)/(0.03) = 3000 \text{ V}; \quad V_2 = -9000 \text{ V}.$$

73E

(a) The charge q_α of a α particle is +2e and its mass $m_\alpha = 4m_p$; e = 1.6 X 10^{-19} C and $m_p = 1.67 \times 10^{-27}$ kg is the mass of a proton. Let V be the accelerating potential and K be kinetic energy. Then, by the work-energy theorem and Eq.4,

$$W = \Delta K_\alpha = q_\alpha V = (2e)V = 2(1.6 \times 10^{-19} \text{ C})(10^6 \text{ V}) = 3.2 \times 10^{-13} \text{ J}.$$

(b) Similarly, $K_p = eV = 1.6 \times 10^{-13}$ J.

(c) Evidently $K_\alpha = 2K_p$, and therefore

$$K_p = \frac{1}{2}m_p v_p^2 = \frac{1}{2}K_\alpha = \frac{1}{2}(\frac{1}{2}m_\alpha v_\alpha^2) = \frac{1}{2}[\frac{1}{2}(4m_p)v_\alpha^2] \rightarrow v_p = v_\alpha\sqrt{2}.$$

Thus the proton, due to its greater charge/mass ratio, reaches the greater final speed.

75P

(a) The potential of the shell is $V = q/4\pi\epsilon_0 r$ and the electric field near the shell's surface is $E = q/4\pi\epsilon_0 r^2$, so that E = V/r. For E < 10^8 V/m, it is therefore necessary that

$$r > V/E = (9 \times 10^6)/(10^8) = 0.09 \text{ m} = 9 \text{ cm}.$$

(b) The work done in bringing up to the machine a charge Q is W = QV. Therefore, the power P supplied must be

$$P = \frac{dW}{dt} = V\frac{dQ}{dt} = (9 \times 10^6)(3 \times 10^{-4}) = 2700 \text{ W} = 2.7 \text{ kW}.$$

(c) If the surface charge density is σ and x denotes a length of the belt, then since $Q = \sigma A = \sigma(wx)$, then $dQ/dt = \sigma(dA/dt) = \sigma d(wx)/dt = \sigma(wv)$; hence,

$$\sigma = \frac{dQ/dt}{wv} = \frac{3 \times 10^{-4}}{(0.50)(30)} = 2 \times 10^{-5} \text{ C/m}^2 = 20 \text{ } \mu\text{C/m}^2.$$

3E

By Eq.1,

$$q = CV = (25 \times 10^{-6} \text{ F})(120 \text{ V}) = 3.00 \times 10^{-3} \text{ C} = 3.00 \text{ mC}.$$

5E

(a) The area of the plates is $A = \pi r^2 = \pi(0.082 \text{ m})^2 = 2.112 \times 10^{-2} \text{ m}^2$. The capacitance is given by Eq.9:

$$C = \varepsilon_0 A/d = (8.85 \times 10^{-12} \text{ F/m})(0.02112 \text{ m}^2)/(0.0013 \text{ m}) = 1.44 \times 10^{-10} \text{ F} = 144 \text{ pF}.$$

(b) By Eq.1,

$$q = CV = (1.44 \times 10^{-10} \text{ F})(120 \text{ V}) = 1.73 \times 10^{-8} \text{ C} = 17.3 \text{ nC}.$$

7E

This is a cylindrical capacitor to which Eq.14 applies; working in SI units, we have

$$C = \frac{1}{2}(4\pi\varepsilon_0)\frac{L}{\ln(b/a)} = \frac{1}{2}(8.99 \times 10^9)^{-1}\frac{(0.024)}{\ln(0.009/0.0008)} = 5.51 \times 10^{-13} \text{ F} = 0.551 \text{ pF}.$$

11P

Suppose that one plate of the capacitor is removed. Then, the charge on the other plate will move so as to distribute itself uniformly on both sides (rather than one side) of the plate, so that there is now $q/2$ on each side of the plate. The electric field set up by this isolated plate is, by Eq.12 of Chapter 25,

$$E_{isol} = \sigma_{isol}/\varepsilon_0 = (\tfrac{1}{2}q/A)/\varepsilon_0 = q/2A\varepsilon_0,$$

where q is the total charge on the plate. The electric field between the plates of the assembled capacitor is, from Eq.7,

$$E_{cap} = q/A\varepsilon_0.$$

Hence, the electric field acting on a plate due to the presence of the other is

$$E = E_{cap} - E_{isol} = q/A\varepsilon_0 - q/2A\varepsilon_0 = q/2A\varepsilon_0.$$

Thus, the force exerted by one plate on the other is

$$F = qE = q^2/2A\varepsilon_0.$$

15P

The new capacitance in terms of the new radius is given by Eq.18: $C* = 4\pi\epsilon_0 R*$. But the new radius $R*$ in terms of the old radius R is found from the addition of volumes:

$$4\pi R*^3/3 = 2(4\pi R^3/3) \rightarrow R* = 2^{1/3}R,$$

so that

$$C* = 4\pi\epsilon_0 R* = 4\pi\epsilon_0(2^{1/3}R) = 5.04\pi\epsilon_0 R.$$

17P

Consider the forces acting on a small area of the charged bubble's surface. These forces are due to the

(i)   gas pressure p_g acting outward,

(ii)   atmospheric pressure p acting inward,

(iii) electrostatic stress acting outward (see Problem 12).

Thus, for equilibrium, noting that the area cancels,

$$p_g + \frac{1}{2}\epsilon_0 E^2 = p.$$

It is assumed that

$$p_g = p(V_0/V).$$

The electric field due to the spherically symmetric distribution of charge on the bubble is given by

$$E = q/4\pi\epsilon_0 R^2.$$

Substituting E and p_g into the equilibrium condition equation and multiplying by V gives

$$pV_0 + \frac{1}{2}\epsilon_0 E^2 V = pV,$$

$$p(\frac{4}{3}\pi R_0^3) + \frac{1}{2}\epsilon_0(\frac{q^2}{16\pi^2\epsilon_0^2 R^4})(\frac{4}{3}\pi R^3) = p(\frac{4}{3}\pi R^3),$$

$$pR_0^3 + \frac{q^2}{32\pi^2\epsilon_0 R} = pR^3,$$

$$q^2 = 32\pi^2\epsilon_0 pR(R^3 - R_0^3).$$

18E

The desired equivalent capacitance is

$$C_{eq} = q/V = (1\ C)/(110\ V) = 9.091 \times 10^{-3}\ F = 9091\ \mu F.$$

For capacitors connected in parallel, Eq.19 states that

$$C_{eq} = \Sigma C_n.$$

Thus, for N capacitors connected in parallel C_{eq} = NC, C the capacitance of each; hence,

$$N = C_{eq}/C = (9091 \ \mu F)/(1 \ \mu F) = 9091.$$

30P

The capacitance of a parallel-plate (flat) capacitor is $C = \varepsilon_0 A/d$ (Eq.9) where d is the plate separation. If the plates of the upper capacitor are a distance D apart, then the plates of the lower capacitor must be separated by [a - (b + D)]. Hence, the equivalent capacitance as given by Eq.20 leads to

$$\frac{1}{C_{eq}} = \frac{1}{C_1} + \frac{1}{C_2},$$

$$\frac{1}{C_{eq}} = \frac{1}{\varepsilon_0 A/D} + \frac{1}{\varepsilon_0 A/(a - b - D)} = \frac{1}{\varepsilon_0 A}(D + a - b - D) = \frac{a - b}{\varepsilon_0 A},$$

$$C_{eq} = \varepsilon_0 A/(a - b),$$

and this last result is independent of D, and therefore is independent of the position of the center section.

32P

(a) Let starred quantities be the final values (switches closed). The final polarities must be the same since the capacitors are then connected in parallel. By the conservation of charge,

$$-q_1 + q_2 = q_1^* + q_2^*.$$

But we are also given that

$$q_1/C_1 = q_2/C_2 = 100 \ V.$$

Since $C_1 = 10^{-6}$ F and $C_2 = 3 \times 10^{-6}$ F, the equation above gives $q_1 = 10^{-4}$ C, $q_2 = 3 \times 10^{-4}$ C. Substituting these results into the first equation yields

$$q_1^* + q_2^* = 2 \times 10^{-4}.$$

However, in the final arrangement we also have

$$V^* = q_1^*/C_1 = q_2^*/C_2,$$

so that

$$q_2^* = 3q_1^*,$$

just as $q_2 = 3q_1$. Solving the two equations for q_1^* and q_2^* yields

$$q_1^* = 0.5 \times 10^{-4} \text{ C}, \quad q_2^* = 1.5 \times 10^{-4} \text{ C}.$$

Therefore

$$V_{ef} = V^* = q_2^*/C_2 = (1.5 \times 10^{-4})/(3 \times 10^{-6}) = 50 \text{ V}.$$

(b) From (a), $q_1^* = 50 \text{ } \mu\text{C}$.

(c) Also from (a), $q_2^* = 150 \text{ } \mu\text{C}$.

33P

The initial charges are given to the left
and the final charges are to the right of
each capacitor in the sketch. Initially,

$$q = C_1 V_0.$$

By the conservation of charge applied to
conductor X,

$$-q = -q_1 - q_3 \rightarrow q_1 + q_3 = C_1 V_0.$$

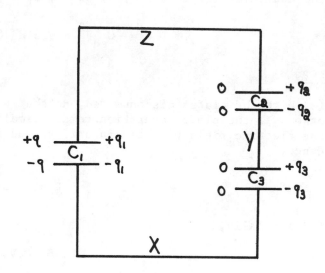

For conductor Y,

$$0 = -q_2 + q_3 \rightarrow q_2 = q_3.$$

Applying conservation of charge to conductor
Z leads to no new information. Now, electrostatic fields are conservative; this means
that the total drop in potential around the complete circuit must be zero. Since the q's
above are considered positive, this indicates that

$$0 = \frac{q_2}{C_2} + \frac{q_3}{C_3} - \frac{q_1}{C_1}.$$

We now have three equations in the three new charges as unknowns. The solutions are
found to be

$$q_1 = \frac{C_1 C_2 + C_1 C_3}{C_1 C_2 + C_1 C_3 + C_2 C_3} C_1 V_0,$$

$$q_2 = q_3 = \frac{C_2 C_3}{C_1 C_2 + C_1 C_3 + C_2 C_3} C_1 V_0.$$

38E

In SI units, the energy to be stored is

$$U = 10 \text{ kW} \cdot \text{h} = 10(10^3 \text{ J/s})(3600 \text{ s}) = 3.6 \times 10^7 \text{ J}.$$

By Eq.22, we have

$$C = 2U/V^2 = 2(3.6 \times 10^7 \text{ J})/(10^3 \text{ V})^2 = 72 \text{ F}.$$

41E

The equivalent capacitance, by Eq.19, is $C_{eq} = 6 \text{ μF}$. Hence,

$$U = \frac{1}{2}C_{eq}V^2 = \frac{1}{2}(6 \times 10^{-6} \text{ F})(300 \text{ V})^2 = 0.27 \text{ J}.$$

49P

(a) With the plates disconnected, no charge can be transferred from one plate to the other. If the plate separation remains small compared with the dimensions of the plates, the electric field will remain uniform and independent of the distance between the plates. Hence,

$$V_f = E_f d_f = E(2d) = 2(Ed) = 2V.$$

(b) Initially,

$$U_i = \frac{1}{2}C_i V_i^2 = \frac{1}{2}(\epsilon_0 A/d)V^2.$$

Both the capacitance and the potential difference change as the plates are pulled apart; thus,

$$U_f = \frac{1}{2}C_f V_f^2 = \frac{1}{2}(\epsilon_0 A/d_f)(2V)^2 = \frac{1}{2}(\epsilon_0 A/2d)(2V)^2 = \epsilon_0 AV^2/d.$$

(c) The work W required to pull the plates apart is, noting that $U_f = 2U_i$,

$$W = U_f - U_i = U_i = \frac{1}{2}(\epsilon_0 A/d)V^2.$$

50P

The energy stored in an electric field occupying a volume V is

$$U = \int u_E dV = \int \frac{1}{2}\epsilon_0 E^2 dV,$$

where u_E is the electric field energy density. From Eq.12,

$$E = \frac{q}{2\pi\epsilon_0 L}\frac{1}{r},$$

between the plates of a coaxial cylindrical capacitor of length L and charge q; r is measured from the axis. The volume element dV is the volume contained in a very thin cylindrical shell of radius r, thickness dr and length L, so that

$$dV = 2\pi rLdr.$$

Hence, the energy stored within a region between the inner shell of radius a and a cylindrical surface of radius s is

$$U_s = \int_a^s \frac{1}{2}\epsilon_0 \left(\frac{q}{2\pi\epsilon_0 Lr}\right)^2 2\pi rLdr = \frac{q^2}{4\pi\epsilon_0 L}\int_a^s dr/r = \frac{q^2}{4\pi\epsilon_0 L}\ln\left(\frac{s}{a}\right).$$

Using the expression for the capacitance C given in Eq.14, the energy stored between the plates is

$$U = q^2/2C = \frac{1}{2}\frac{q^2}{2\pi\epsilon_0 L}\ln\left(\frac{b}{a}\right).$$

Thus, the value of s for which $U_s = U/2$ is found from

$$\frac{q^2}{4\pi\epsilon_0 L}\ln\left(\frac{s}{a}\right) = \frac{1}{2}\frac{q^2}{4\pi\epsilon_0 L}\ln\left(\frac{b}{a}\right) \rightarrow s = \sqrt{(ab)}.$$

53E

Adding a dielectric to fill completely the air space between capacitor plates results in a new capacitance given by Eq.26:

$$C = \kappa C_{air}.$$

The stored energy is, by Eq.22,

$$U = \frac{1}{2}CV^2 = \frac{1}{2}(\kappa C_{air})V^2,$$

$$7.4 \times 10^{-6} = \frac{1}{2}\kappa(7.4 \times 10^{-12})(652)^2 \rightarrow \kappa = 4.70.$$

In Table 2, we see that pyrex has this value of dielectric constant.

56E

Eq.14 gives the capacitance of a cylindrical air capacitor; with Eq.26 for the effect of the dielectric we get

$$C = \kappa(2\pi\epsilon_0 L)/\ln\left(\frac{b}{a}\right).$$

For L = 1 m, this becomes

$$C = (2.6)2\pi(8.85 \times 10^{-12})(1)/\ln\left(\frac{0.6}{0.1}\right) = 80.7 \times 10^{-12} \text{ F} = 80.7 \text{ pF},$$

or 80.7 pF/m of cable.

57P

The capacitance of the capacitor is

$$C = 7 \times 10^{-8} = \kappa\varepsilon_0 A/d = 2.8\varepsilon_0 A/d.$$

The dielectric strength is the maximum potential gradient possible without breakdown, so that

$$(V/d)_{max} = 4000/d = 18 \times 10^6,$$

and therefore $d = 2.222 \times 10^{-4}$ m $= 0.2222$ mm. Then, from the first equation,

$$A = (7 \times 10^{-8})(2.222 \times 10^{-4})/2.8\varepsilon_0 = 0.628 \text{ m}^2.$$

64P

Let q, $-q$ be the charges on the plates separated by a distance d. If V is the potential difference across the plates, then

$$C = q/V.$$

The uniform electric fields in the dielectrics are $\sigma/\kappa\varepsilon_0$, so that

$$E_1 = \frac{q/A}{\kappa_1\varepsilon_0}, \quad E_2 = \frac{q/A}{\kappa_2\varepsilon_0}.$$

Therefore,

$$C = \frac{q}{E_1(d/2) + E_2(d/2)} = \frac{2q}{d}[\frac{q/A}{\kappa_1\varepsilon_0} + \frac{q/A}{\kappa_2\varepsilon_0}]^{-1} = \frac{2A\varepsilon_0}{d}\frac{\kappa_1\kappa_2}{\kappa_1 + \kappa_2}.$$

For an air capacitor, $\kappa_1 = \kappa_2 = 1$, and the equation above reduces to $C = \varepsilon_0 A/d$, as expected. If $\kappa_1 = \kappa_2 = \kappa$, then, as anticipated, the equation becomes $C = \kappa\varepsilon_0 A/d$.

66E

(a) The free charge is $q = CV = (100 \times 10^{-12})(50) = 5 \times 10^{-9}$ C. By Eq.32, in the dielectric,

$$E = q/\kappa\varepsilon_0 A = (5 \times 10^{-9})/(5.4)(8.85 \times 10^{-12})(10^{-2}) = 1.05 \times 10^4 \text{ V/m}.$$

(c) By Eq.33,

$$q' = q(1 - \frac{1}{\kappa}) = (5.0)(1 - \frac{1}{5.4}) = 4.07 \text{ nC}.$$

71P

Let the electric field in the gaps be E and the field in the dielectric be E_d. Now $E_d = E/\kappa$ and therefore the potential difference across the plates is

$$V = xE + bE_d + (d - b - x)E,$$

$$V = E(d - b) + E_d b,$$

$$V = E(d - b + \frac{b}{\kappa}).$$

But, for a parallel-plate capacitor,

$$E = \sigma/\varepsilon_0 = q/A\varepsilon_0,$$

A the area of each plate. Thus,

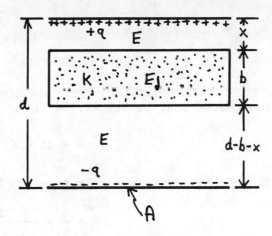

$$V = \frac{q}{A\varepsilon_0}[d + b(\frac{1}{\kappa} - 1)] = \frac{q}{A\varepsilon_0 \kappa}[\kappa d - b(\kappa - 1)],$$

and therefore the capacitance is

$$C = \frac{q}{V} = \frac{\kappa \varepsilon_0 A}{\kappa d - b(\kappa - 1)}.$$

The cases $b = 0$ and $\kappa = 1$ each correspond to an air capacitor for which $C = \varepsilon_0 A/d$ by the formula above, as expected. If $b = d$, the dielectric fills entirely the space between the plates, giving $C = \kappa \varepsilon_0 A/d$, again as anticipated.

CHAPTER 28

1E

(a) The charge that passes is, by Eq.1,

$$q = it = (5 \text{ A})(240 \text{ s}) = 1200 \text{ C}.$$

(b) The number of electrons is given by Eq.8 in Chapter 23:

$$n = \frac{q}{e} = \frac{1200 \text{ C}}{1.6 \times 10^{-19} \text{ C}} = 7.5 \times 10^{21}.$$

3P

The potential of a conducting sphere carrying a charge q is

$$V = \frac{1}{4\pi\epsilon_0} \frac{q}{R}.$$

Therefore, if q changes V will also, and at the rate

$$\frac{dV}{dt} = \frac{1/R}{4\pi\epsilon_0} \frac{dq}{dt} = \frac{1}{4\pi\epsilon_0 R} i = \frac{1}{4\pi\epsilon_0 R}(i_{in} - i_{out}),$$

i being the net current. If i does not vary with time, then

$$\frac{\Delta V}{\Delta t} = \frac{1}{4\pi\epsilon_0 R} i,$$

$$\frac{1000}{\Delta t} = \frac{(8.99 \times 10^9)}{(0.10)}(1.000002 - 1) \rightarrow \Delta t = 5.56 \times 10^{-3} \text{ s} = 5.56 \text{ ms}.$$

5E

(a) Apply Eq.7. That equation was derived for electrons, magnitude of charge = e, in a wire. For other charged particles, replace e with their charge q; hence we have (since $1 \text{ m}^3 = 10^6 \text{ cm}^3$),

$$J = nqv_d = n[2e]v_d = (2 \times 10^{14} \text{ m}^{-3})[2(1.6 \times 10^{-19} \text{ C})](10^5 \text{ m/s}) = 6.4 \text{ A/m}^2.$$

Since the charges are positive, J is directed parallel to v_d, i.e., to the north.

(b) We are not told the cross-sectional area of the beam, and therefore we cannot find the current i = JA.

7E

The cross-sectional area of a circular cylinder of diameter D is $A = \pi D^2/4$. Therefore D must satisfy the relation

$$J = \frac{i}{A} = \frac{i}{\pi D^2/4},$$

$$440 \text{ A/cm}^2 = \frac{0.5 \text{ A}}{\pi D^2/4} \rightarrow D = 3.80 \text{ X } 10^{-2} \text{ cm} = 0.380 \text{ mm}.$$

14P

(a) The current carried is $i = 0.25 \text{ } \mu\text{A} = 2.5 \text{ X } 10^{-7}$ C/s. Since each particle carries a charge of 2e, the number n that strike in 3 s is given by

$$n = \frac{q}{2e} = \frac{it}{2e} = \frac{(2.5 \text{ X } 10^{-7})(3)}{(2)(1.6 \text{ X } 10^{-19})} = 2.34 \text{ X } 10^{12}.$$

(b) Let the number of particles in a length L of the beam be called N. The current is

$$i = \frac{q}{t} = \frac{N(2e)}{t} = \frac{N(2e)}{L/v} = \frac{2evN}{L} \rightarrow N = \frac{iL}{2ev}.$$

To find v, note that the mass of an alpha particle is $m = 4m_p$, where m_p is the proton mass. Thus, in SI units,

$$K = \frac{1}{2}mv^2,$$

$$(20)(1.6 \text{ X } 10^{-13}) = \frac{1}{2}[4(1.67 \text{ X } 10^{-27})]v^2 \rightarrow v = 3.095 \text{ X } 10^7 \text{ m/s}.$$

(This is a nonrelativistic calculation.) Substituting this and the other numerical data into the equation above for N yields

$$N = \frac{(2.5 \text{ X } 10^{-7})(0.2)}{2(1.6 \text{ X } 10^{-19})(3.095 \text{ X } 10^7)} = 5.05 \text{ X } 10^3.$$

(c) The required potential V is given by

$$K = qV.$$

Expressing the charge q in terms of e gives

$$20 \text{ MeV} = (2e)V \rightarrow V = 10 \text{ MV}.$$

17E

Apply Eq.15; substitute the data in SI units to obtain

$$R = \rho\frac{L}{A} = \rho\frac{L}{\pi D^2/4},$$

$$50 \text{ X } 10^{-3} = \rho\frac{2}{\pi(0.001)^2/4} \rightarrow \rho = 1.96 \text{ X } 10^{-8} \text{ } \Omega\cdot\text{m}.$$

18E

From Eq.8,

$$V = iR = (50 \times 10^{-3} \text{ A})(2 \times 10^3 \text{ } \Omega) = 100 \text{ V}.$$

19E

The length of the wire forming the coil is

$$L = (250)2\pi(0.12 \text{ m}) = 188.50 \text{ m}.$$

The cross-sectional area of the wire is

$$A = \pi(0.0013 \text{ m})^2/4 = 1.327 \times 10^{-6} \text{ m}^2.$$

From Table 1, the resistivity of copper is found to be $\rho = 1.69 \times 10^{-8}$ $\Omega \cdot$m. Eq.15 now gives for the resistance

$$R = \rho \frac{L}{A} = (1.69 \times 10^{-8} \text{ } \Omega \cdot \text{m}) \frac{188.50 \text{ m}}{1.327 \times 10^{-6} \text{ m}^2} = 2.40 \text{ } \Omega.$$

27E

In order that the density of the material not be altered by the stretching, the volume V of the wire must be unchanged, since the mass remains unaffected. If the length increases by a factor of 3, then since V = AL, the cross-sectional area must decrease by a factor of 3. Hence, the new resistance R is

$$R = \rho \frac{L}{A} = \rho \frac{3L_0}{A_0/3} = 9(\rho \frac{L_0}{A_0}) = 9(6) = 54 \text{ } \Omega.$$

30P

Since the material of the two conductors is the same, we have $\rho_A = \rho_B$. Also, it is given that $L_A = L_B$. Therefore, since $R = \rho L/A$ for each, we have for the ratio of resistances,

$$\frac{R_A}{R_B} = \frac{A_B}{A_A} = \frac{\pi(d_o^2 - d_i^2)/4}{\pi d^2/4} = \frac{d_o^2 - d_i^2}{d^2} = \frac{2^2 - 1^2}{1^2} = 3.0.$$

37P

The cross-sectional area of the wire is

$$A = \pi r^2 = \pi(0.3 \times 10^{-3} \text{ m})^2 = 2.827 \times 10^{-7} \text{ m}^2.$$

The current i must be

$$i = JA = (1.4 \times 10^4 \text{ A/m}^2)(2.827 \times 10^{-7} \text{ m}^2) = 3.958 \times 10^{-3} \text{ A}.$$

Therefore the resistance is

$$R = V/i = (115 \text{ V})/(3.958 \text{ X } 10^{-3} \text{ A}) = 2.906 \text{ X } 10^4 \text{ } \Omega.$$

Hence, by Eq.15,

$$\rho = RA/L = (2.906 \text{ X } 10^4 \text{ } \Omega)(2.827 \text{ X } 10^{-7} \text{ m}^2)/(10 \text{ m}) = 8.22 \text{ X } 10^{-4} \text{ } \Omega \cdot \text{m}.$$

42P

(a) Since $R = \rho L/A$ we have, assuming small changes,

$$\Delta R = \frac{\partial R}{\partial \rho}(\Delta \rho) + \frac{\partial R}{\partial L}(\Delta L) + \frac{\partial R}{\partial A}(\Delta A).$$

Taking the derivatives we obtain

$$\frac{\partial R}{\partial \rho} = L/A = R/\rho,$$

$$\frac{\partial R}{\partial L} = \rho/A = R/L,$$

$$\frac{\partial R}{\partial A} = -\rho L/A^2 = -R/A.$$

Substituting these into the expression for ΔR gives

$$\frac{\Delta R}{R} = \frac{\Delta \rho}{\rho} + \frac{\Delta L}{L} - \frac{\Delta A}{A}.$$

Let β be the coefficient of thermal expansion (called α in Chapter 19), so that

$$\Delta L = \beta L(\Delta T), \quad \Delta A = 2\beta A(\Delta T),$$

where ΔT is the change in temperature. Therefore, the percent change in length is $\beta(\Delta T)$ = $(1.7 \text{ X } 10^{-5})(1) = 0.0017\%$, and the percent change in area is twice this, or 0.0034%. The percent change in R is, from the above,

$$\frac{\Delta R}{R} = (\alpha + \beta - 2\beta)\Delta T = (\alpha - \beta)\Delta T = 0.428\%,$$

since $\alpha = 4.3 \text{ X } 10^{-3}/°\text{C}$ for copper.

(b) The relative change in resistivity is much larger than those in length and area, and has the greatest effect on the resistance.

43P

(a) Consider a thin slice perpendicular to the axis of the cone, the slice having a thickness dx and being situated at a distance x from the narrow end of the cone (see the sketch on p.228). The cross-sectional area of this slice is $A = \pi r^2$ and its resistance dR is

$$dR = \rho dL/A = \rho dx/\pi r^2.$$

Let θ be the semi-angle of the cone from which the resistor was cut. Let the vertex of the cone be at a distance d from the narrow end of the resistor, as shown. Then,

$$\tan\theta = \frac{a}{d} = \frac{r}{x + d} = \frac{b}{L + d}.$$

From the first equation,

$$\frac{a}{d} = \frac{r}{x + d},$$

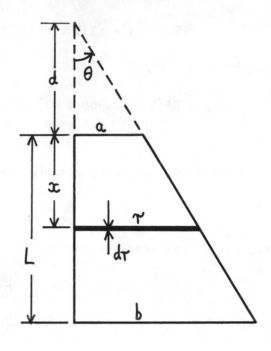

we find that

$$d = \frac{xa}{r - a}.$$

Substituting this into the second equation

$$\frac{r}{x + d} = \frac{b}{L + d},$$

gives

$$r = (\frac{b - a}{L})x + a.$$

Thus, the resistance R of the object is

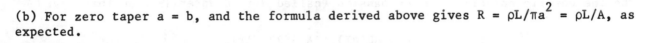

$$R = \int dR = (\rho/\pi)\int_0^L [(\frac{b - a}{L})x + a]^{-2}dx = \rho L/\pi ab.$$

(b) For zero taper a = b, and the formula derived above gives $R = \rho L/\pi a^2 = \rho L/A$, as expected.

<u>46E</u>

Eq.20 yields

$$P = iV = (7 \times 10^{-3} \text{ A})(80 \times 10^3 \text{ V}) = 560 \text{ W}.$$

<u>49E</u>

(a) By Eq.22,

$$P = V^2/R = (120 \text{ V})^2/(14 \ \Omega) = 1030 \text{ W} = 1.03 \text{ kW}.$$

(b) The heat energy produced in 5 hours is

$$Q = E = Pt = (1.03 \text{ kW})(5 \text{ h}) = 5.15 \text{ kW·h}.$$

Hence the cost is (5.15 kW·h)(5¢/kW·h) = 25.75¢.

56P

(a) Since the power dissipated is $P = iV$, the current i must be

$$i = \frac{P}{V} = \frac{1250}{115} = 10.9 \text{ A}.$$

(b) By Ohm's law $V = iR$ so that

$$R = \frac{V}{i} = \frac{115}{10.9} = 10.6 \ \Omega.$$

(c) In one hour the heater generates

$$Q = Pt = (1250 \text{ J/s})(3600 \text{ s}) = 4.5 \times 10^6 \text{ J} = 4.5 \text{ MJ}.$$

58P

For the two temperatures 200°C and 800°C, Eq.22 gives

$$P_8 = V_8^2/R_8, \quad P_2 = V_2^2/R_2.$$

But $V_2 = V_8$ (= 110 V) so that

$$\frac{P_8}{P_2} = \frac{R_2}{R_8}.$$

But $R = \rho L/A$; if we can ignore changes in L and A with temperature, then

$$\frac{R_2}{R_8} = \frac{\rho_2}{\rho_8} \rightarrow \frac{P_8}{P_2} = \frac{\rho_2}{\rho_8}.$$

By Eq.16,

$$\rho_2 - \rho_8 = \rho_8 \alpha (T_2 - T_8),$$

$$\frac{\rho_2}{\rho_8} = 1 + \alpha(T_2 - T_8) = 1 + (4 \times 10^{-4})(200 - 800) = 0.76.$$

Therefore, by the third equation above,

$$P_2 = P_8 (\frac{\rho_8}{\rho_2}) = \frac{500}{0.76} = 658 \text{ W}.$$

60P

(a) The charge q accelerated during each pulse is, in SI units,

$$q = iT = (0.5)(0.1 \times 10^{-6}) = 5 \times 10^{-8} \text{ C}.$$

Each electron carries charge 1.6×10^{-19} C in magnitude. Therefore, the number n of electrons accelerated is

$$n = (5 \times 10^{-8})/(1.6 \times 10^{-19}) = 3.125 \times 10^{11}.$$

(b) In one second a total charge

$$Q = (500 \text{ pulses})(5 \times 10^{-8} \text{ C/pulse}) = 25 \times 10^{-6} \text{ C},$$

is accelerated. The average current is

$$\bar{i} = \frac{Q}{t} = 25 \times 10^{-6} \text{ A} = 25 \text{ μA}.$$

(c) The accelerating voltage must be

$$V = K/e = (50 \text{ MeV})/(1 \text{ e}) = 5 \times 10^{7} \text{ V}.$$

The power output is P = iV; per pulse this is

$$P = iV = (0.5)(5 \times 10^{7}) = 2.5 \times 10^{7} \text{ W}.$$

Over one second the average power is

$$\bar{P} = \bar{i}V = (25 \times 10^{-6})(5 \times 10^{7}) = 1250 \text{ W}.$$

CHAPTER 29

2E

If q is the charge that moves through the circuit in time t, then the energy delivered by the battery is, by Eq.1,

$$E = W = q\varepsilon.$$

But $q = it$, so we have

$$E = it\varepsilon = (5 \text{ C/s})(360 \text{ s})(6 \text{ J/C}) = 10800 \text{ J}.$$

4P

The work done in moving a charge q through a potential difference V is $W = qV$. This is the available energy E, so that $E = qV$. In terms of power, $E = Pt$. Combining these relations gives

$$Pt = qV,$$

$$t = \frac{qV}{P} = \frac{[(120 \text{ C/s})(3600 \text{ s})](12 \text{ V})}{100 \text{ J/s}} = 5.184 \times 10^4 \text{ s} = \frac{5.184 \times 10^4}{3600} = 14.4 \text{ h}.$$

8E

(a) The net emf in the circuit is $\varepsilon_1 - \varepsilon_2 = 6$ V counterclockwise. Therefore,

$$i = (\varepsilon_1 - \varepsilon_2)/(R_1 + R_2) = (6 \text{ V})/(12 \text{ }\Omega) = 0.5 \text{ A}.$$

(b) The power dissipated in the resistors is

$$P_1 = i^2 R_1 = (0.5)^2(4) = 1 \text{ W}; \quad P_2 = i^2 R_2 = (0.5)^2(8) = 2 \text{ W}.$$

(c) Battery 1 supplies energy at the rate $i\varepsilon_1 = (0.5)(12) = 6$ W. The emf of battery 2 opposes the current, so it absorbs energy and at the rate $i\varepsilon_2 = (0.5)(6) = 3$ W. Hence the batteries together supply $6 - 3 = 3$ W, equal to the total power dissipated in the resistors.

11E

(a) The potential between A and B is $V = V_A - V_B > 0$, since energy is removed from this circuit section. Since $P = iV$, we have $V = (50 \text{ W})/(1 \text{ A}) = 50 \text{ V}$.

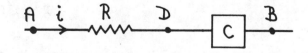

(b) Consider a point D between resistor R and element C; we have

$$V_A - V_D = iR = 2 \text{ V},$$

$$V_A - V_B = (V_A - V_D) + (V_D - V_B),$$

$$50 = 2 + (V_D - V_B) \rightarrow V_D - V_B = 48 \text{ V}.$$

(c) Since $V_D > V_B$, the negative terminal must be at B.

20P

(a) Let the points a and b denote the two terminals of battery 1. The current i flows clockwise in the diagram and

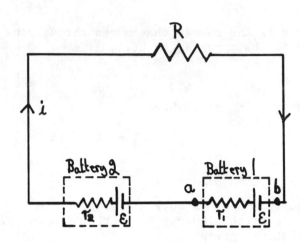

$$V_a + \varepsilon - ir_2 - iR = V_b,$$

$$V_a - V_b = i(r_2 + R) - \varepsilon.$$

If $V_a - V_b = 0$, so that the potential drop is zero across the terminals of battery 1, then the last equation gives

$$i(r_2 + R) = \varepsilon.$$

By the loop theorem,

$$i = 2\varepsilon/(r_1 + r_2 + R),$$

so that the battery condition becomes

$$\frac{2\varepsilon}{r_1 + r_2 + R}(r_2 + R) = \varepsilon \quad \rightarrow. \quad R = r_1 - r_2.$$

(b) Since we must have $R > 0$, it follows that $r_1 > r_2$. But we took the potential drop to be zero across battery 1; now we see that this means across the battery with the larger internal resistance.

22P

(a) The current in the circuit is

$$i = \frac{\varepsilon}{r + R},$$

and therefore the rate of thermal energy delivery is

$$P = i^2 R = \frac{\varepsilon^2 R}{(r + R)^2}.$$

To find the value of R that maximizes P, set dP/dR equal to zero and solve for R:

$$\frac{dP}{dR} = 0 = \varepsilon^2 \frac{(r - R)}{(r + R)^3} \quad \rightarrow \quad R = r.$$

(b) The maximum power at which thermal energy is dissipated is

$$P(R=r) = i^2 r = \varepsilon^2 r/(r + r)^2 = \varepsilon^2/4r.$$

<u>26E</u>

The equivalent resistance is given by Eq.20:

$$\frac{1}{R_{eq}} = \frac{1}{R} + \frac{1}{R} + \frac{1}{R} + \frac{1}{R} = \frac{4}{R} \rightarrow R_{eq} = R/4.$$

Thus, the current through the battery is

$$i = \frac{\varepsilon}{R_{eq}} = \frac{\varepsilon}{R/4} = 4\varepsilon/R = 4(25 \text{ V})/(18 \text{ }\Omega) = 5.56 \text{ A.}$$

<u>29E</u>

Apply the loop theorem, in the counter-clockwise direction to the top and bottom loops to get

$$\varepsilon_1 - \varepsilon_2 - \varepsilon_3 - i_2 R_2 = 0,$$

$$-i_1 R_1 + \varepsilon_2 = 0.$$

Substitution of the given data into these equations gives

$$6 - 5 - 4 - i_2(50) = 0 \rightarrow i_2 = -0.06 \text{ A;}$$

$$-i_1(100) + 5 = 0 \rightarrow i_1 = 0.05 \text{ A.}$$

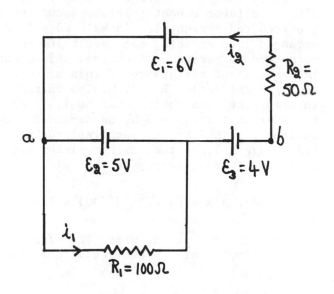

Note that i_2 actually flows in the direction opposite to that shown on the sketch. To obtain the potential between points a and b, start at a and proceed directly across to b:

$$V_b + 4 + 5 = V_a \rightarrow V_b - V_a = -9 \text{ V.}$$

<u>33E</u>

(a) The brighter bulb indicates a higher value of $P = i^2 R$. In parallel, $i_2 = \varepsilon/R_2$ so that $i_2^2 R_2 = \varepsilon^2/R_2$. Similarly, $i_1^2 R_1 = \varepsilon^2/R_1$. Since $R_2 < R_1$, we have $\varepsilon^2/R_2 > \varepsilon^2/R_1$, so the R_2 bulb is brighter.

(b) When connected in series, the currents through the bulbs are the same; therefore $i^2 R_1 > i^2 R_2$ since $R_1 > R_2$, and therefore R_1 is now the brighter bulb.

<u>35E</u>

The length of the composite resistor is still ℓ, the length of any single resistor, but the area through which current can flow is now 9 times the area A of a single wire. The new resistance is therefore, by Eq.15 in Chapter 28,

$$R = \frac{\rho\ell}{9A} = \frac{\rho\ell}{9\pi d^2/4}.$$

234

The cross-sectional area of the equivalent single wire is $\pi D^2/4$, but its length still is ℓ. For this wire to have the same resistance as the composite given above, the diameter D must be chosen to satisfy the condition that Eq. 15 of Chapter 28 yield the same value of resistance for each; that is,

$$\frac{\rho \ell}{9\pi d^2/4} = \frac{\rho \ell}{\pi D^2/4} \rightarrow D = 3d.$$

40P

Let there be n resistors. They cannot be connected either all in series or all in parallel, since the equivalent resistance in series would be nR and in parallel would be R/n (R is the resistance of a single resistor), and these equal R only for n = 1. But a single resistor cannot tolerate a current greater than $\sqrt{(P/R)} = \sqrt{(1\ W/10\ \Omega)} = 1/\sqrt{10}$ A. The required current is $\sqrt{(5\ W/10\ \Omega)} = 1/\sqrt{2}$ A. Suppose instead that the n resistors are arranged into $\sqrt{n}$ sets, each set being $\sqrt{n}$ resistors connected in parallel, the sets being connected in series. (Of course, this means that n must have an integral square root.) The equivalent resistance of this arrangement is $R_{eq} = \sqrt{n}(R/\sqrt{n}) = R = 10\ \Omega$. The current entering the combination must be i = $1/\sqrt{2}$ A in order that $i^2 R_{eq} = 5$ W, as demanded. The current passing through each resistor is $i/\sqrt{n} = 1/\sqrt{(2n)}$, so that each resistor will dissipate thermal energy at the rate

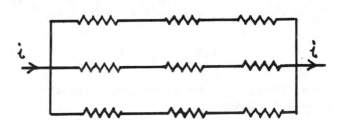

$$P = [i/\sqrt{n}]^2 R = [1/\sqrt{(2n)}]^2 (10) = 5/n.$$

This must be 1 W at the most. Therefore 5/n < 1, so the minimum n allowed is n = 9 (integral value of $\sqrt{n}$ required). Thus, there are 9 resistors, connected as shown.

47P

Let the currents be directed as shown, with the currents already labelled to satisfy the junction rule. Applying the loop rule to I,

$$i_1(2R) - i_2(R) = 0 \rightarrow i_2 = 2i_1.$$

Applied to II the loop rule gives

$$(i_1 + i_3)R - (i_2 - i_3)R = 0,$$

$$i_2 - i_1 = 2i_3.$$

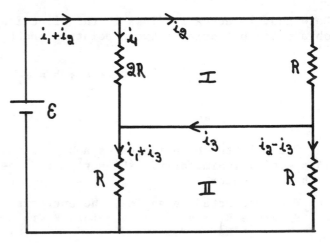

Eliminating i_1 between the resulting current equations above gives $i_2 = 4i_3$. But, the loop rule applied to the outside loop yields

$$\varepsilon - i_2 R - (i_2 - i_3)R = 0.$$

Substituting $i_2 = 4i_3$ into this equation gives $i_3 = \varepsilon/7R$.

50P

(a) By the loop rule applied to the left
and right branches,

$$\varepsilon_1 - i_3 R_3 - i_1 R_1 = 0,$$

$$\varepsilon_2 + i_2 R_2 - i_1 R_1 = 0,$$

assuming that the currents are directed as
shown. The junction rule further provides
that

$$i_3 = i_1 + i_2.$$

These three equations can be solved for the
currents; the results are

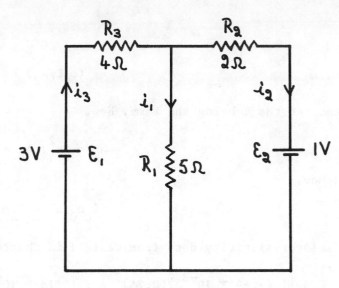

$$r^2 i_1 = \varepsilon_1 R_2 + \varepsilon_2 R_3,$$

$$r^2 i_2 = \varepsilon_1 R_1 - \varepsilon_2 (R_1 + R_3),$$

$$r^2 i_3 = \varepsilon_1 (R_1 + R_2) - \varepsilon_2 R_1,$$

in which

$$r^2 = R_1 R_2 + R_1 R_3 + R_2 R_3.$$

Substituting the numerical data gives

$$i_1 = 5/19 \text{ A}; \quad i_2 = 3/19 \text{ A}; \quad i_3 = 8/19 \text{ A}.$$

The rates at which thermal energy appears in these resistors are

$$P_1 = i_1^2 R_1 = 0.346 \text{ W}; \quad P_2 = i_2^2 R_2 = 0.050 \text{ W}; \quad P_3 = i_3^2 R_3 = 0.709 \text{ W}.$$

(b) The powers supplied by the batteries are

$$P_{b1} = i_3 \varepsilon_1 = 1.263 \text{ W}; \quad P_{b2} = -i_2 \varepsilon_2 = -0.158 \text{ W}.$$

The result for battery 2 is negative since the current i_2 flows through the battery in
the direction opposite to the battery's emf.

(c) Energy is supplied to the circuit by battery 1 and is dissipated as thermal energy in
the three resistors and stored chemically in battery 2. As expected then, we have that
1.263 W = 0.346 W + 0.050 W + 0.709 W + 0.158 W.

54P

(a) The copper wire and aluminum jacket are connected in parallel, so that the potential
drop V = iR is the same for each; that is,

$$i_C R_C = i_A R_A,$$

$$i_C (\rho_C L/A_C) = i_A (\rho_A L/A_A),$$

the lengths L being the same. Now,

$$A_C = \pi a^2; \ A_A = \pi(b^2 - a^2).$$

Hence,

$$i_C \rho_C / a^2 = i_A \rho_A / (b^2 - a^2).$$

Taking resistivity data from Table 1 in Chapter 28,

$$i_C (1.69 \times 10^{-8})/(0.25)^2 = i_A (2.75 \times 10^{-8})/(0.38^2 - 0.25^2) \rightarrow i_C = 1.242 i_A.$$

But,

$$i = i_C + i_A,$$

$$2 = 1.242 i_A + i_A \rightarrow i_A = 0.892 \ A.$$

Therefore, the current in the copper wire is $i_C = i - i_A = 2 - 0.892 = 1.108$ A.

(b) The potential drops across the ends of the wire and the jacket are identical. We evaluate this by working with the wire:

$$V_C = i_C \rho_C L/A_C,$$

$$L = \frac{V_C (\pi a^2)}{i_C \rho_C} = \frac{(12)\pi(0.25 \times 10^{-3})^2}{(1.108)(1.69 \times 10^{-8})} = 126 \ m.$$

56P

The method is illustrated by solving part (a) for the equivalent resistance of a face diagonal. Connect a battery between points B and C, the points between which the equivalent resistance is desired. This equivalent resistance is defined by

$$\varepsilon = IR_{eq}.$$

By the junction rule,

$$I = i_1 + i_2 + i_3,$$

and by the loop rule,

$$\varepsilon - i_1 R - i_5 R = 0.$$

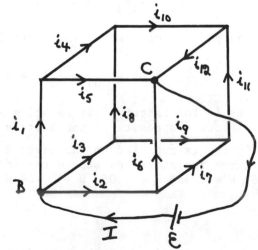

Therefore,

$$IR_{eq} = (i_1 + i_5)R.$$

Now write down all possible loop rule equations (excepting those that are redundant). Since the resistances are identical, the R in the equations will cancel out, leaving

$$i_1 + i_5 = i_2 + i_6; \quad i_7 + i_{11} = i_6 - i_{12};$$

$$i_3 + i_9 = i_7 + i_2; \quad i_8 + i_{10} = i_9 + i_{11};$$

$$i_4 + i_1 = i_3 + i_8; \quad i_4 + i_{10} = i_5 - i_{12}.$$

The junction rule equations are

$$i_1 = i_4 + i_5; \quad i_{10} + i_{11} = i_{12};$$

$$i_{10} = i_4 + i_8; \quad i_2 = i_6 + i_7;$$

$$i_3 = i_8 + i_9; \quad i_{11} = i_7 + i_9.$$

Solving these twelve equations for the twelve currents gives

$$i_1 = i_2 = i_5 = i_6 = i,$$

$$i_3 = i_{12} = \frac{2}{3} i,$$

$$i_4 = i_7 = 0,$$

$$i_8 = i_9 = i_{10} = i_{11} = \frac{1}{3} i.$$

Thus $I = i_1 + i_2 + i_3 = 8i/3$, and $i_1 + i_5 = 2i$. Substituting these into the equation at the top of this page gives

$$2iR = (\frac{8i}{3})R_{eq} \rightarrow R_{eq} = 3R/4.$$

58P

With $R_V = \infty$, it follows that $i_V = 0$ and

$$i = i_1 = \frac{\varepsilon}{r + R_1 + R_2} = 4.6154 \text{ mA}.$$

Hence, the voltmeter reading is $iR_1 = 1.1538$ V. When R_V is finite,

$$i_1 R_1 = i_V R_V = \text{voltmeter reading.}$$

But in this case we also have

$$\varepsilon - (r + R_2)i - i_1 R_1 = 0,$$

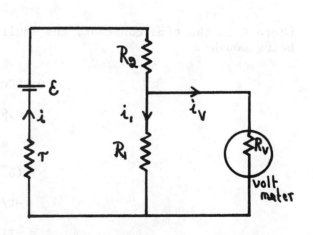

$$i = i_1 + i_V.$$

These three equations yield

$$i_1 R_1 = \frac{R_1 R_V \varepsilon}{rR_V + R_2 R_V + R_1 R_V + R_1 R_2 + rR_1} = 1.1194 \text{ V.}$$

Thus, the percent error is

$$\frac{1.1538 - 1.1194}{1.1538}(100) = 2.98\%.$$

63P

Ignore the battery and R_0 and focus on the rhombus. By the loop rule,

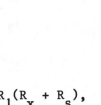

$$i(R_1 + R_2) = i*(R_x + R_s).$$

Also $V_{ac} = V_{bc}$, so that

$$iR_1 = i*R_s.$$

Eliminate i between these two equations to obtain

$$i*R_s(R_1 + R_2) = i*R_1(R_x + R_s),$$

$$R_x = R_s(R_2/R_1).$$

66E

During charging, the charge across the capacitor is given by Eq.29:

$$q = C\varepsilon(1 - e^{-t/RC}) = C\varepsilon(1 - e^{-t/\tau}),$$

where τ is the time constant. The equilibrium charge is $q(t=\infty) = C\varepsilon$. Hence, for the time being sought,

$$0.99C\varepsilon = C\varepsilon(1 - e^{-t/\tau}),$$

$$0.99 = 1 - e^{-t/\tau},$$

$$e^{-t/\tau} = 0.01,$$

$$\ln(e^{-t/\tau}) = \ln(0.01),$$

$$-t/\tau = \ln(0.01),$$

$$t = -\tau\ln(0.01) = 4.605\tau,$$

meaning that 4.605 time constants must elapse.

69P

(a) The charge on the capacitor is given by Eq.34 and therefore the potential difference across the capacitor is

$$V = q/C = (q_0/C)e^{-t/\tau} = V_0 e^{-t/\tau}.$$

When $t = 0$, $V = 100$ V so that, by the equation above, we have $V_0 = 100$ V also. Hence,

$$V = 100e^{-t/\tau}.$$

Now, when $t = 10$ s, $V = 1$ V. Substituting these data into the equation above gives

$$1 = 100e^{-10/\tau},$$

$$\ln 1 = 0 = \ln 100 - 10/\tau,$$

$$\tau = 10/\ln 100 = 2.17 \text{ s}.$$

(b) At $t = 17$ s,

$$V = 100e^{-t/\tau} = 100e^{-17/2.17} = 3.96 \times 10^{-2} \text{ V} = 39.6 \text{ mV}.$$

71P

(a) The energy stored in the capacitor is

$$U_C = q^2/2C.$$

At $t = 0$, $U_C = 0.5$ J, so that

$$0.5 = q_0^2/2(10^{-6}) \rightarrow q_0 = 10^{-3} \text{ C}.$$

(b) Use Eq.34 to calculate the current i:

$$i = \frac{dq}{dt} = -(q_0/\tau)e^{-t/\tau}.$$

Therefore, disregarding the sign,

$$i(0) = q_0/\tau = q_0/RC = \frac{10^{-3}}{(10^6)(10^{-6})} = 10^{-3} \text{ A}.$$

(c) The potential change across the capacitor is

$$V_C = q/C = (q_0/C)e^{-t/\tau}.$$

But, from (b), the time constant is evidently $\tau = RC = 1$ s, $q_0/C = (10^{-3})/(10^{-6}) = 10^3$ V. All this gives, in volts,

$$V_C = 1000e^{-t}.$$

From the loop rule

$$V_R = -V_C = -1000e^{-t},$$

also in volts.

(d) The rate of generation of thermal energy in the resistor is

$$P = i^2R = [(-q_0/\tau)e^{-t/\tau}]^2R = (q_0^2R/\tau^2)e^{-2t/\tau} = e^{-2t},$$

in watts; to evaluate P numerically, we used results from other parts of the problem.

72P

(a) The time constant is $\tau = RC = (3 \times 10^6)(10^{-6}) = 3$ s. Also,

$$q = C\varepsilon(1 - e^{-t/\tau}),$$
$$i = (\varepsilon/R)e^{-t/\tau}.$$

Hence, at $t = 1$ s,

$$\frac{dq}{dt} = i = (4)(e^{-1/3})/(3 \times 10^6) = 9.5538 \times 10^{-7} \text{ A}.$$

(b) The rate at which energy is being stored in the capacitor is

$$dU_C/dt = d(q^2/2C)/dt = \frac{q}{C}\frac{dq}{dt} = iq/C.$$

Now $q(1) = (10^{-6})(4)(1 - e^{-1/3}) = 1.1339 \times 10^{-6}$ C; $i(1)$ is found in (a) and therefore

$$dU_C/dt = (1.1339 \times 10^{-6})(9.5538 \times 10^{-7})/(10^{-6}) = 1.0833 \times 10^{-6} \text{ W}.$$

(c) For the resistor,

$$dU_R/dt = i^2R = (9.5538 \times 10^{-7})^2(3 \times 10^6) = 2.7383 \times 10^{-6} \text{ W}.$$

(d) Finally, the battery delivers energy at the rate

$$dU_B/dt = i\varepsilon = (9.5538 \times 10^{-7})(4) = 3.8215 \times 10^{-6} \text{ W}.$$

Note that, within round-off error,

$$dU_B/dt = dU_C/dt + dU_R/dt,$$

as required by the loop rule.

78P

(a) At $t = 0$ the capacitor exerts no influence and therefore the loop rule and junction rule require that

$$\epsilon - i_1 R - i_2 R = 0,$$

$$-i_3 R + i_2 R = 0,$$

$$i_1 = i_2 + i_3.$$

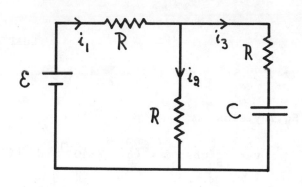

Solving for the currents gives

$$i_2 = i_3 = \epsilon/3R = 5.48 \times 10^{-4} \text{ A},$$

$$i_1 = 2\epsilon/3R = 1.10 \times 10^{-3} \text{ A}.$$

At $t = \infty$, the capacitor prevents the flow of current in its branch of the circuit, so that $i_3 = 0$. In the other branch, now just a series circuit,

$$i_1 = i_2 = i = \epsilon/2R = 8.22 \times 10^{-4} \text{ A}.$$

(b) At intermediate values of t,

$$i_1 = i_2 + i_3,$$

$$\epsilon - i_1 R - i_2 R = 0,$$

$$-\frac{q}{C} - i_3 R + i_2 R = 0.$$

Use the first two equations to eliminate i_2 from the third equation to get

$$(\frac{3R}{2})\frac{dq}{dt} + (\frac{1}{C})q = \epsilon/2,$$

since $i_3 = dq/dt$. The solution of this equation, for $q(0) = 0$ is, and this can be verified by direct substitution,

$$q = \frac{1}{2}C\epsilon(1 - e^{-2t/3RC}).$$

Therefore,

$$i_3 = (\epsilon/3R)e^{-2t/3RC}.$$

Using the third equation in (b) above, the potential drop across the resistor is found to be

$$V_R = i_2 R = \frac{\epsilon}{6}(3 - e^{-2t/3RC}).$$

(c) From (b), $V_R(0) = \epsilon/3 = 400$ V and $V_R(\infty)$ = $\epsilon/2 = 600$ V.

(d) The time constant is $3RC/2 = 7.1$ s. After many time constants i_3 is close to zero.

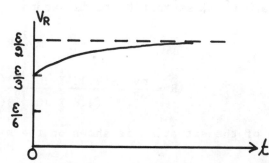

CHAPTER 30

4E

(a) Eq.7 yields

$$v = F_B/qB\sin\phi = (6.5 \times 10^{-17} \text{ N})/[(1.6 \times 10^{-19} \text{ C})(2.6 \times 10^{-3} \text{ T})\sin 23°],$$

$$v = 4.00 \times 10^5 \text{ m/s} = 400 \text{ km/s}.$$

(b) Since $v \ll c$, we use the classical formula for kinetic energy,

$$K = \tfrac{1}{2}mv^2 = \tfrac{1}{2}\frac{(1.67 \times 10^{-27} \text{ kg})(4.00 \times 10^5 \text{ m/s})^2}{1.6 \times 10^{-19} \text{ J/eV}} = 835 \text{ eV}.$$

7P

(a) The charge on the electron is $q = -e$, with $e = 1.6 \times 10^{-19}$ C. Therefore $\vec{F} = q\vec{v} \times \vec{B} = -e\vec{v} \times \vec{B} = e\vec{B} \times \vec{v}$, and this is directed to the east.

(b) In magnitude, since $\phi = 90°$,

$$F = evB\sin 90° = evB = ma,$$

$$a = (\tfrac{e}{m})vB.$$

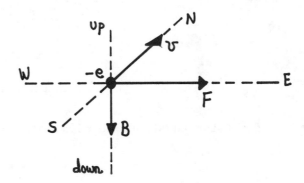

The kinetic energy K of the electron is employed to calculate the speed of the electron. Using the non-relativistic formula,

$$K = 12 \text{ keV} = (12 \times 10^3 \text{ eV})(1.6 \times 10^{-19} \text{ J/eV}) = 1.92 \times 10^{-15} \text{ J},$$

$$v = \sqrt{[2K/m]} = \sqrt{[2(1.92 \times 10^{-15})/(9.11 \times 10^{-31})]} = 6.492 \times 10^7 \text{ m/s}.$$

Thus,

$$a = (\tfrac{e}{m})vB = (\frac{1.6 \times 10^{-19}}{9.11 \times 10^{-31}})(6.492 \times 10^7)(5.5 \times 10^{-5}) = 6.27 \times 10^{14} \text{ m/s}^2.$$

(c) The electrons follow a circular path at the constant speed **v**. For uniform circular motion, $a = v^2/R$, so that Newton's second law gives

$$mv^2/R = evB,$$

$$R = \frac{mv}{eB} = \frac{(9.11 \times 10^{-31})(6.492 \times 10^7)}{(1.6 \times 10^{-19})(5.5 \times 10^{-5})} = 6.7207 \text{ m}.$$

A sketch of the situation is shown on the next page. The deflection sought is Δ. From the

242

sketch we see that

$$\tan\theta = \frac{0.2}{6.7207} = 0.02976.$$

Also (see Appendix G),

$$R^2 = (R + \Delta*)^2 + \Delta^2 - 2\Delta(R + \Delta*)\cos\theta.$$

Now $\Delta* = R(\sec\theta - 1)$, and therefore the equation above becomes

$$\Delta^2 - 2R\Delta + R^2\tan^2\theta = 0.$$

Numerically, in meters,

$$\Delta^2 - 13.4414\Delta + 0.040003 = 0,$$

$$\Delta = 0.002977; \quad 2R - 0.002977.$$

Thus the required deflection is 2.98 mm. The other solution above is the deflection from the same N-S line to the other side of the upper semicircular path. (The electrons will strike the screen and so will not actually complete this part of the trajectory.)

10E

(a) The velocity filter condition is given in Eq.10. First, calculate the speed v:

$$v = \sqrt{[\frac{2K}{m}]} = \sqrt{[\frac{2(10 \text{ keV})(1.6 \times 10^{-16} \text{ J/keV})}{9.11 \times 10^{-31} \text{ kg}}]} = 5.927 \times 10^7 \text{ m/s}.$$

We now have

$$B = \frac{E}{v} = \frac{10^4 \text{ V/m}}{5.927 \times 10^7 \text{ m/s}} = 1.69 \times 10^{-4} \text{ T}.$$

See Fig.9 for the directions of E, B and v.

(b) A proton has a positive charge. Hence, the electric and magnetic forces will be reversed in direction relative to the forces on the electron, but they will still allow a proton to pass through undeflected provided that its speed is that of the electron in (a) above.

12P

By Eq.10, the needed magnetic field is $B = E/v$. But $mv^2/2 = qV_{acc}$; also $E = V_{pl}/d$. In SI units, the result of combining these equations is

$$B = \frac{V_{pl}/d}{\sqrt{[2qV_{acc}/m]}} = \frac{(100)/(0.020)}{\sqrt{[2(1.6 \times 10^{-19})(10^3)/(9.11 \times 10^{-31})]}} = 2.67 \times 10^{-4} \text{ T}.$$

19E

Rearrange Eq.16 to solve for the magnetic field:

$$B = \frac{mv}{qr} = \frac{(9.11 \times 10^{-31} \text{ kg})(1.3 \times 10^6 \text{ m/s})}{(1.6 \times 10^{-19} \text{ C})(0.35 \text{ m})} = 2.11 \times 10^{-5} \text{ T} = 21.1 \text{ } \mu\text{T.}$$

23E

(a) The speed follows from $K = qV$ so we have, in SI units,

$$v = \sqrt{[\frac{2qV}{m}]} = \sqrt{[\frac{2(1.6 \times 10^{-19})(350)}{9.11 \times 10^{-31}}]} = 1.11 \times 10^7 \text{ m/s.}$$

(b) By Eq.16,

$$r = \frac{mv}{qB} = \frac{(9.11 \times 10^{-31})(1.11 \times 10^7)}{(1.6 \times 10^{-19})(0.200)} = 3.16 \times 10^{-4} \text{ m} = 0.316 \text{ mm.}$$

27E

If the radius of the circular path of the electrons after they emerge from the tube is less than d, then the electrons will not reach the plate. For radius d exactly, Newton's second law gives

$$evB = mv^2/d.$$

Since the kinetic energy is $K = mv^2/2$, this equation in terms of K rather than v is

$$e\sqrt{(2K/m)}B = 2K/d.$$

This can be solved for B; a stronger field gives a smaller radius, so the requirement is

$$B > \sqrt{(2Km/e^2d^2).}$$

The field must be arranged so that it is perpendicular to the velocity of the emergent electrons ($\sin 90° = 1$), as has been assumed in writing the equations above.

31P

The ion enters the spectrometer with a speed v related to the accelerating potential V by $W = K = qV$, so that

$$\frac{1}{2}mv^2 = qV.$$

Inside the instrument the ion undergoes uniform circular motion with the speed v not changing. By Newton's second law,

$$mv^2/r = qvB.$$

But, from the first equation above, $v^2 = 2qV/m$. Also $r = x/2$ (see Fig.33). Substituting these gives

$$\frac{m\sqrt{(2qV/m)}}{x/2} = qB \quad \rightarrow \quad m = B^2qx^2/8V.$$

33P

(a) Apply the result of Problem 31. Note that x = 2r = 2.0 m. Solving the equation in Problem 31 for B and substituting the data in SI units gives

$$B = (\frac{1}{x})\sqrt{[\frac{8mV}{q}]} = (\frac{1}{2})\sqrt{[\frac{8(3.92 \times 10^{-25})(10^5)}{3.2 \times 10^{-19}}]} = 0.495 \text{ T.}$$

(b) The number N of uranium ions collected in one hour is

$$N = \frac{10^{-4} \text{ kg}}{3.92 \times 10^{-25} \text{ kg}} = 2.551 \times 10^{20}.$$

Since each ion carries charge 2e, the current is

$$i = \frac{Q}{t} = \frac{N(2e)}{t} = \frac{(2.551 \times 10^{20})(3.2 \times 10^{-19} \text{ C})}{3600 \text{ s}} = 2.27 \times 10^{-2} \text{ A.}$$

(c) Each ion delivers kinetic energy K = qV and this will be converted to thermal energy in the cup. Hence, in one hour the thermal energy delivered is

$$E = NK = NqV = (2.551 \times 10^{20})(3.2 \times 10^{-19} \text{ C})(10^5 \text{ V}) = 8.16 \times 10^6 \text{ J.}$$

35P

(a) See Sample Problem 4. The period is

$$T = \frac{2\pi r}{v\sin\phi} = \frac{2\pi}{v\sin\phi}(\frac{mv\sin\phi}{qB}) = \frac{2\pi m}{qB}.$$

A positron is a positive electron so that, in SI units,

$$T = \frac{2\pi(9.11 \times 10^{-31})}{(1.6 \times 10^{-19})(0.1)} = 3.58 \times 10^{-10} \text{ s.}$$

(b) The pitch p is p = (vcosϕ)T, so that we must first find v from the kinetic energy:

$$v = \sqrt{[\frac{2K}{m}]} = \sqrt{[\frac{2(2 \times 10^3)(1.6 \times 10^{-19})}{9.11 \times 10^{-31}}]} = 2.651 \times 10^7 \text{ m/s.}$$

Therefore,

$$p = (v\cos\phi)T = (2.651 \times 10^7)\cos89°(3.58 \times 10^{-10}) = 1.66 \times 10^{-4} \text{ m} = 0.166 \text{ mm.}$$

(c) The radius is

$$r = \frac{mv\sin\phi}{qB} = \frac{(9.11 \times 10^{-31})(2.651 \times 10^7)\sin89°}{(1.6 \times 10^{-19})(0.10)} = 1.51 \times 10^{-3} \text{ m} = 1.51 \text{ mm.}$$

246

37P

(a) Write Eq.16 in the form

$$r = \frac{p}{qB},$$

where p is the relativistic momentum:

$$p = \frac{mv}{\sqrt{[1 - v^2/c^2]}}.$$

Substituting this into the equation for r yields, after a little rearrangement,

$$\frac{v}{\sqrt{[1 - v^2/c^2]}} = \frac{qBr}{m},$$

$$v = \frac{qBr/m}{\sqrt{[1 + (qBr/mc)^2]}}.$$

Inserting the numerical data gives

$$\frac{qBr}{mc} = \frac{(1.6 \times 10^{-19})(41 \times 10^{-6})(6.37 \times 10^6)}{(1.67 \times 10^{-27})(3 \times 10^8)} = 83.408,$$

$$\frac{qBr}{m} = 2.5022 \times 10^{10},$$

$$v = 2.99974 \times 10^8 \text{ m/s}.$$

(c) The arrangement of the vectors is shown on the sketch. In actuality, it is the magnetic south pole (where the field lines enter the earth) that is in the geographic northern hemisphere. Also, the geographic and magnetic poles do not coincide: see Fig.7 in Chapter 34.

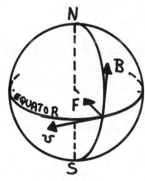

42P

From Sample Problem 5, the final deuteron energy is 17 MeV and the dee radius is 0.53 m. Since there are two "kicks" per revolution, the number of revolutions made by a deuteron is (17 MeV)/2(80 keV) = 106. Over each of these accelerations, as the deuteron passes from one dee to the other, its speed increases from v to v* and the radius of its semi-circular path increases from r to r*. These quantities are related by the work-energy theorem (Eq.16 in Chapter 7):

$$\frac{1}{2}mv^2 + qV = \frac{1}{2}mv*^2,$$

$$\frac{1}{2}m(r\omega)^2 + qV = 2\pi^2 m\nu^2 r^2 + qV = 2\pi^2 m\nu^2 r*^2,$$

$$r*^2 = r^2(1 + qV/2\pi^2 m\nu^2 r^2).$$

If we can write r* = r + Δr with Δr/r << 1, then (see Appendix G),

$$r*^2 \simeq r^2(1 + 2\Delta r/r).$$

Comparison of the last two equations suggests that

$$\Delta r = \frac{qV}{4\pi^2 m\nu^2} \frac{1}{r}.$$

We see that Δr is largest at small r and decreases toward the edge of the dees; see Fig. 16. Thus the deuteron spends more of the 106 revolutions at radii greater than the average, (0.53 m)/2, than at smaller radii. As an actual average radius then, let's pick (0.53 m)$(\sqrt{2}/2)$, for which the circumference of the corresponding circle is $2\pi(0.37$ m) = 2.3 m. Since the deuteron makes 106 revolutions, the path traveled must be approximately (106)(2.3 m) = 240 m. This answer is approximate because we did not make a careful calculation of the actual average radius.

44E

Apply Eq.24. The magnitude of the force is

$$F_B = iLB\sin\phi = (5000 \text{ A})(100 \text{ m})(60 \times 10^{-6} \text{ T})\sin70° = 28.2 \text{ N}.$$

The direction, found by applying the right-hand rule to the vector product, is horizontal and to the west.

46P

To remove the tension in the leads the force F_B exerted by the magnetic field on the wire of length L must equal mg in magnitude, but be directed vertically up. By the right-hand rule, the current must pass from left to right. Then,

$$F_B = iLB\sin90° = iLB = mg,$$

$$i = \frac{mg}{LB} = \frac{(0.013 \text{ kg})(9.8 \text{ m/s}^2)}{(0.62 \text{ m})(0.44 \text{ T})} = 0.467 \text{ A}.$$

52P

Let B make an angle θ with the vertical. The magnetic force is $F_B = iLB$, since B is at 90° to the horizontal rod. For the rod to be on the verge of sliding, apply Eq.1 of Chapter 6 to get

$$iLB\cos\theta - \mu N = 0,$$

where μ is the coefficient of static friction. In the vertical direction,

$$N + iLB\sin\theta - mg = 0.$$

Eliminating the normal force N between these equations gives

$$iLB\cos\theta - \mu(mg - iLB\sin\theta) = 0,$$

$$B = \frac{\mu mg/iL}{\cos\theta + \mu\sin\theta}.$$

To find the minimum field needed, set $dB/d\theta = 0$; this gives

$$\theta = \tan^{-1}\mu = \tan^{-1}(0.6) = 31°.$$

Numerically, $\mu mg/iL = (0.6)(1)(9.8)/(50)(1) = 0.1176$ T. Using the equation for B above, with the value of θ just determined, gives $B_{min} = 0.101$ T.

<u>54E</u>

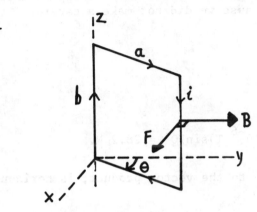 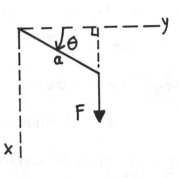

Let the hinge lie along the z axis. It is evident that the only force that can exert a torque along the hinge is the one exerted on the side of the rectangle opposite the hinge. (The forces on the 5 cm sides are in the z direction and, by the cross-product rule, their torques will be perpendicular to z.) The magnitude F of this force is

$$F = ibB\sin90° = ibB.$$

From the sketch, the appropriate moment arm is seen to be $a\cos\theta$, so that the torque for N loops will be

$$\tau = N(ibB)(a\cos\theta) = (20)(0.1 \text{ A})(0.1 \text{ m})(0.5 \text{ T})(0.05 \text{ m})\cos30° = 4.33 \times 10^{-3} \text{ N·m}.$$

From the right-hand rule, it is seen that this is directed in the -z direction, that is, down.

<u>56P</u>

If N closed loops are formed from the wire of length L, the circumference of each loop is L/N, the radius of each loop is L/2πN, and therefore the area of each is

$$A = \pi(L/2\pi N)^2 = L^2/4\pi N^2.$$

For maximum torque, orient the plane of the loops parallel to the magnetic field lines, so that θ in Eq.27 is 90°. By Eq.27, then,

$$\tau = NiAB = Ni(L^2/4\pi N^2)B = iL^2B/4\pi N.$$

Since N appears in the denominator, the maximum torque is that for N = 1: $\tau_{max} = iL^2B/4\pi$.

60P

(a) The deflection of a galvanometer is proportional to the current passing through it. If the device obeys Ohm's law, this current is proportional to the voltage across the instrument. Hence, the deflection is proportional to the voltage also. Thus, if the total resistance of the galvanometer plus auxiliary resistor is R,

$$V = iR,$$

$$1 = (0.00162)R \rightarrow R = 617.3 \ \Omega.$$

Therefore, connect the auxiliary resistor r in series with the galvanometer, choosing r = 617.3 − 75.3 = 542 Ω.

(b) In this case, insert the auxiliary resistor in parallel with the galvanometer, so that a current i = 0.00162 A flows through the galvanometer when it is attached to a circuit branch in which the current I = 0.050 A, as shown. For a parallel connection,

$$i_a r_a = i r_G;$$

furthermore,

$$I = i + i_a.$$

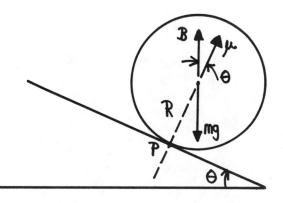

Eliminating i_a gives

$$r_a = \frac{i}{I - i} r_G = \frac{0.00162}{0.05 - 0.00162}(75.3) = 2.52 \ \Omega.$$

61P

If the cylinder rolls, the instantaneous axis of rotation will pass through P, the point on the rim in contact with the plane (see Fig. 3c in Chapter 12). Neither the normal force nor the force of friction (not shown) will exert any torque about P since their lines of action pass through P. The torque due to gravity is

$$\tau_g = mgr = mgR\sin\theta,$$

directed into the page. The torque due to the magnetic field on the current loop is

$$\tau_m = \mu B\sin\theta = NiAB\sin\theta = Ni2RLB\sin\theta,$$

directed out of the page. For no rotation the net torque must vanish, so that

$$Ni2RLBsin\theta = mgRsin\theta,$$

$$i = \frac{mg}{2NBL} = \frac{(0.25)(9.8)}{2(10)(0.5)(0.1)} = 2.45 \text{ A.}$$

62E

(a) The magnitude of the magnetic dipole moment of a current loop is $\mu = NiA$, so that

$$i = \mu/NA = (2.3 \text{ A} \cdot m^2)/(160)[\pi(0.019 \text{ m})^2] = 12.7 \text{ A.}$$

(b) From Eq.29 we find the maximum torque by setting $\theta = 90°$ to obtain

$$\tau = \mu B = (2.3 \text{ A} \cdot m^2)(0.035 \text{ T}) = 0.0805 \text{ N} \cdot m.$$

CHAPTER 31

3E

The magnetic field due to the current in the wire, at a point a distance r from the wire, is, by Eq.8,

$$B = \mu_0 i/2\pi r.$$

Set $r = 20$ ft $= 6.096$ m, so that

$$B = (4\pi \times 10^{-7} \text{ T} \cdot \text{m/A})(100 \text{ A})/2\pi(6.096 \text{ m}) = 3.28 \times 10^{-6} \text{ T} = 3.28 \text{ } \mu\text{T}.$$

This will affect the compass reading.

7E

(a) The field B_w due to the wire must be directed opposite to the earth's field; i.e., to the south; it also must equal the earth's field in magnitude. If we call the earth's field B_e, then we must have, by Eq.8,

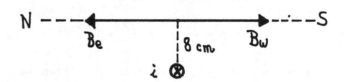

$$i = 2\pi r B_w/\mu_0 = 2\pi r B_e/\mu_0,$$

$$i = 2\pi(0.08 \text{ m})(39 \times 10^{-6} \text{ T})/(4\pi \times 10^{-7} \text{ T} \cdot \text{m/A}) = 15.6 \text{ A}.$$

(b) By the right-hand rule, the current must be directed into the paper, i.e., flowing from west to east.

15P

By the Biot-Savart law, the contribution to the magnetic field B from the short section ds of the wire is, from Eq.7,

$$d\vec{B} = (\mu_0 i/4\pi) \frac{d\vec{s} \times \vec{r}}{r^3}.$$

Now sections he and fg give nothing since $d\vec{s} \times \vec{r} = 0$ there. Along fe,

$$B_2 = (\mu_0 i/4\pi)\int\frac{r\sin(\vec{r},d\vec{s})ds}{r^3},$$

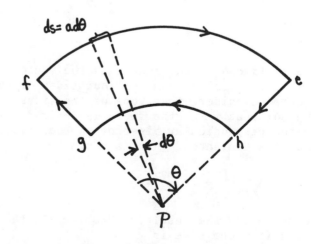

$$B_2 = (\mu_0 i/4\pi)\int\frac{a(ad\theta)\sin90°}{a^3} = (\mu_0 i/4\pi a)\int_0^\theta d\theta = \mu_0 i\theta/4\pi a.$$

Similarly, the contribution from section gh is $B_1 = (\mu_0 i\theta/4\pi b)$. Also, note that $B_1 > B_2$.

(This is because a > b.) Furthermore, B_1 is out of and B_2 into, the page. Hence, the magnitude of the field at P is

$$B = B_1 - B_2 = (\mu_0 i \theta/4\pi)(\frac{1}{b} - \frac{1}{a}),$$

directed out of the page.

17P

Let the wire rest along the x axis, its midpoint at the origin. Since the wire is of finite length, the Biot-Savart law is called for. All elements of the wire give rise to a field that is directed into the paper. Therefore, the strength of the total field is found from

$$B = (\mu_0 i/4\pi) \int_{x=-\frac{1}{2}L}^{x=+\frac{1}{2}L} (ds) r \sin\theta/r^3.$$

But the variables are related by

$$\sin\theta = \frac{R}{r}; \quad r^2 = x^2 + R^2.$$

Therefore, the integral can be written as

$$B = (\mu_0 iR/4\pi) \int_{-\frac{1}{2}L}^{+\frac{1}{2}L} (x^2 + R^2)^{-3/2} dx,$$

$$B = (\mu_0 i/2\pi R) \frac{L}{\sqrt{(L^2 + 4R^2)}}.$$

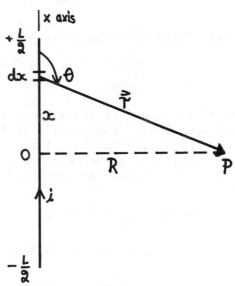

If R << L, ignore the R^2 in the denominator to obtain $B = \mu_0 i/2\pi R$, which is the field due to a very long wire; from points close to the wire, the wire appears to be very long.

20P

The field due to the square is the vector sum of the fields from the four sides of the square. Consider, then, one of the sides. The point at which the field is desired lies on the perpendicular bisector to that side, and at a distance R that is given by

$$R = \sqrt{(4x^2 + a^2)}/2.$$

Hence, using the result of Problem 17, the field from one side is

$$B = (\mu_0 i/2\pi R) \frac{a}{\sqrt{(a^2 + 4R^2)}},$$

putting a for L as the length of each side segment. Substituting for R from the first

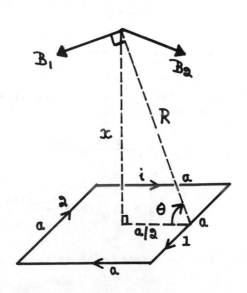

equation gives

$$B = (\mu_0 i/\pi)\frac{a}{\sqrt{(4x^2 + a^2)}\sqrt{(4x^2 + 2a^2)}}.$$

The direction of this field is perpendicular to the plane that contains the side under consideration and the perpendicular bisector of that side. The component of this field that is perpendicular to the normal of the square itself will be cancelled by the analogous component of the field due to the opposite side of the square (see the sketch on p.252). Thus, the total field will be

$$B_T = 4B\cos\theta = 4B\frac{a/2}{R} = \frac{4aB}{\sqrt{(4x^2 + a^2)}} = (4\mu_0 i/\pi)\frac{a^2}{(4x^2 + a^2)\sqrt{(4x^2 + 2a^2)}}.$$

As expected, at x = 0 (the center of the square), this result reduces to that of Problem 18.

21P

Use the result of Problem 18 for the square, but put a = L/4 to get

$$B_{sq} = 8\sqrt{2}(\mu_0 i/\pi L) = 11.31(\mu_0 i/\pi L).$$

For the circle, use Eq.24 with z = 0, to obtain

$$B_{ci} = \mu_0 i/2R.$$

Since R = L/2π, this becomes

$$B_{ci} = \pi^2(\mu_0 i/\pi L) = 9.87(\mu_0 i/\pi L).$$

Thus, as asserted, the square yields the larger magnetic field at its center.

23P

B at Q is zero since, in using the law of Biot-Savart, $\vec{ds}$ is parallel to $\vec{r}$, so that ds X r = 0. As for B at P, use the result of Problem 22 with L = D = a to find

$$B_P = \frac{\mu_0 i}{4\pi a}\frac{a}{\sqrt{(a^2 + a^2)}} = \sqrt{2}\mu_0 i/8\pi a.$$

27E

(a) See Section 4. For parallel currents, the fields midway between the wires are in opposite directions and sum to zero. Thus there is no possible answer.

(b) For antiparallel currents, the fields due to the wires at a point midway between them are equal and parallel. Hence, the field due to one wire is 150 μT. By Eq.8, the current in the wire must be

$$i = 2\pi rB/\mu_0 = 2\pi(0.04 \text{ m})(150 \text{ X } 10^{-6} \text{ T})/(4\pi \text{ X } 10^{-7} \text{ T·m/A}) = 30.0 \text{ A}.$$

40E

(a) The net current enclosed by the dotted path is 2.0 A, out of the page. However, a net current out of the page gives rise to a field directed in the counterclockwise sense (the right-hand rule applied to Ampere's law). Since the dotted path is traversed clockwise, the term $\vec{B} \cdot d\vec{s}$ gives rise to a factor $\cos 180° = -1$ in the integral, so that the integral yields $-2\mu_0$, T·m.

(b) The net current enclosed by the dashed path is zero (4.0 A in, 4.0 A out). Hence the value of the line integral is zero.

45P

Apply Ampere's law by integrating around the dashed rectangle shown on the sketch. Since there is no current cutting the area bounded by this rectangle, it is anticipated that

$$\oint \vec{B} \cdot d\vec{s} = 0.$$

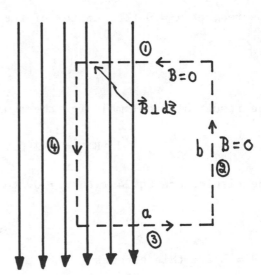

Now examine the sides of the rectangle in turn. On side 2, $\int \vec{B} \cdot d\vec{s} = 0$ since B = 0 on this side. On sides 1 and 3 the integral is zero either because B = 0 or B is at a 90° angle to ds. On side 4, however,

$$\int \vec{B} \cdot d\vec{s} = B_4 b,$$

where B_4 is the value of the magnetic field along side 4. Thus, adding up the separate values of the integral along the four sides gives

$$\oint \vec{B} \cdot d\vec{s} = B_4 b \neq 0.$$

This contradicts the expectation from Ampere's law expressed in the first equation. We conclude that the geometry pictured for the magnetic field must be in error. The lines actually bulge out and their density decreases gradually near the edges of the magnet.

50P

(a) Let P represent a point in the hole. By superposition, the field $\vec{B}$ at P = (field $\vec{B}_1$ at P due to a wire of radius a and current I containing no hole) − (field $\vec{B}_2$ at P due to the hole if it carried a uniform current of current density $I/\pi a^2$ in the same direction as i). That is,

$$\vec{B} = \vec{B}_1 - \vec{B}_2.$$

The condition of equal current densities is

$$\frac{i}{\pi a^2 - \pi b^2} = \frac{I}{\pi a^2}.$$

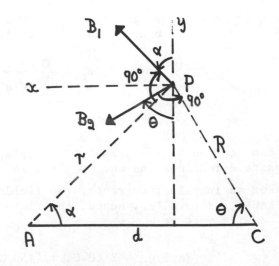

$$B_1 = \mu_0 Ir/2\pi a^2; \quad B_2 = \mu_0 IR/2\pi a^2.$$

Let A be the center of the wire and C the center of the hole. Then, from the sketch, we see that

$$B_y = B_1\cos\alpha + B_2\cos\theta = (\mu_0 I/2\pi a^2)(r\cos\alpha + R\cos\theta) = \mu_0 Id/2\pi a^2,$$

$$B_x = B_1\sin\alpha - B_2\sin\theta = (\mu_0 I/2\pi a^2)(r\sin\alpha - R\sin\theta) = 0.$$

But, from the current density condition,

$$I = (\frac{a^2}{a^2 - b^2})i,$$

and therefore

$$B = (\mu_0 i/2\pi)\frac{d}{a^2 - b^2},$$

perpendicular to AC and independent of the position of P within the hole.

(b) If b = 0, the result in (a) reduces to Eq.18 (with a different notation), the field inside a current carrying wire. With d = 0, we get B = 0. That is, B = 0 inside the hole if it is centrally located. This can be verified easily by Ampere's law.

51P

(a)

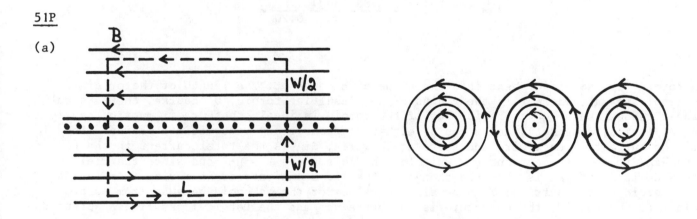

Consider a large number of long wires carrying identical currents and arranged in a plane as in the sketch above. By the right-hand rule, the fields above the plane all are to the left and below the plane all are directed to the right. Between the wires the field due to neighboring wires are oppositely pointing. It seems reasonable that as the wires are brought closer together the already opposing fields between them will vanish (cancel), yielding the pattern shown above on the left.

(b) Selecting as a path of integration a rectangle of length L and width W aligned as shown, perpendicular to the current sheet which evenly divides the rectangle, we see that B is at 90° to ds along the sides W, so the line integral in Ampere's law gives zero on the sides. Along the top of length L, however,

$$\int \vec{B}\cdot d\vec{s} = \int B ds \cos 0° = B\int ds = BL.$$

Since both of the sides (length L, parallel to the sheet) are at the same distance (W/2) from the sheet, B has the same magnitude at each of them, so that Ampere's law gives

$$\oint \vec{B} \cdot d\vec{s} = 2BL = \mu_0 i_{enc} = \mu_0 (\lambda L) \;\rightarrow\; B = \tfrac{1}{2}\mu_0 \lambda.$$

54E

Eq.21 applies; the number of turns per unit length (SI units, of course) is

$$n = \frac{200}{0.25 \text{ m}} = 800 \text{ m}^{-1},$$

so that

$$B = \mu_0 i_0 n = (4\pi \times 10^{-7} \text{ T·m/A})(0.30 \text{ A})(800 \text{ m}^{-1}) = 3.02 \times 10^{-4} \text{ T} = 0.302 \text{ mT}.$$

56E

(a) Eq.22 applies to the toroid. The inner radius is r = 15 cm = 0.15 m, so we have (in SI units),

$$B = \frac{\mu_0 i_0 N}{2\pi} \frac{1}{r} = \frac{(4\pi \times 10^{-7})(0.80)(500)}{2\pi} \frac{1}{0.15} = 5.33 \times 10^{-4} \text{ T} = 533 \text{ μT}.$$

(b) The outer radius is 15 + 5 = 20 cm = 0.20 m; at this location,

$$B = \frac{(4\pi \times 10^{-7})(0.80)(500)}{2\pi} \frac{1}{0.20} = 400 \text{ μT}.$$

59P

A toroid is a solenoid bent into the shape of a doughnut. The length of the original solenoid is $2\pi r$, where r is the radius of the resulting toroid. Of course, there is an inner radius and an outer radius but if the toroid is ideal, these radii must be nearly equal (skinny doughnut), and r can be either of them or any intermediate radius. Since i_0 is the current in each wire and there are N wires, Ni_0 is the total current in the toroid and $Ni_0/2\pi r = \lambda$, the current per unit length. This gives $B = \mu_0 \lambda$ and since the field outside is zero, $\mu_0 \lambda$ is also the change in the field encountered in moving from inside the toroid to the outside. This equals the value for the solenoid because from points close to the toroid, the curvature is not perceived and the toroid looks like a solenoid.

62P

(a) The force due to the magnetic field must be directed to the center of curvature; by the right-hand rule, then, the particle must carry negative charge.

(b) The changing radius of curvature of the path is given by Eq.16 of Chapter 30:

$$R = \frac{mv}{qB},$$

where q is the magnitude of the charge. Now, the magnetic field does zero work on the particle; by the work-energy theorem, its kinetic energy must remain constant. Thus, its speed v does not change. Therefore, at two points 1 and 2 on the trajectory, RB = mv/q = constant, so that

$$R_1 B_1 = R_2 B_2.$$

But, for a toroid, by Eq.22,

$$B = (\mu_0 i_0 N / 2\pi)\frac{1}{r},$$

where r is the distance of the particle from the axis of the toroid. Hence,

$$R_1 / r_1 = R_2 / r_2,$$

$$11/125 = R_2/110 \rightarrow R_2 = 9.68 \text{ cm}.$$

63E

See Section 9 in Chapter 30. We find that

$$\mu = NiA = (200)(0.30 \text{ A})[\pi(0.10 \text{ m})^2/4] = 0.471 \text{ A} \cdot \text{m}^2.$$

66E

(a) The magnetic moment is

$$\mu = NiA = (300)(4.0 \text{ A})[\pi(0.05 \text{ m})^2/4] = 2.36 \text{ A} \cdot \text{m}^2.$$

(b) Use Eq.25; note that in Eq.25, the authors use z, not R, for distance from the loop. We have, in SI units,

$$z = [\frac{\mu_0}{2\pi} \frac{\mu}{B}]^{1/3} = [\frac{4\pi \times 10^{-7}}{2\pi} \frac{2.36}{5 \times 10^{-6}}]^{1/3} = 0.455 \text{ m} = 45.5 \text{ cm}.$$

74P

(a) Consider a narrow ring of radius r and width dr; it carries charge

$$dq = \frac{q}{\pi R^2}(2\pi r dr),$$

i.e., the charge per unit area of the disk times the area of the ring. In time $T = 2\pi/\omega$ all this charge passes any fixed point near the ring, so that the equivalent current is

$$di = \frac{dq}{T} = \frac{2\pi q r dr/\pi R^2}{2\pi/\omega} = q\omega r dr/\pi R^2.$$

By Eq.24, with z = 0 (note the difference in notation), this ring sets up a field dB at the center of the disk given by

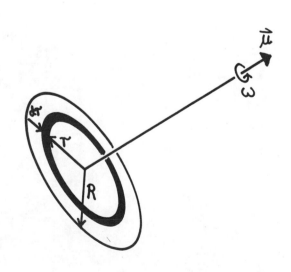

$$dB = \frac{\mu_0 di}{2r} = \frac{\mu_0}{2r}(q\omega r dr/\pi R^2).$$

Thus, the total field is

$$B = \int dB = (\frac{\mu_0 q\omega}{2\pi R^2})\int_0^R dr = \mu_0 q\omega/2\pi R.$$

(b) The dipole moment is

$$\mu = \int A di = \int_0^R (\pi r^2)(\frac{q\omega r dr}{\pi R^2}) = (\frac{q\omega}{R^2})\int_0^R r^3 dr = q\omega R^2/4.$$

CHAPTER 32

3E

At any instant, the flux through the antenna is

$$\Phi_B = \int \vec{B} \cdot d\vec{A} = \int BdA\cos 0° = \int BdA = B\int dA = BA = B(\pi R^2),$$

R being the radius of the antenna. The flux changes as B changes (R stays fixed) so that, disregarding the sign in Lenz's law (the direction, or "sense" of the emf is not of importance), we have, by Eq.7,

$$\varepsilon = N[d\Phi_B/dt] = N[d(B\pi R^2)/dt] = N\pi R^2[dB/dt],$$

$$\varepsilon = (1)\pi(0.055 \text{ m})^2(0.16 \text{ T/s}) = 1.52 \times 10^{-3} \text{ V} = 1.52 \text{ mV}.$$

10P

(a) By Faraday's law, Eq.7, the current i induced in the coil is

$$i = \frac{\varepsilon}{R} = (\frac{N}{R})d\Phi_B/dt.$$

Now $\Phi_B = \int \vec{B} \cdot d\vec{A}$, the integral taken over the cross-section of the coil, or any area that is bounded by the coil, and B is the magnetic field due to the solenoid. Since $B = \mu_0 n i_s$ inside the solenoid (i_s is the current in the solenoid) and $B = 0$ outside the solenoid, the integral for the flux between coil and solenoid will be zero, so that

$$\Phi_B = \int (\mu_0 n i_s)dA + 0 = \mu_0 n i_s A_s,$$
$$\text{area of}$$
$$\text{solenoid}$$

where A_s is the cross-sectional area of the solenoid. Thus,

$$i = (\frac{N}{R})d(\mu_0 n i_s A_s)/dt = (\frac{N}{R})\mu_0 n A_s d(i_s)/dt = (\frac{N}{R})\mu_0 n A_s 2i_0/t.$$

Numerically, $N = 120$, $R = 5.3$ Ω, $\mu_0 = 4\pi \times 10^{-7}$ T·m/A, $n = 22,000$ m^{-1}, $A_s = \pi(0.016 \text{ m})^2$, $i_0 = 1.5$ A and $t = 0.05$ s; these give $i = 30.2$ mA.

(b) As the magnetic field changes, electric fields appear and these affect the conduction electrons in the coil.

14P

Thermal energy is generated at the rate $\varepsilon^2/R = P$. The resistance of the loop is, in SI units (see Eq.15 and Table 1 in Chapter 28),

$$R = \rho L/A = (1.69 \times 10^{-8})\frac{(0.50)}{\pi(10^{-3})^2/4} = 1.0759 \times 10^{-2} \text{ } \Omega.$$

The emf is found from Faraday's law, Eq.7 with $N = 1$:

$$\varepsilon = d\Phi_B/dt = \pi r^2(dB/dt) = \pi(L/2\pi)^2(dB/dt) = L^2(dB/dt)/4\pi,$$

$$\varepsilon = (0.5)^2(10^{-2})/4\pi = 1.9894 \times 10^{-4} \text{ V.}$$

Therefore,

$$P = \varepsilon^2/R = (1.9894 \times 10^{-4})^2/(1.0759 \times 10^{-2}) = 3.68 \ \mu\text{W.}$$

16P

(a) Far away from the larger loop, the field may be taken, across the area of the small loop, as approximately uniform and equal to the value on the axis. From Section 7 in Chapter 31, writing x for z, this field is

$$B = \mu_0 iR^2/2x^3,$$

so that the flux through the small loop is

$$\Phi_B = B(\pi r^2) = \mu_0 iR^2 \pi r^2/2x^3.$$

(b) For a single loop, the emf is, from Eq.6,

$$\varepsilon = d\Phi_B/dt = (\mu_0 iR^2 \pi r^2/2)\frac{d(1/x^3)}{dt}.$$

But,

$$d(1/x^3)dt = -3x^{-4}dx/dt = -3v/x^4.$$

Hence, except for sign,

$$\varepsilon = \frac{3}{2}\mu_0 i\pi R^2 r^2 v/x^4.$$

(c) The field due to the current i in the larger loop, near its axis, is directed away from the larger loop toward the smaller loop. Therefore, as the small loop moves away from the larger, the small loop will see a steadily decreasing upward directed flux cutting across it. To oppose this, the induced emf will seek to set up an induced current the magnetic field of which will be directed upward also, in the area bounded by the small loop. By the right-hand rule, this will require a counterclockwise current as seen looking down from above the small loop (i.e., in the same direction as the current i in the large loop).

17P

(a) By Faraday's law,

$$i = \varepsilon/R = (-1/R)d\Phi_B/dt.$$

But i = dq/dt and therefore

$$\frac{dq}{dt} = -\frac{1}{R}d\Phi_B/dt,$$

$$\int_0^q \frac{dq}{dt}\, dt = -\frac{1}{R}\int_{\Phi_B(0)}^{\Phi_B(t)} (d\Phi_B/dt)dt,$$

$$q = -\frac{1}{R}[\Phi_B(t) - \Phi_B(0)] = \frac{1}{R}[\Phi_B(0) - \Phi_B(t)].$$

The limits on the integrals are the values of charge and flux at times t and zero.

(b) If the flux at time t and the flux at time t = 0 both are zero, the equation derived in (a) implies that the net charge through the circuit is zero. If the flux changed during the time interval, this result indicates that equal amounts of charge flowed "clockwise" through the circuit for part of the time, and "counterclockwise" for the remaining time, giving a net charge flow of zero. This means that induced currents moving in different directions could have existed during the entire time interval, although the current could be zero for part of the time interval also.

19P

(a) The flux, at any time, is (with L = length of each side of the square)

$$\Phi_B = B(L^2/2),$$

only half of the square being in the field. Since dB/dt = -0.87 T/s, we have, by Faraday's law,

$$\varepsilon = -d\Phi_B/dt = -(L^2/2)(dB/dt) = -(2^2/2)(-0.87) = 1.74 \text{ V}.$$

Now the field B inside the square is directed out of the page and is decreasing in strength. The induced current will flow so as to produce an upward directed flux inside the square. To do this a counterclockwise current is needed. But this is the direction of the current set up by the battery. Thus the induced emf must be in the same direction as the battery emf, for a total emf of 20 + 1.74 = 21.74 V.

(b) From (a), the current is counterclockwise.

21P

(a) First, calculate the flux through the interior of a single wire due to the wire itself. Choose dA in the same direction as B as it cuts the surface, which is a rectangle with one side on the axis of the wire and extending to the surface. From Eq.18 in Chapter 31, the field inside the wire is

$$B = \mu_0 ir/2\pi R^2.$$

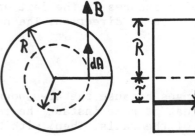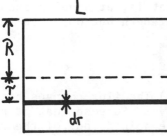

Thus,

$$\Phi_B = \int \vec{B}\cdot d\vec{A} = \int B(dA)\cos 0° = \int B dA.$$

Now B varies across the surface in the radial, or r, direction, but not over the shaded strip above, which is very thin in the radial direction, so that all points on the strip are essentially at the same distance from the axis of the wire. Hence,

$$\Phi_B = \int_0^R (\mu_0 ir/2\pi R^2) L dr = \mu_0 iL/4\pi.$$

Now call the radius of each wire b and let a be the distance between their axes. The flux inside each wire due to its own current is

$$\Phi_i = \mu_0 iL/4\pi.$$

The flux outside, out to a distance r = a from the axis is

$$\Phi_o = \int B dA = \int_b^a (\mu_0 i/2\pi r) L dr = (\mu_0 iL/2\pi) \ln\left(\frac{b}{a}\right),$$

choosing dA in the same direction as B. The wires and currents being identical, the fields due to the two wires, in the space between them, are in the same direction since their currents are antiparallel. Hence, the total flux per meter is

$$\Phi = 2\Phi_o + 2\Phi_i = (2\mu_0 i/4\pi)[1 + 2\ln\left(\frac{b}{a}\right)].$$

Numerically, a/b = 20/1.25 = 16, i = 10 A, μ_0 = 4π X 10^{-7} T·m/A; these give Φ = 13.09 μWb/m.

(b) Consider one of the wires. Part of its flux for r > b lies inside the other wire; this flux is

$$\Phi* = \int B dA = (\mu_0 iL/2\pi) \ln\left(\frac{a}{a-b}\right).$$

A similar flux lies inside the other wire due to the first wire and therefore the total flux inside the wires is

$$2\Phi_i + 2\Phi* = (2\mu_0 iL/4\pi)[1 + 2\ln\left(\frac{a}{a-b}\right)].$$

Numerically, per meter (i.e., L = 1 m) this is 2.258 μWb/m. Therefore, the fraction that is sought is (2.258/13.09) = 0.1725 = 17.25%.

(c) If the currents are in the same direction, the net magnetic field between the wires is antisymmetric about a line parallel to the axes of the wires and midway between them. That is, at equal distances to the left and right of this line the fields are equal in strength but opposite in direction. The net flux, therefore, vanishes.

25E

(a) The flux changes because the area bounded by rod and rails is increasing as the rod moves. Suppose that at some instant the rod is at a distance x from the right-hand end, the joined end, of the rails. Then, with N = 1, the flux through the area is

$$\Phi_B = BA = BxL,$$

L being the length of the rod. Thus, by Faraday's law,

$$\varepsilon = d\Phi_B/dt = d(BxL)/dt = BL(dx/dt) = BLv,$$

$$\epsilon = (0.35 \text{ T})(0.25 \text{ m})(0.55 \text{ m/s}) = 0.0481 \text{ V} = 48.1 \text{ mV}.$$

(b) The current can be found from Ohm's law (Eq.8 in Chapter 28),

$$i = \epsilon/R = (48.1 \text{ mV})/(18 \text{ }\Omega) = 2.67 \text{ mA}.$$

26E

(a) The emf induced is $\epsilon = d\Phi_B/dt = d(BA)/dt = d(BLx)/dt = BL(dx/dt) = BLv$, where x is the distance of the rod from the closed end of the rails. Thus,

$$\epsilon = BLv = (1.2 \text{ T})(0.1 \text{ m})(5 \text{ m/s}) = 0.60 \text{ V}.$$

(b) The induced current is $i = \epsilon/R = (0.6 \text{ V})/(0.4 \text{ }\Omega) = 1.5$ A. As the rod moves to the left, an increasing upward directed flux becomes contained within the rod-rail circuit. The induced current will flow so that the magnetic field it sets up will point downward in this same region bounded by the circuit. By the right-hand rule, this requires a clockwise flowing induced current.

(c) Thermal energy is generated at the rate $P = i^2R = (1.5 \text{ A})^2(0.4 \text{ }\Omega) = 0.90$ W.

(d) The force F exerted by the external agent must overcome the $i\vec{L} \times \vec{B}$ force exerted by the external magnetic field B on the induced current in the rod. The latter force points to the right; for zero acceleration, set F = iLB exactly (zero net force); then we have F = iLB = (1.5 A)(0.1 m)(1.2 T) = 0.18 N.

(e) By Eq.25 in Chapter 7, the rate that this force F does work is Fv = (0.18 N)(5 m/s) = 0.90 W; this equals (c) by energy conservation.

31P

(a) By Faraday's law $\epsilon = -Nd\Phi_B/dt$ and therefore the flux through the rectangular loop at any time t must be found first. This flux is

$$\Phi_B = \int \vec{B} \cdot d\vec{A} = \int B(dA)\cos\theta = B\cos\theta \int dA = abB\cos\theta.$$

This last result follows because B is uniform and θ, the angle between B and dA, has at any instant the same value at all points on the rectangle. If the loop rotates at the constant rate ν, then $\theta = \omega t = 2\pi\nu t$ and, by Faraday's law,

$$\epsilon = -Nd\Phi_B/dt = -Nd(abB\cos 2\pi\nu t)/dt = 2\pi N\nu abB\sin 2\pi\nu t = (2\pi NabB\nu)\sin 2\pi\nu t = \epsilon_0\sin 2\pi\nu t.$$

(b) If $\epsilon_0 = 150$ V, $\nu = 60$ Hz and B = 0.50 T, then

$$2\pi Nab = \epsilon_0/\nu B = 150/[(60)(0.5)] = 5 \text{ m}^2$$

and any loop built according to the specification Nab = $2.5/\pi$ = 0.796 m^2 will produce the desired effect.

36P

(a) Over the thin rectangular strip, the field due to the wire is uniform (see sketch on p.264), so the flux through this shaded rectangle is

$d\Phi_B = B(xdr) = (\mu_0 i/2\pi r)(xdr) = \mu_0 ixdr/2\pi r.$

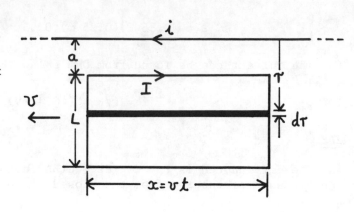

The flux through the entire rectangle swept out by the rod is

$$\Phi_B = \int d\Phi_B = (\mu_0 ix/2\pi)\int_a^{a+L} dr/r,$$

$$\Phi_B = (\mu_0 ix/2\pi)\ln(1 + \frac{L}{a}).$$

Hence,

$$\varepsilon = d\Phi_B/dt = (\mu_0 i/2\pi)\ln(1 + \frac{L}{a})(dx/dt) = (\mu_0 iv/2\pi)\ln(1 + \frac{L}{a}).$$

Numerically, this is

$$\varepsilon = [(4\pi \times 10^{-7})(100)(5)/2\pi]\ln(1 + \frac{10}{1}) = 2.398 \times 10^{-4} \text{ V.}$$

(b) The current is

$$I = \varepsilon/R = (2.398 \times 10^{-4})/(0.4) = 5.995 \times 10^{-4} \text{ A.}$$

(c) The rate of thermal energy generation is

$$P = \varepsilon^2/R = (2.398 \times 10^{-4})^2/(0.4) = 14.376 \times 10^{-8} \text{ W.}$$

(d) The force F is given by

$$F = P/v = (14.376 \times 10^{-8})/5 = 2.875 \times 10^{-8} \text{ N.}$$

(e) By energy conservation, the external agent does work at the rate that thermal energy is generated in the rod.

38P

The field B varies in the y direction but not in the x direction (since it does not depend on x). Thus, at any time the field B is constant over the shaded strip shown, and the flux through this strip is

$$d\Phi_B = B(Ldy) = 4Lt^2 ydy,$$

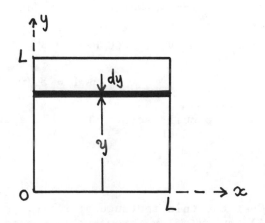

L being the length of each side of the square. Over the entire square, then, the flux is

$$\Phi_B = \int d\Phi_B = 4Lt^2\int_0^L ydy = 2L^3 t^2.$$

Thus, the emf is

$$\varepsilon = d\Phi_B/dt = 4L^3t = 4(0.02)^3(2.5) = 8 \times 10^{-5} \text{ V} = 80 \ \mu\text{V}.$$

The flux inside the square is directed out of the page and is increasing with time. A clockwise current sets up a field that, inside the square, is directed into the paper. This meets the demand of Lenz's law. Thus the sense (direction) of the induced emf is clockwise.

41P

(a) The forces acting on the wire are its weight mg, the normal force N, and the force F exerted by the magnetic field on the current induced in the wire by virtue of its motion induced by gravity. Newton's second law requires that

$$mg\sin\theta - F\cos\theta = ma.$$

But $F = i\ell B = (\varepsilon/R)\ell B$, ℓ the length of the wire rod. Since the angle between dA (the normal to the inclined plane) and B is θ, the angle of the incline, $\varepsilon = B\ell v\cos\theta$, and thus

$$mg\sin\theta - (B^2\ell^2v/R)\cos^2\theta = ma.$$

Initially $v = 0$; as the wire rod accelerates down the plane v will increase until it reaches a value v_t given by

$$v_t = \frac{mgR\sin\theta}{B^2\ell^2\cos^2\theta},$$

at which point, by the preceding equation, we see that ma = 0. Subsequently the rod will slide at the constant speed v_t.

(b) The kinetic and gravitational potential energies are $K = mv^2/2$ and $U = mgy$. Let x be the distance of the rod from the bottom of the incline measured along the incline. Then $y = x\sin\theta$. By the conservation of energy it is expected that

$$\frac{dU}{dt} = \frac{dK}{dt} + P,$$

the last term P being the rate of generation of thermal energy. Since $v = dx/dt$ and the acceleration $a = dv/dt$, this equation becomes

$$mgv\sin\theta = (mv)\frac{dv}{dt} + i^2R,$$

$$mg\sin\theta = ma + (B^2\ell^2v/R)\cos^2\theta,$$

since $i = \varepsilon/R = B\ell v\cos\theta/R$. But this last equation agrees with (a); hence the result in (a) is consistent with energy conservation. After the rod has reached its terminal speed, $dK/dt = 0$ so that $P = dU/dt$, as asserted.

(c) If B is directed down instead of up, the induced current will flow in the direction opposite to that in (a), but $\vec{F} = (-i)\vec{\ell} \times (-\vec{B})$ will be in the same direction as in (a), so the motion of the wire will be unaffected.

43E

(a) See Sample Problem 4. Since r = 2.2 cm < R = 6 cm, Eq.21 applies. We obtain

$$E = \frac{1}{2}(dB/dt)r = \frac{1}{2}(6.5 \times 10^{-3} \text{ T/s})(0.022 \text{ m}) = 71.5 \text{ } \mu\text{V/m}.$$

(b) In this case r = 8.2 cm > R = 6 cm; use Eq.22:

$$E = \frac{1}{2}(dB/dt)R^2(\frac{1}{r}) = \frac{1}{2}(6.5 \times 10^{-3} \text{ T/s})(0.06 \text{ m})^2(\frac{1}{0.082 \text{ m}}) = 143 \text{ } \mu\text{V/m}.$$

CHAPTER 33

<u>1E</u>

Solve Eq.2 for the magnetic flux Φ:

$$\Phi = Li/N = (8.0 \text{ mH})(5.0 \text{ mA})/(400) = 0.100 \text{ }\mu\text{Wb}.$$

<u>8P</u>

The inductance L can be found by combining
Faraday's law with Eq.11 to get

$$-L \frac{di}{dt} = -d\Phi_B/dt,$$

where the flux is calculated from

$$\Phi_B = \int \vec{B} \cdot d\vec{A}.$$

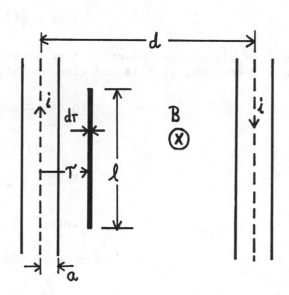

The area of integration for the flux is the
area of the loop formed by imagining two
short additional wires connecting those that
are given, to form a closed circuit. The
lengths of these new wires are very small
compared with those of the original wires,
so their fluxes can be ignored. Then, the
magnetic field B is the vector sum of the fields set up by the two given wires. Both of
these are into the paper and therefore, by Ampere's law (Eq.17 in Chapter 31), we have

$$B = \frac{\mu_0 i}{2\pi r} + \frac{\mu_0 i}{2\pi(d - r)}.$$

B does not vary in the direction parallel to the wires, and therefore for dA take a very
skinny rectangle of length ℓ and width dr; choose the direction of dA as into the paper
(the same direction as B). Then,

$$\Phi_B = \int B(\ell dr)\cos 0° = (\mu_0 i\ell/2\pi)\int_a^{d-a} [\frac{1}{r} + \frac{1}{d - r}]dr = (\mu_0 i\ell/\pi)\ln(\frac{d - a}{a}).$$

Hence,

$$d\Phi_B/dt = (\mu_0\ell/\pi)(\frac{di}{dt})\ln(\frac{d - a}{a}) = L \frac{di}{dt},$$

and therefore, with the flux within the wires not taken into account,

$$L = (\mu_0\ell/\pi)\ln(\frac{d - a}{a}).$$

10E

Since $\varepsilon = -L(di/dt)$, cause the current to change at the rate

$$di/dt = \varepsilon/L = (60 \text{ V})/(12 \text{ H}) = 5 \text{ A/s}.$$

15E

By Eq.18, the current is

$$i = \frac{\varepsilon}{R}(1 - e^{-t/\tau_L}),$$

where $\tau_L = L/R$ is the inductive time constant. The equilibrium value of the current is ε/R and therefore, if t is the time sought,

$$(0.999)\frac{\varepsilon}{R} = \frac{\varepsilon}{R}(1 - e^{-t/\tau_L}),$$

$$0.999 = 1 - e^{-t/\tau_L},$$

$$e^{-t/\tau_L} = 1 - 0.999 = 0.001,$$

$$\ln(e^{-t/\tau_L}) = \ln(0.001),$$

$$-t/\tau_L = -6.908 \rightarrow t = 6.908\tau_L;$$

that is, 6.908 time constants must elapse.

16E

Current decay in an LR circuit is described by Eq.20; substituting the data yields

$$i = i_0 e^{-t/\tau_L} = i_0 e^{-tR/L},$$

$$10 \times 10^{-3} = (1)e^{-(1)R/10},$$

$$\ln(10 \times 10^{-3}) = \ln(1) - R/10,$$

$$-4.6052 = 0 - R/10 \rightarrow R = 46.1 \ \Omega.$$

19E

(a) The equilibrium current is ε/R. To find the required time, then, use

$$i = \frac{\varepsilon}{R}(1 - e^{-t/\tau_L}),$$

with $i = 0.80\varepsilon/R$: this gives

$$0.80 = 1 - e^{-t/\tau_L},$$

$$e^{-t/\tau_L} = 0.20,$$

$$t = -\tau_L \ln(0.02) = 1.609\tau_L.$$

The numerical value of the time constant is

$$\tau_L = L/R = (6.3 \times 10^{-6} \text{ H})/(1.2 \times 10^3 \ \Omega) = 5.25 \times 10^{-9} \text{ s} = 5.25 \text{ ns}.$$

Therefore, t = 8.45 ns.

(b) Set $t = \tau_L$ in the first equation to get

$$i = (\frac{14}{1200})(1 - e^{-1}) = 7.37 \text{ mA}.$$

25P

(a) The inductor "breaks" the right-hand branch: $i_1 = i_2 = i$. But,

$$i = \frac{\varepsilon}{R_1 + R_2} = \frac{100}{10 + 20} = 3.33 \text{ A}.$$

(b) Now the inductor has no effect and therefore, by the loop rule,

$$\varepsilon - i_1 R_1 - i_1 R_{eq} = 0,$$

where, by Eq.20 in Chapter 29,

$$\frac{1}{R_{eq}} = \frac{1}{R_2} + \frac{1}{R_3},$$

giving $R_{eq} = 12 \ \Omega$. Thus $i_1 = (100 \text{ V})/(22 \ \Omega) = 4.55$ A. Further,

$$\varepsilon - i_1 R_1 - i_2 R_2 = 0,$$

$$100 = (4.545)(10) + i_2(20) \rightarrow i_2 = 2.73 \text{ A}.$$

(c) The left-hand branch is now broken so that $i_1 = 0$. The current through R_2 equals the current through R_3 since the remaining elements form a series circuit. The initial value of this current equals the current through R_3 a long time after S was originally closed. From (b) this is $i_2 = 4.545 - 2.727 = 1.82$ A.

(d) There are now no sources of emf in the circuit and hence all currents vanish.

30E

Combine Eqs.24 and 26, using the fact that $U_B = u_B V$. Hence, we have

$$\frac{1}{2}Li^2 = u_B V,$$

$$i = \sqrt{[2u_B V/L]} = \sqrt{[2(70 \text{ J/m}^3)(0.020 \text{ m}^3)/(0.090 \text{ H})]} = 5.58 \text{ A}.$$

35P

(a) The current, from Eq.18, is

$$i = \frac{\varepsilon}{R}(1 - e^{-t/\tau_L}) = \frac{50}{10,000}(1 - e^{-t/\tau_L}) = 0.005(1 - e^{-t/\tau_L}).$$

At $t = 5$ ms, $i = 2$ mA; hence, inserting these values we get

$$0.002 = 0.005(1 - e^{-0.005/\tau_L}),$$

$$\tau_L = -\frac{0.005}{\ln(0.6)} = 9.79 \times 10^{-3} = L/R = L/10^4 \rightarrow L = 97.9 \text{ H}.$$

(b) By Eq.24, the stored energy is

$$U_B = \tfrac{1}{2}Li^2 = \tfrac{1}{2}(97.9)(0.002)^2 = 1.96 \times 10^{-4} \text{ J} = 196 \text{ }\mu\text{J}.$$

37P

(a) The energy density is given by Eq.26; the magnetic field of a toroid is found in Eq. 22 in Chapter 31; hence,

$$u_B = B^2/2\mu_0 = (\mu_0 iN/2\pi r)^2/2\mu_0 = \mu_0 i^2 N^2/8\pi^2 r^2.$$

(b) Since u_B depends on r, take as a volume element the volume between two coaxial circular cylinders, radii r and r + dr, the axes of which coincide with the axis of the toroid. That is, $dV = 2\pi r h dr$. The stored energy will be

$$U_B = \int u_B dV = \int_a^b (\mu_0 i^2 N^2/8\pi^2 r^2)2\pi r h dr = (\mu_0 i^2 N^2 h/4\pi)\int_a^b dr/r = (\tfrac{1}{4\pi})(\mu_0 i^2 N^2 h)\ln(\tfrac{b}{a}),$$

$$U_B = (\tfrac{1}{4\pi})(4\pi \times 10^{-7})(0.50)^2(1250)^2(0.013)\ln(\tfrac{95}{52}) = 3.06 \times 10^{-4} \text{ J}.$$

(c) Using the expression for the inductance L from Eq.7, the energy can also be found by invoking Eq.24:

$$U_B = \tfrac{1}{2}Li^2 = \tfrac{1}{2}[\frac{\mu_0 N^2 h}{2\pi}\ln(\tfrac{b}{a})]i^2,$$

and this is seen to agree with the expression in (b).

39P

Suppose the switch to have been in position a for a time T before being thrown to b. The energy stored in the inductor the instant the switch is thrown is

$$U_B(T) = Li_T^2/2,$$

where

$$i_T = \frac{\varepsilon}{R}(1 - e^{-T/\tau_L}),$$

and $\tau_L = L/R$. With the switch in position **b** the current in the circuit becomes

$$i = i_T e^{-(t - T)/\tau_L},$$

time still being measured from the instant the switch was closed on **a**. The thermal energy generated in the resistor after the switch is thrown to **b**, over all subsequent time, is

$$E = \int_T^\infty i^2 R \, dt = i_T^2 R \int_T^\infty e^{-2(t - T)/\tau_L} dt = i_T^2 R \tau_L/2 = i_T^2 RL/2R = Li_T^2/2,$$

proving the assertion.

40E

(a) For the solenoid, Eq.21 of Chapter 31 yields

$$B = \mu_0 i_0 n = (4\pi \times 10^{-7} \text{ T·m/A})(6.6 \text{ A})(\frac{950}{0.85 \text{ m}}) = 9.2695 \times 10^{-3} \text{ T}.$$

Eq.26 now gives

$$u_B = B^2/2\mu_0 = (9.2695 \times 10^{-3} \text{ T})^2/2(4\pi \times 10^{-7} \text{ T·m/A}) = 34.2 \text{ J/m}^3.$$

(b) The solenoid's volume is $V = (17 \times 10^{-4} \text{ m}^2)(0.85 \text{ m}) = 1.445 \times 10^{-3} \text{ m}^3$. Therefore, the contained energy is $U_B = u_B V = (34.2 \text{ J/m}^3)(0.001445 \text{ m}^3) = 49.4 \text{ mJ}$.

47E

(a) The mutual inductance M is given from Eq.34:

$$\varepsilon_1 = M(di_2/dt),$$

$$25 \text{ mV} = M(15 \text{ A/s}) \rightarrow M = 1.67 \text{ mH}.$$

(b) The flux linkage is, by Eqs.29 and 32,

$$N_2 \Phi_{21} = M_{21} i_1 = M i_1,$$

$$N_2 \Phi_{21} = (1.67 \text{ mH})(3.6 \text{ A}) = 6.01 \text{ mWb}.$$

49P

(a) Connect the coils to a battery. The resistors in the circuit sketch shown are the resistances of the coils. Now apply the loop rule to obtain

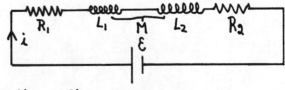

$$\varepsilon - iR_1 - L_1 \frac{di}{dt} \pm M \frac{di}{dt} - L_2 \frac{di}{dt} \pm M \frac{di}{dt} - iR_2 = 0,$$

$$\epsilon = i(R_1 + R_2) + (L_1 + L_2 \pm 2M)\frac{di}{dt},$$

and therefore the second term in parenthesis is the equivalent inductance, for the analogous equation for a circuit containing only one inductor would be

$$\epsilon = iR_{eq} + (L_{eq})\frac{di}{dt}.$$

(b) If the coils are wound so that their fluxes are in the same direction, use the plus sign, whereas if the windings of one are reversed so that the fluxes tend to cancel, use the negative sign.

52P

By Eq.33, $\epsilon_2 = -M(di_1/dt)$. Compute ϵ_2 by Faraday's law. The flux linking the inner with the outer solenoid is

$$\Phi_{21} = \int \vec{B}_1 \cdot d\vec{A},$$

where B_1 is the field set up by a current i_1 in the inner solenoid, and the integral is over the cross-sectional area of the outer solenoid. But $B_1 = \mu_0 n_1 i_1$ inside solenoid 1 and zero outside. Thus there is no contribution to the integral from the area between the solenoids (and therefore the size of solenoid 2 does not enter), so that

$$\Phi_{21} = B_1(\pi R_1^2) = \mu_0 n_1 i_1 \pi R_1^2.$$

Since there are $n_2\ell$ turns of solenoid 2 in a length ℓ, Faraday's law gives

$$\epsilon_2 = -(n_2\ell)(d\Phi_{21}/dt) = -\mu_0 n_1 n_2 \pi \ell R_1^2 (di_1/dt) = -M(di_1/dt) \rightarrow M = \mu_0 n_1 n_2 \pi R_1^2 \ell.$$

CHAPTER 34

5P

(a) In a time $t = 2\pi/\omega$, one period of rotation, all of the charge q on the ring passes any fixed point near the ring; thus the equivalent current i is

$$i = \frac{q}{t} = \frac{q}{2\pi/\omega} = \frac{\omega q}{2\pi}.$$

Therefore, the magnetic moment is

$$\mu = NiA = (1)(\frac{\omega q}{2\pi})(\pi r^2) = \tfrac{1}{2}\omega q r^2.$$

(b) By the right-hand rule, the magnetic moment vector $\vec{\mu}$ is parallel to the angular velocity vector $\vec{\omega}$.

10E

See Sample Problem 3. The magnitude B of the earth's magnetic field and its horizontal component B_h are related by

$$B_h = B\cos\phi_i,$$

where ϕ_i is the inclination; see Fig.10. Hence,

$$B = B_h/\cos\phi_i = (16\ \mu T)/\cos73° = 54.7\ \mu T.$$

15P

From Problem 13,

$$B = \frac{\mu_0\mu}{4\pi r^3}\sqrt{[1 + 3\sin^2\lambda_m]}.$$

At the surface of the earth r = R, where R is the radius of the earth. At altitude h, set r = R + h, the distance of that point from the earth's center. The requirement is that

$$\frac{\mu_0\mu}{4\pi(R + h)^3}\sqrt{[1 + 3\sin^2\lambda_m]} = \frac{1}{2}\frac{\mu_0\mu}{4\pi R^3}\sqrt{[1 + 3\sin^2\lambda_m]},$$

$$\frac{1}{(R + h)^3} = \frac{1}{2}\frac{1}{R^3},$$

$$R + h = 2^{1/3}R,$$

$$h = (2^{1/3} - 1)R = (2^{1/3} - 1)(6370\ km) = 1660\ km.$$

19E

The magnetization M and magnetic moment μ of the rod are related by

$$M = \mu/V,$$

where V is the volume of the rod. The rod, being a circular cylinder, its volume is

$$V = \pi r^2 h = \pi(0.5 \text{ cm})^2(5 \text{ cm}) = 3.927 \text{ cm}^3 = 3.927 \times 10^{-6} \text{ m}^3.$$

Hence,

$$\mu = MV = (5.3 \times 10^3 \text{ A/m})(3.927 \times 10^{-6} \text{ m}^3) = 0.0208 \text{ A·m}^2 = 20.8 \text{ mJ/T}.$$

21P

Examine the maximum value of B/T in the range of observation (note that T here stands for temperature; it also is the abbreviation for tesla). This will be $(0.5 \text{ T})/(10 \text{ K}) =$ 0.05 T/K. This is very close to the origin of Fig.11, well within the range in which Curie's law is obeyed, and therefore Curie's law will be found to be valid in this test.

24P

(a) By Eq.9, the magnetic dipole moment of an orbiting charge can be written as

$$\mu = \frac{evr}{2},$$

assuming that the magnitude of the charge is e. Now, for circular motion perpendicular to a uniform magnetic field, Eq.16 in Chapter 30 gives for the radius of the orbit,

$$r = \frac{mv}{eB},$$

so that

$$\mu = \left(\frac{ev}{2}\right)\left(\frac{mv}{eB}\right) = \frac{1}{B}\left(\frac{1}{2}mv^2\right) = K_e/B.$$

An electron circulates clockwise about a magnetic field that is directed into the paper (to choose an example). The resulting angular velocity vector $\vec{\omega}$ therefore is into the paper also. But the electron charge is negative, so that $\vec{\mu}$ is antiparallel to $\vec{\omega}$ and therefore is directed out of the paper, or antiparallel to $\vec{B}$.

(b) The charge cancels out in the calculation of μ in (a). Thus, for a positive ion, the same relation holds: $\mu = K_i/B$. A positive ion circulates counterclockwise in a magnetic field pointing into the paper, so $\vec{\omega}$ is directed out of the paper. Since the ion carries positive charge, $\vec{\mu}$ is parallel to $\vec{\omega}$, and therefore antiparallel to $\vec{B}$, just as for the electron.

(c) The directions of the dipole moments due to electrons and ions are the same direction, by (b) above, so that

$$\mu = N_e\mu_e + N_i\mu_i = N_eK_e/B + N_iK_i/B = \frac{1}{B}[N_eK_e + N_iK_i],$$

where N_e, N_i are the numbers of electrons and ions. Now these numbers are equal, so put $N_e = N_i = N$. Therefore, the magnetization becomes

$$M = \frac{\mu}{V} = \frac{1}{B}(\frac{N}{V})[K_e + K_i],$$

$$M = (\frac{1}{1.2 \text{ T}})(5.3 \times 10^{21} \text{ m}^{-3})[6.2 \times 10^{-20} \text{ J} + 7.6 \times 10^{-21} \text{ J}] = 307 \text{ A/m}.$$

27P

As the magnetic field is introduced and B increases, a counterclockwise electric field is induced; by Eq.21 in Chapter 32, this field is given by

$$E = (r/2)(dB/dt).$$

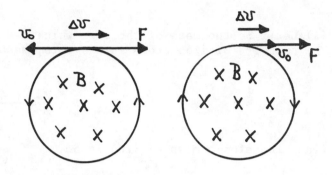

Thus, the electrons feel an electric force, directed as shown. Suppose that the magnetic field increases by an amount B in time T. Then, for each electron,

$$\Delta v = aT = (\frac{F}{m})T = (\frac{eE}{m})T = \frac{e}{m}(\frac{r}{2}\frac{B}{T})T = \frac{erB}{2m}.$$

The new velocities are

$$v = v_0 \pm \frac{erB}{2m},$$

+ for the clockwise circulating electron and − for the other. Dividing v by r, r assumed not to change, gives

$$\omega = \omega_0 \pm \frac{eB}{2m}.$$

The changed angular velocity leads to an increase or decrease in the orbital magnetic moment. Also, the existence of the diamagnetic effect in a constant magnetic field can be "explained" by noting that the circulating electron still cuts through lines of magnetic flux.

30E

The saturation magnetization, corresponding to complete alignment of all the dipoles, is given by

$$M_{max} = \mu N/V;$$

(see p.780 of HR). Set $V = 1 \text{ m}^3$. The mass of nickel in 1 m^3 is $(8.90 \text{ g/cm}^3)(10^6 \text{ cm}^3) = 8.90 \times 10^6$ g. Hence, there are

$$n = (8.90 \times 10^6 \text{ g})/(58.71 \text{ g/mol}) = 1.5159 \times 10^5 \text{ mol},$$

of nickel in 1 m^3 of nickel. By Eq.2 of Chapter 21, the number of atoms is

$$N = nN_A = (1.5159 \times 10^5 \text{ mol})(6.02 \times 10^{23} \text{ mol}^{-1}) = 9.126 \times 10^{28}.$$

Hence, by the first equation,

$$\mu = M_{max}V/N = (4.7 \text{ X } 10^5 \text{ A/m})(1 \text{ m}^3)/(9.126 \text{ X } 10^{28}) = 5.15 \text{ X } 10^{-24} \text{ A} \cdot \text{m}^2.$$

32P

(a) Let M be the mass of the core which has a radius r. The mass of an iron atom is m. The number N of iron atoms in the core, assumed to be pure iron, is N = M/m. But M = ρV, so that

$$N = \frac{M}{m} = \frac{\rho(4\pi r^3/3)}{m} = \frac{\rho V}{m}.$$

Since the atomic mass of iron is 56,

$$m = 56u = 56(1.66 \text{ X } 10^{-27} \text{ kg}).$$

Thus, if μ is the magnetic moment of a single iron atom, Nμ = the magnetic moment of the core, and therefore

$$\frac{\rho(4\pi r^3/3)}{56u}(\mu) = 8 \text{ X } 10^{22},$$

$$\frac{(14 \text{ X } 10^3)(4\pi r^3/3)}{(56)(1.66 \text{ X } 10^{-27})}(2.1 \text{ X } 10^{-23}) = 8 \text{ X } 10^{22} \rightarrow r = 1.82 \text{ X } 10^5 \text{ m} = 182 \text{ km}.$$

(b) The fraction f of the volume of the earth, radius R, that would be occupied by the hypothetical iron core is

$$f = (\frac{r}{R})^3 = (\frac{182}{6370})^3 = 2.33 \text{ X } 10^{-5}.$$

34P

(a) The field in a toroid is B = $(\mu_0 iN)/2\pi r$, where N is the total number of turns. This is a nonuniform field, but assume the field to be approximately uniform and equal to the actual value at the middle of the toroid tube. It is desired that this field B_0 equal 0.2 mT. Thus,

$$B_0 = \mu_0 iN/2\pi r,$$

$$2 \text{ X } 10^{-4} = (4\pi \text{ X } 10^{-7})(i)(400)/2\pi(0.055) \rightarrow i = 0.14 \text{ A}.$$

(b) With the iron present inside the tube, the field is $B_M + B_0 = 801B_0$. Let A be the cross-sectional area of the toroid tube. By Problem 17 in Chapter 32, the charge induced in a coil of N_c turns and resistance R_c is

$$q = N_c[\Phi_B(\text{final}) - \Phi_B(\text{initial})]/R_c = N_c(B_0 + B_M)A/R_c,$$

$$q = (50)(801)(2 \text{ X } 10^{-4})\pi(0.005)^2/(8) = 7.86 \text{ X } 10^{-5} \text{ C} = 78.6 \text{ } \mu\text{C}.$$

CHAPTER 35

4E

(a) The total energy that can be stored in the capacitor is given by

$$U_E = Q^2/2C = (2.9 \times 10^{-6} \text{ C})^2/[2(3.6 \times 10^{-6} \text{ F})] = 1.17 \times 10^{-6} \text{ J} = 1.17 \text{ }\mu\text{J}.$$

This equals the total energy in the circuit.

(b) Since $U_E = U_B$,

$$U_B = \frac{1}{2}Li^2,$$

$$1.17 \times 10^{-6} \text{ J} = \frac{1}{2}(75 \times 10^{-3} \text{ H})i^2 \rightarrow i = 5.59 \times 10^{-3} \text{ A} = 5.59 \text{ mA}.$$

10E

The time required is T/4, where T is the period of oscillation; see Fig.2. Hence, by Eq.12,

$$t = \frac{T}{4} = \frac{2\pi/\omega}{4} = \frac{2\pi/[1/\sqrt{(LC)}]}{4},$$

$$t = \frac{\pi}{2}\sqrt{(LC)} = \frac{\pi}{2}\sqrt{[(0.05 \text{ H})(4 \times 10^{-6} \text{ F})]} = 7.02 \times 10^{-4} \text{ s} = 0.702 \text{ ms}.$$

11E

With S_1 closed and the others open, we have an RC circuit, so that $\tau_C = RC$. With S_2 closed and the others open we have an LR circuit, for which $\tau_L = L/R$. With S_3 closed and the others open we have an LC circuit; for an LC circuit, Eq.12 gives

$$T = 2\pi/\omega = 2\pi\sqrt{[LC]} = 2\pi[\sqrt{(\frac{L}{R})}\sqrt{(RC)}] = 2\pi\sqrt{[\tau_L\tau_C]}.$$

20P

(a) If $U_E = U_B/2$, then since

$$U_E + U_B = \frac{1}{2}Q^2/C,$$

it follows that

$$U_E = \frac{1}{3}(\frac{1}{2}Q^2/C),$$

$$q^2/2C = \frac{1}{3}(Q^2/2C) \rightarrow q = Q/\sqrt{3}.$$

(b) The charge as a function of time, with $q = Q$ at $t = 0$, is

$$q = Q\cos\omega t.$$

Substituting the value of q from (a) yields

$$1/\sqrt{3} = \cos\omega t,$$

$$t = \frac{1}{\omega}\cos^{-1}(1/\sqrt{3}) = \frac{1}{\omega}(0.9553 \text{ rad}).$$

But $\omega = 2\pi/T$, T the period. Hence,

$$t = \frac{0.9553}{2\pi} T = 0.152T.$$

23P

(a) The frequencies that can be tuned are $\nu = 1/2\pi\sqrt{(LC)}$; thus

$$\frac{\nu_{max}}{\nu_{min}} = \frac{1/\sqrt{(LC_{min})}}{1/\sqrt{(LC_{max})}} = \sqrt{(C_{max}/C_{min})} = \sqrt{(365/10)} = 6.04.$$

(b) Let C be the added capacitance. For capacitors in parallel, the equivalent capacitance C* is the sum of the individual capacitances, and therefore C* ranges between (C + 10) and (C + 365) pF. Hence,

$$\frac{1.60}{0.54} = \sqrt{(\frac{365 + C}{10 + C})} \rightarrow C = 35.63 \text{ pF}.$$

The inductance must be given by

$$\nu = \frac{\omega}{2\pi} = (\frac{1}{2\pi})\ 1/\sqrt{[(L)(C + 10)]} = 1.6 \times 10^{6} \rightarrow L = 2.2 \times 10^{-4} \text{ H} = 0.22 \text{ mH},$$

using the value of C (expressed in farads) found above.

25P

(a) The charge q as a function of time is

$$q = Q\sin\omega t,$$

choosing sine rather than cosine so that q(0) = 0 as required. The current i is

$$i = \frac{dq}{dt} = \omega Q\cos\omega t.$$

Thus the maximum current is $I = \omega Q$ and therefore

$$Q = I/\omega = I\sqrt{(LC)} = (2)\sqrt{[(3 \times 10^{-3})(2.7 \times 10^{-6})]} = 1.8 \times 10^{-4} \text{ C}.$$

(b) The energy stored in the capacitor at any time t is

$$U = q^2/2C = (Q^2/2C)\sin^2\omega t.$$

The rate of increase of this energy is just

$$\frac{dU}{dt} = (Q^2/2C)(2\omega)(\sin\omega t)(\cos\omega t) = (\omega Q^2/2C)\sin 2\omega t.$$

This reaches its greatest value when $\sin 2\omega t = 1$; i.e.,

$$2\omega t = 2(\frac{2\pi}{T})t = \frac{\pi}{2} \rightarrow t = T/8.$$

(c) From (b), the greatest rate of increase is

$$(\frac{dU}{dt})_{max} = \omega Q^2/2C.$$

Numerically, $\omega = 1/\sqrt{(LC)} = (9 \times 10^{-5})^{-1}$, and $Q = 1.8 \times 10^{-4}$ C, from (a). Also, $C = 2.7$ μF. These give $(dU/dt)_{max} = 66.7$ W.

29P

The energy needed to raise the 100 μF capacitor to 300 V is $C_1V^2/2 = 4.5$ J. Initially, the 900 μF capacitor has an energy $C_9V^2/2 = 4.5$ J also, since V = 100 V. First, transfer all of this energy to the inductor by leaving S_1 open, closing S_2 and waiting a time $T_9/4$, where T_9 is the period of the LC_9 system. Then, before the energy starts to flow back to C_9, open S_2 and close S_1 simultaneously. Wait a time $T_1/4$ for the energy to flow from the inductor to C_1. Finally, open S_1 to prevent any energy from flowing back to the inductor. The periods T_1 and T_9 are

$$T_1 = 2\pi/\omega_1 = 2\pi\sqrt{(LC_1)} = 0.199 \text{ s}; \quad T_9 = 2\pi\sqrt{(LC_9)} = 0.596 \text{ s}.$$

30E

The decay of charge is described by Eq.18; ignoring the oscillations we have

$$q = Qe^{-Rt/2L}.$$

The time t required for 50 cycles is

$$t = 50T = 50(2\pi/\omega) = 100\pi\sqrt{[LC]} = 100\pi\sqrt{[(0.220 \text{ H})(12 \times 10^{-6} \text{ F})]} = 0.5104 \text{ s}.$$

Now set $q = 0.99Q$:

$$0.99Q = Qe^{-Rt/2L},$$

$$R = -[\frac{2L}{t}]\ln(0.99) = -[\frac{2(0.22 \text{ H})}{0.5104 \text{ s}}]\ln(0.99) = 8.66 \text{ m}\Omega.$$

32P

The charge q on the capacitor as a function of time t in a damped LC, or LCR circuit, is, by Eq.18 (with $\phi = 0$),

$$q = Qe^{-Rt/2L}\cos\omega' t.$$

The energy present in the capacitor is

$$U = q^2/2C = (Q^2 e^{-Rt/L}/2C)\cos^2\omega't.$$

The maximum energy is

$$U_{max} = Q^2 e^{-Rt/L}/2C.$$

At $t = 0$ this is $U_{max}(0) = Q^2/2C$. We want to find the time t such that

$$U_{max} = (\tfrac{1}{2})U_{max}(0),$$

$$Q^2 e^{-Rt/L}/2C = (\tfrac{1}{2})Q^2/2C,$$

$$e^{-Rt/L} = \frac{1}{2} \rightarrow t = (\tfrac{L}{R})\ln 2.$$

35P

The charge q on the capacitor is, by Eq.18 with $\phi = 0$,

$$q = Qe^{-Rt/2L}\cos\omega't.$$

This function may be considered to be a harmonic oscillation with a time-dependent amplitude Q*; that is, $q = Q*\cos\omega't$. If the damping is small, then just as for the undamped LC system, the total energy in the system during one oscillation is $U = Q*^2/2C$, to a high degree of approximation. The energy loss per cycle is

$$\Delta U \simeq \frac{dU}{dt}(\Delta t),$$

where $\Delta t = 2\pi/\omega' \simeq 2\pi/\omega$. But $\omega = 1/\sqrt{(LC)}$ so that

$$\frac{\Delta U}{U} = \frac{\frac{d}{dt}(Q*^2/2C)}{Q*^2/2C}(\frac{2\pi}{\omega}) = \frac{2\pi R}{\omega L},$$

where we used $Q* = Qe^{-Rt/2L}$ in taking the derivative in the numerator above.

CHAPTER 36

3E

(a) The current amplitude is given by Eq.18: $I_L = V_L/X_L$, so we need the inductive reactance X_L to get I_L. From Eq.16,

$$X_L = \omega L = 2\pi\nu L = 2\pi(10^3 \text{ Hz})(0.050 \text{ H}) = 314.2 \ \Omega.$$

Since the circuit contains only the battery and the inductor we expect, from the loop rule, that $V_L = \varepsilon_m$. Hence,

$$I_L = V_L/X_L = \varepsilon_m/X_L = (30 \text{ V})/(314.2 \ \Omega) = 95.5 \text{ mA}.$$

(b) The frequency being greater by a factor of eight over that in (a), the current amplitude, being inversely proportional to the frequency, is smaller by a factor of eight; i.e., $I_L = (95.5 \text{ mA})/8 = 11.9 \text{ mA}.$

7E

(a) Set Eqs.9 and 16 equal and solve for the frequency, as requested. We have

$$\frac{1}{\omega C} = \omega L,$$

$$\omega = 2\pi\nu = \frac{1}{\sqrt{LC}},$$

$$\nu = (\frac{1}{2\pi})\frac{1}{\sqrt{LC}} = (\frac{1}{2\pi})\frac{1}{\sqrt{[(0.006 \text{ H})(10 \times 10^{-6} \text{ F})]}} = 650 \text{ Hz}.$$

(b) Evaluate either X_C or X_L since at this frequency they are equal; if we pick X_L,

$$X_L = \omega L = 2\pi\nu L = 2\pi(650 \text{ s}^{-1})(0.006 \text{ H}) = 24.5 \ \Omega.$$

(c) We see that the expression for ω in (a) is identical to Eq.12 in Chapter 35, thus proving the assertion.

10P

(a) The generator emf reaches a maximum when $(\omega t - \pi/4) = \pi/2$, since $\sin\pi/2 = 1$, the maximum possible value. Hence, the time at which this occurs is

$$t = 3\pi/4\omega = 3\pi/4(350 \text{ s}^{-1}) = 6.73 \text{ ms}.$$

(b) Similarly, the current reaches a maximum when

$$\omega t - 3\pi/4 = \pi/2 \rightarrow t = 5\pi/4\omega = 5\pi/4(350 \text{ s}^{-1}) = 11.2 \text{ ms}.$$

(c) Comparing (a) and (b), we see that the current lags the emf, so the circuit is inductive, with the element being an inductor; see Fig.5(c).

(d) Besides the generator, the circuit contains only an inductor; by the loop rule, then we have $V_L = \varepsilon_m$. By Eqs.16 and 18,

$$L = X_L/\omega = V_L/I_L\omega = \varepsilon_m/I_L\omega = (30 \text{ V})/(0.620 \text{ A})(350 \text{ s}^{-1}) = 138 \text{ mH.}$$

15E

(a) The capacitive reactance, by Eq.9, is

$$X_C = \frac{1}{\omega C} = \frac{1}{2\pi\nu C} = \frac{1}{2\pi(60)(70 \times 10^{-6})} = 37.9 \ \Omega.$$

The inductive reactance is unchanged: $X_L = 87 \ \Omega$. The new impedance, by Eq.23, is

$$Z = \sqrt{[R^2 + (X_L - X_C)^2]} = \sqrt{[(160)^2 + (87 - 37.9)^2]} = 167 \ \Omega.$$

The current amplitude is found from Eq.24:

$$I = \varepsilon_m/Z = 36/167 = 0.216 \text{ A.}$$

Finally, from Eq.26, we get for the phase angle,

$$\phi = \tan^{-1}[(X_L - X_C)/R] = \tan^{-1}[(87 - 37.9)/160] = 17.1°.$$

(b) To draw the phasor diagram, calculate the quantities:

$$V_R = IR = (0.216)(160) = 34.6 \text{ V,}$$

$$V_L = IX_L = (0.216)(87) = 18.8 \text{ V;}$$

$$V_C = IX_C = (0.216)(37.9) = 8.19 \text{ V.}$$

Note that $X_L > X_C$, so that ε_m leads I.

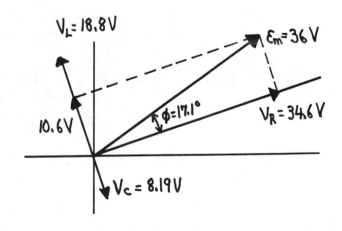

18P

The amplitude desired is $V_L = IX_L$. At resonance, $\omega = 1/\sqrt{(LC)}$, so that

$$X_L = \omega L = L/\sqrt{(LC)} = \sqrt{(L/C)} = \sqrt{(1/10^{-6})} = 1000 \ \Omega.$$

Also at resonance $X_L = X_C$ so that

$$I = \varepsilon_m/Z = \varepsilon_m/R = 10/10 = 1 \text{ A.}$$

Hence,

$$V_L = IX_L = (1)(1000) = 1000 \text{ V.}$$

21P

(a) The current amplitude is, by Eq.25,

$$I = \epsilon_m/\sqrt{[R^2 + (\omega L - 1/\omega C)^2]}.$$

To find the maximum value of I, set $dI/d\omega = 0$:

$$-\frac{dI}{d\omega} = \frac{\epsilon_m}{2}[R^2 + (\omega L - 1/\omega C)^2]^{-3/2}(2)(\omega L - \frac{1}{\omega C})(L + \frac{1}{\omega^2 C}) = 0,$$

$$(\omega L - \frac{1}{\omega C}) = 0,$$

$$\omega = 1/\sqrt{[LC]} = \omega_0 = 1/\sqrt{[(1)(20 \times 10^{-6})]} = 223.61 \text{ rad/s}.$$

(b) The maximum value of the current is ϵ_m/R, deduced from (a) with $\omega = 1/\sqrt{(LC)}$; i.e., I = 30/5 = 6.0 A.

(c) To find ω_1 and ω_2, put $I = \epsilon_m/2R$ into the first equation in (a) to get

$$\epsilon_m/2R = \epsilon_m/\sqrt{[R^2 + (\omega L - 1/\omega C)^2]},$$

$$[R^2 + (\omega L - 1/\omega C)^2] = 4R^2,$$

$$\omega^2 L^2 - 2L/C + 1/\omega^2 C^2 = 3R^2,$$

$$\omega^4 L^2 C^2 - 2LC\omega^2 + 1 = 3R^2 C^2 \omega^2,$$

$$(L^2 C^2)\omega^4 - (3R^2 C^2 + 2LC)\omega^2 + 1 = 0.$$

Substituting the values of L, C, R gives

$$(4 \times 10^{-5})\omega^4 - (4.003)\omega^2 + 10^5 = 0.$$

Treat this as a quadratic equation with the unknown $x = \omega^2$. The quadratic formula (see Appendix G) gives for the solutions

$$\omega^2 = 5.197435 \times 10^4; \quad 4.810065 \times 10^4.$$

Taking the positive square roots ($\omega > 0$) yields

$$\omega_1 = 227.98 \text{ rad/s}; \quad \omega_2 = 219.32 \text{ rad/s}.$$

(d) Taking ω_0 from (a), it is found that

$$(\omega_1 - \omega_2)/\omega_0 = 0.0387.$$

27P

See Problem 21 above; from the second equation in part (c) there, we have

$$\omega L - \frac{1}{\omega C} = R\sqrt{3}.$$

Now set $\omega = \omega_0 + \delta\omega = \omega_0(1 + \delta\omega/\omega_0)$. This should yield an accurate result, treating $\delta\omega/\omega_0 \ll 1$, if the half-width is small compared to the resonant frequency (sharp peak). To this same degree of approximation (see Appendix G, Binomial theorem) we have

$$\frac{1}{\omega} = \frac{1}{\omega_0}(1 - \frac{\delta\omega}{\omega_0}).$$

The first equation above now becomes

$$\omega_0(1 + \frac{\delta\omega}{\omega_0})L - \frac{1}{\omega_0 C}(1 - \frac{\delta\omega}{\omega_0}) = R\sqrt{3},$$

$$(\omega_0 L - \frac{1}{\omega_0 C}) + \frac{\delta\omega}{\omega_0}(\omega_0 L + \frac{1}{\omega_0 C}) = R\sqrt{3}.$$

But $\omega_0^2 = 1/LC$; hence the first term in the equation above vanishes identically; also,

$$\omega_0 L + \frac{1}{\omega_0 C} = \frac{\omega_0^2 LC + 1}{\omega_0 C} = \frac{2}{\omega_0 C} = 2\sqrt{(\frac{L}{C})}.$$

Thus,

$$\frac{\delta\omega}{\omega_0} = \frac{R}{2}\sqrt{(3C/L)}.$$

The half-width is $\Delta\omega = 2\delta\omega$, so that

$$\frac{\Delta\omega}{\omega_0} = R\sqrt{(\frac{3C}{L})}.$$

For the values of L, C, R from Problem 21, this formula gives $\Delta\omega/\omega_0 = 0.0387$, the same as in Problem 21.

30E

The average power produced by an alternating current is given by Eq.29: $P_{av} = I_{rms}^2 R$. But $I_{rms} = I/\sqrt{2}$, by Eq.30, so that

$$P_{av} = I^2 R/2.$$

For the direct current $P = i^2 R$. Setting the powers equal gives

$$I^2 R/2 = i^2 R,$$

$$i = I/\sqrt{2} = (2.6 \text{ A})/\sqrt{2} = 1.84 \text{ A}.$$

34E

(a) By Eq.23,

$$Z = \sqrt{[R^2 + (X_L - X_C)^2]} = \sqrt{[12^2 + (1.3 - 0)^2]} = 12.1 \; \Omega.$$

(b) Combining Eqs.31 and 32 leads to

$$P_{av} = (\varepsilon^2_{rms}/Z)\cos\phi.$$

But, from p.822,

$$\cos\phi = R/Z = 12/12.1 = 0.9917,$$

so that

$$P_{av} = (120^2/12.1)(0.9917) = 1.18 \text{ kW.}$$

40P

(a) Writing $i = I\sin(\omega t - \phi)$, Eq.2, we see that the power factor is $\cos\phi = \cos(-42°) = 0.743$.

(b) Since $\phi < 0$, $\omega t - \phi > \omega t$, so that the current leads the emf.

(c) With $\phi = -42°$, $\tan\phi = -0.900$. But

$$\tan\phi = (X_L - X_C)/R = -0.9 < 0,$$

so that $X_C > X_L$, and the circuit is capacitive.

(d) At resonance, $X_C = X_L$, giving $\tan\phi = 0$, $\phi = 0°$. Since $\phi = -42° \neq 0°$, the circuit is not in resonance.

(e) If the box contained a resistor only, ε and i would be in phase, or $\phi = 0°$. A capacitor only or an inductor only would lead to currents depending exponentially on the time. If there is no resistor but only a capacitor and an inductor (LC circuit), the current would be out of phase by 90° (set R = 0 in the equation for $\tan\phi$). The remaining possibilities, then, are LR, CR, LCR. But the circuit was found to be capacitive; hence, there must be a capacitor. But an inductor is not necessary, for a pair of values of R,C can be found to yield the given current amplitude and phase angle with L = 0. The circuit must, therefore, contain a capacitor and a resistor, but may or may not contain an inductor.

(f) By Eqs.30 and 33, the average power is

$$P_{av} = \frac{1}{2}\varepsilon_m I\cos\phi = \frac{1}{2}(75)(1.2)(0.743) = 33.4 \text{ W.}$$

(g) Although the power factor $\cos\phi$ depends on the values of L, C, R, ω through

$$\tan\phi = (X_L - X_C)/R = (\omega L - \frac{1}{\omega C})/R,$$

the value of ϕ has been given and does not have to be computed; the value of ω is not needed therefore. The power above is an average over many cycles and $\langle\sin\omega t \cdot \sin(\omega t - \phi)\rangle = \frac{1}{2}\cos\phi$, independent of ω when the average is so taken.

41P

(a) The average power dissipated is, by Eq.33,

$$P_{av} = \epsilon_{rms} I_{rms} \cos\phi.$$

But, by Eq.30, this becomes

$$P_{av} = (\tfrac{1}{2})\epsilon_m I \cos\phi.$$

By Eq.24, $I = \epsilon_m/Z$; also $\cos\phi = R/Z$; Z is given in Eq.25. Putting all this together gives

$$P_{av} = (\epsilon_m^2 R/2)\frac{1}{Z^2} = (\epsilon_m^2 R/2)\frac{1}{R^2 + (\omega L - 1/\omega C)^2}.$$

To find the desired value of C, set $dP_{av}/dC = 0$; we get

$$dP_{av}/dC = -(\epsilon_m^2 R/2)[R^2 + (\omega L - 1/\omega C)^2]^{-2}[2(\omega L - \frac{1}{\omega C})(\frac{1}{\omega C^2})] = 0,$$

$$(\omega L - \frac{1}{\omega C}) = 0 \;\rightarrow\; C = 1/\omega^2 L = 117 \;\mu F,$$

since $\omega = 2\pi(60 \; s^{-1})$ and $L = 0.060$ H.

(b) Rewrite the expression for P_{av} in the form

$$P_{av} = (\epsilon_m^2 R/2)[\frac{\omega^2 C^2}{(\omega RC)^2 + (\omega^2 LC - 1)^2}].$$

From this it is clear that a minimum value of $P_{av} = 0$ is attained when $C = 0$.

(c) Substitute the expression for C from (a) into the expression for P_{av} itself gives for the maximum,

$$P_{av} = \epsilon_m^2/2R = (30)^2/2(5) = 90 \; W.$$

From (b), the minimum value is $P_{av} = 0$.

(d) For $C = 1/L\omega^2$, the impedance reduces to $Z = R$ and therefore $\cos\phi = R/Z = 1$, so that $\phi = 0°$. With $C = 0$, the impedance is $Z = 0$, so the phase angle is undefined. However, for purely inductive circuits $\phi = 90°$.

(e) The power factors are $\cos\phi$; from (d) these are 1 and zero.

45E

(a) Use Eq.36:

$$V_s = V_p(N_s/N_p) = (120 \; V)(10/500) = 2.4 \; V.$$

(b) By Ohm's law,

$$I_s = V_s/R_s = (2.4 \; V)/(15 \; \Omega) = 0.16 \; A.$$

Now use Eq.37,

$$I_p = I_s(N_s/N_p) = (0.16 \text{ A})(10/500) = 0.0032 \text{ A.}$$

48P

The resistance added to the amplifier by virtue of the added transformer is not R, but rather, by Eq.38,

$$r_t = (N_p/N_s)^2 R,$$

and thus the equivalent resistance of the amplifier is $r + r_t$. The average power delivered to R is

$$P_{av} = I_s^2 R = (I_p N_p/N_s)^2 R.$$

But $I_p = \varepsilon_{rms}/(r + r_t)$, so that

$$P_{av} = \varepsilon_{rms}^2 (N_p/N_s)^2 R \frac{1}{[r + (N_p/N_s)^2 R]^2}.$$

If we write $x = (N_p/N_s)^2$, then as far as dependence on x is concerned,

$$P_{av} = \frac{x}{(r + xR)^2}.$$

To find the value of x leading to a maximum P_{av}, set $dP_{av}/dx = 0$:

$$dP_{av}/dx = \frac{r - xR}{(r + xR)^3} = 0,$$

$$x = \frac{r}{R} = (N_p/N_s)^2 = 1000/10 \rightarrow N_p/N_s = 10.$$

CHAPTER 37

3E

There are different expressions for B at points with r < R and points at r > R. The maximum value of the field is at r = R. This maximum, which can be found by setting r = R into either of the two expressions referred to above, is

$$B_{max} = \frac{1}{2}\mu_0\epsilon_0 R(dE/dt).$$

There will be a point at r < R at which $B = B_{max}/2$. To find it, set $B = B_{max}/2$ using the appropriate expression for B given in Sample Problem 1. We get

$$\frac{1}{2}\mu_0\epsilon_0 r(dE/dt) = \frac{1}{2}[\frac{1}{2}\mu_0\epsilon_0 R(dE/dt)],$$

$$r = \frac{1}{2}R = \frac{1}{2}(55 \text{ mm}) = 27.5 \text{ mm}.$$

Now do the analogous calculation using the expression for B at points exterior to R:

$$\frac{1}{2r}(\mu_0\epsilon_0 R^2)(dE/dt) = \frac{1}{2}[\frac{1}{2}\mu_0\epsilon_0 R(dE/dt)],$$

$$r = 2R = 2(55 \text{ mm}) = 110 \text{ mm}.$$

6E

The displacement current i_d is, by Eq.11,

$$i_d = \epsilon_0 A \frac{dE}{dt}.$$

Now let x be the separation between the plates. Thus E = V/x, V the potential difference across the plates. Then, if the plates are fixed in position, so that x does not change,

$$i_d = \epsilon_0 A \frac{d(V/x)}{dt} = \frac{\epsilon_0 A}{x} \frac{dV}{dt}.$$

But for a parallel-plate capacitor, with $\kappa = 1$, we have $C = \epsilon_0 A/x$, giving

$$i_d = C \frac{dV}{dt}.$$

8E

The displacement current density J_d is defined in a manner analogous to the conduction current density J: see Eq.4 in Chapter 28; i.e., $J_d = i_d/A$. But Eq.11 gives the expression for i_d, so we have directly

$$J_d = \frac{1}{A}[i_d] = \frac{1}{A}[\epsilon_0 A(\frac{dE}{dt})] = \epsilon_0(\frac{dE}{dt}).$$

14P

(a) By Problem 6,

$$i_d = C\frac{dV}{dt} = C\frac{d}{dt}(V_m\sin\omega t) = \omega CV_m\cos\omega t.$$

The maximum displacement current is, therefore,

$$i_{d,max} = \omega CV_m = 2\pi\nu CV_m = (2\pi)(50\ s^{-1})(10^{-10}\ F)(1.74\ X\ 10^5\ V) = 5.47\ mA.$$

(b) From (a) it is apparent that i_d is proportional to V_m, and therefore a large value of V_m yields a more easily detectable and measureable displacement current.

15P

(a) The displacement current i_d in the gap between the plates is equal to the conduction current i in the wires; hence,

$$i_{max} = i_{d,max} = 7.6\ \mu A.$$

(b) By definition, $i_d = \varepsilon_0(d\Phi_E/dt)$, so that

$$(d\Phi_E/dt)_{max} = i_{d,max}/\varepsilon_0 = 7.6\ X\ 10^{-6}/8.85\ X\ 10^{-12} = 8.59\ X\ 10^5\ V\cdot m/s.$$

(c) By Problem 6,

$$i_d = \frac{\varepsilon_0 A}{d}\frac{dV}{dt} = \frac{\varepsilon_0 A}{d}\frac{d\varepsilon}{dt} = \frac{\varepsilon_0 A}{d}\varepsilon_m\omega\cos\omega t,$$

the potential across the capacitor, except for sign, being equal to the battery emf, by the loop rule. The plate separation, then, must be

$$d = \varepsilon_0 A\varepsilon_m\omega/i_{d,max} = \frac{(8.85\ X\ 10^{-12})[\pi(0.18)^2](220)(130)}{7.6\ X\ 10^{-6}} = 3.39\ mm.$$

(d) In the gap between the plates the conduction current i = 0. The Ampere-Maxwell law, Eq.6, becomes

$$\oint\vec{B}\cdot d\vec{s} = \mu_0 I_d,$$

where I_d is the displacement current flowing through the area bounded by the circular path of integration of radius r. Then, if the current density J_d is uniform, we have

$$I_d = J_d(\pi r^2) = (i_d/A)(\pi r^2) = i_d r^2/R^2.$$

Substituting this into the Ampere-Maxwell law gives

$$B(2\pi r) = \mu_0 i_d r^2/R^2,$$

$$B_{max} = \mu_0 i_{d,max} r/2\pi R^2 = (4\pi \times 10^{-7})(7.6 \times 10^{-6})(0.11)/2\pi(0.18)^2 = 5.16 \text{ pT}.$$

20P

(a) Since $i = dq/dt$, the charge $q(t)$ on the faces will be

$$q = \int i\,dt = \int \alpha t\,dt = \alpha t^2/2.$$

[With $q(0) = 0$, the constant of integration is zero.]

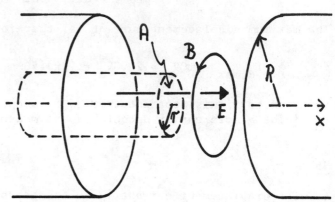

(b) Use the can-shaped Gaussian surface shown. The charges on the rod faces do not contribute to the electric field in the rod, so set $E = 0$ in the rod. In the gap E is parallel to the x axis, and under these conditions Gauss' law gives

$$\varepsilon_0 EA = qA/\pi R^2,$$

$$E = q/\pi\varepsilon_0 R^2 = \alpha t^2/2\pi\varepsilon_0 R^2,$$

using the result from (a) for $q(t)$.

(c) From Fig.1, the lines of B are concentric circles, centered on and perpendicular to the x axis.

(d) Choosing as a path of integration a circle of radius r that coincides with a line of B, we have $\int \vec{B} \cdot \vec{ds} = B(2\pi r)$. In the gap $i = 0$; hence, by the Ampere-Maxwell law,

$$2\pi r B = \mu_0 \varepsilon_0 (d\Phi_E/dt) = \mu_0 \varepsilon_0 \frac{d}{dt}(EA) = \mu_0 \varepsilon_0 (\pi r^2)\frac{dE}{dt}.$$

From (b),

$$dE/dt = \alpha t/\pi\varepsilon_0 R^2,$$

so that

$$B = \mu_0 \varepsilon_0 (\frac{r}{2})(\alpha t/\pi\varepsilon_0 R^2) = \mu_0 r\alpha t/2\pi R^2.$$

CHAPTER 38

2E

(a) Since the speed of light is $c = 3 \times 10^8$ m/s $= 3 \times 10^5$ km/s,

$$t = \frac{d}{c} = \frac{150 \text{ km}}{3 \times 10^5 \text{ km/s}} = 5 \times 10^{-4} \text{ s} = 0.5 \text{ ms}.$$

(b) As seen from the earth, the moon is opposite the sun at full moon. Hence,

$$t = (1.5 \times 10^8 \text{ km} + 7.6 \times 10^5 \text{ km})/(3 \times 10^5 \text{ km/s}) = 502.5 \text{ s} = 8.375 \text{ min}.$$

(c) If d is the earth–Saturn distance,

$$t = 2d/c = (2.6 \times 10^9 \text{ km})/(3 \times 10^5 \text{ km/s}) = 8670 \text{ s} = 2.41 \text{ h}.$$

(d) By the definition of light-year, light took 6500 years to reach the earth after the explosion. Thus, the explosion occured in 1054 A.D. – 6500 = 5446 B.C.

7P

(a) Let the true period of revolution of one of Jupiter's satellites be T and the observed period be T*. The latter may be determined by measuring the time between two successive passages of the satellite into eclipse behind Jupiter. If this observation is made when the earth is near x, it is expected that T = T* since the earth, moving at a tangent to the earth–Jupiter line, does not substantially alter its distance to Jupiter in the time T. However, when the earth is near y, it is moving almost directly away from Jupiter, so that T*(y) will differ from T by vT/c approximately, where v is the speed of the earth in its orbit. Hence, as the earth moves from point x to point y, T* will increase steadily. (Typical values of T for Jupiter's satellites are a few days.)

(b) With T determined (when the earth was near x), a prediction can be made, based on the speed of light being infinite, of the number of times a particular satellite enters Jupiter's shadow before the earth reaches y.   The time of occurence of the shadow entry that occurs closest to the moment the earth reaches y can also be computed. However, this particular entry will in fact be observed at a time R/c, approximately, later than is predicted on the basis of an infinite speed of light. Hence, if R (the radius of the orbit of the earth about the sun) can be determined, then c can be calculated.

16E

The energy is

$$E = Pt = (100 \times 10^{12} \text{ J/s})(10^{-9} \text{ s}) = 10^5 \text{ J} = 0.1 \text{ MJ}.$$

20E

The area illuminated on the moon is a circle with diameter $D = r\theta$, where r is the earth–moon distance. Thus, the area A that is illuminated is

$$A = \pi D^2/4 = \pi[r\theta]^2/4 = \pi[(3.8 \times 10^5 \text{ km})(0.88 \times 10^{-6})]^2/4 = 0.0878 \text{ km}^2.$$

26P

The energy density due to an electric field is

$$u_E = \tfrac{1}{2}\epsilon_0 E^2,$$

provided that the field travels in a vacuum. Using Eq.1 for a traveling wave, this becomes

$$u_E = \tfrac{1}{2}\epsilon_0 E_m^2 \sin^2(kx - \omega t).$$

But, by Eq.5, $E_m = cB_m$ and therefore, referring to Eq.2,

$$u_E = \tfrac{1}{2}\epsilon_0 c^2 B_m^2 \sin^2(kx - \omega t) = \tfrac{1}{2}\epsilon_0 c^2 B^2.$$

Finally, invoke the relation $c^2 = 1/\mu_0\epsilon_0$ to obtain

$$u_E = \tfrac{1}{2}\epsilon_0 \left(\frac{1}{\mu_0\epsilon_0}\right) B^2 = B^2/2\mu_0 = u_B.$$

In a traveling electromagnetic wave the electric and magnetic fields are in phase, so there does not seem to be any need to resort to a time average.

29P

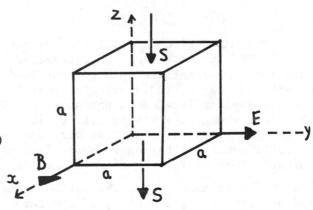

(a) Since $\vec{S} = (\vec{E} \times \vec{B})/\mu_0$, S is directed in the negative z direction, by the right-hand rule. Thus, the only faces across which S cuts are the two that are parallel to the x,y plane. Energy at the rate EBa^2/μ_0 "enters" the top face and "leaves" through the bottom face (a^2 is the area of each face) according to the Poynting vector point of view. If the fields E and B do not depend on time (i.e., are static) however, this picture of energy "flow" does not really correspond to reality.

(b) Since energy appears to enter and leave the cube at the same rate, the net energy in the cube does not change, even in the Poynting scenario. Hence, the Poynting vector picture makes sense when the energy flow through a closed surface is examined, but not always if only part of such a surface is considered.

32P

(a) The airplane receives practically a plane wave. Since $E_{rms} = E_m/\sqrt{2}$ (see Eq. 30 in Chapter 36), Eq.29 for the intensity can be written (and we work, as always in EM problems, in SI units),

$$I = E_m^2/2\mu_0 c,$$

$$E_m = \sqrt{[2\mu_0 cI]} = \sqrt{[2(4\pi \times 10^{-7})(3 \times 10^8)(10 \times 10^{-6})]} = 86.8 \text{ mV/m}.$$

(b) By Eq.26,

$$B_m = E_m/c = (86.8 \text{ X } 10^{-3})/(3 \text{ X } 10^8) = 289 \text{ pT}.$$

(c) Since the transmitted wave is spherical (it appears to be a plane wave from the airplane because that object intercepts only a very small fraction of the wavefront), Eq.24 of Chapter 18 applies:

$$P = 4\pi r^2 I = 4\pi (10 \text{ X } 10^3)^2 (10 \text{ X } 10^{-6}) = 12.6 \text{ kW}.$$

35P

(a) Sighting along the resistor in the direction of the current flow, E points directly away from the observer, while B is directed transverse and at a right angle to the resistor axis in the clockwise sense. By the right-hand rule, $\vec{E} \text{ X } \vec{B}$, and therefore $\vec{S}$, are pointed radially inward.

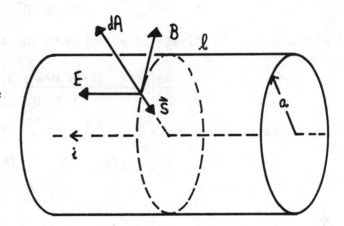

(b) Since E and B are at 90° to each other,

$$S = \frac{EB}{\mu_0} = \frac{(V/\ell)B}{\mu_0} = \frac{i}{\mu_0 \ell} (R)B,$$

$$S = \frac{i}{\mu_0 \ell} (\rho\ell/\pi a^2)(\mu_0 i/2\pi a) = \frac{i^2 \rho}{2\pi^2 a^3}.$$

The integral desired is $\int \vec{S} \cdot d\vec{A}$ over the cylindrical surface of the wire, considering a piece of length ℓ. The angle between S and the element of area dA is 180°, and the magnitude of S is the same at all points on the surface. Thus,

$$\int \vec{S} \cdot d\vec{A} = -SA = -(\frac{i^2 \rho}{2\pi^2 a^3})(2\pi a\ell) = -i^2(\rho\ell/\pi a^2) = -i^2 R.$$

37E

See Sample Problem 2, where it is shown that the pressure is 2S/c. The intensity S is given by

$$S = I = P/A = (1.5 \text{ X } 10^9 \text{ W})/(10^{-6} \text{ m}^2) = 1.5 \text{ X } 10^{15} \text{ W/m}^2.$$

Hence the pressure Pr is

$$Pr = 2S/c = 2(1.5 \text{ X } 10^{15} \text{ W/m}^2)/(3 \text{ X } 10^8 \text{ m/s}) = 10 \text{ MPa}.$$

40E

The pressure exerted on an absorbing surface is Pr = S/c = I/c (see Sample Problem 2). The intensity I for a spherical wave is, from Eq.24 of Chapter 18, $I = P/4\pi r^2$. Thus we have

$$Pr = P/4\pi r^2 c = (500 \text{ W})/[4\pi(1.5 \text{ m})^2(3 \text{ X } 10^8 \text{ m/s})] = 5.89 \text{ X } 10^{-8} \text{ N/m}^2.$$

41P

(a) The frequency is $\nu = c/\lambda = (3 \times 10^8 \text{ m/s})/(3.0 \text{ m}) = 100 \text{ Mhz}$.

(b) $\vec{E} \times \vec{B} = \mu_0 \vec{S}$ must be in the +x direction. By the right-hand rule, if E lies along the y axis, B must be directed along the z axis. By Eq.26, $B_m = E_m/c = (300 \text{ V/m})/(3 \times 10^8 \text{ m/s}) = 1.0 \ \mu T$.

(c) The angular frequency is $\omega = 2\pi\nu = 6.28 \times 10^8 \text{ rad/s}$; the angular wave number (see Eq.5 in Chapter 17) is $k = 2\pi/\lambda = 2.1 \text{ m}^{-1}$.

(d) The time-averaged Poynting vector may be written (see Problem 22)

$$\overline{S} = \overline{E^2}/\mu_0 c = E_m^2/2\mu_0 c = \frac{(300 \text{ V/m})^2}{2(4\pi \times 10^{-7} \text{ T} \cdot \text{m/A})(3 \times 10^8 \text{ m/s})} = 119 \text{ W/m}^2.$$

(e) The rate of delivery of momentum to the absorbing sheet is (see Sample Problem 2),

$$\frac{dp}{dt} = \frac{SA}{c} = \frac{(119 \text{ W/m}^2)(2 \text{ m}^2)}{3 \times 10^8 \text{ m/s}} = 7.93 \times 10^{-7} \text{ N}.$$

Hence, the pressure Pr is

$$Pr = \frac{F}{A} = \frac{(dp/dt)}{A} = \frac{S}{c} = \frac{119 \text{ W/m}^2}{3 \times 10^8 \text{ m/s}} = 3.97 \times 10^{-7} \text{ Pa}.$$

45P

Let f be the fraction of the incident beam that is reflected. The radiation pressure due to the part of the beam energy that is absorbed is, by Sample Problem 2,

$$p_a = \frac{(1 - f)S}{c}.$$

S is the magnitude of the Poynting vector of the incident beam. The radiation pressure due to the reflected part of the incident beam is

$$p_r = \frac{2(fS)}{c}.$$

The factor of 2 occurs because, on reflection, the momentum p of the beam is reversed, in which case $\Delta p = 2p$. The total radiation pressure is

$$p(\text{rad}) = p_a + p_r = \frac{(1 + f)S}{c}.$$

For a plane wave with energy flux S, an amount of energy SAt crosses an area A normal to the beam in time t. But in this same time the wave travels a distance ct. Hence, SAt is the energy contained in a cylindrical volume of base area A and length ct, so that the energy density u in the wave is

$$u = \frac{SAt}{Act} = \frac{S}{c}.$$

In terms of the energy density u, then, the radiation pressure found above is

$$p(\text{rad}) = (1 + f)u = u + fu.$$

The first term on the right is the energy density of the incident beam and the second is the energy density of the reflected beam. Since energy is a scalar, the total radiation energy density just outside the surface is u + fu, rather than u − fu, even though the incident and reflected beams are moving in opposite directions. Thus, the assertion of the problem has been established.

47P

The momentum carried off by the laser beam in time t is, by Eq.33, $p = U/c$, since the process is just the reverse of absorption. U is the energy carried by the beam that is emitted in the same time t. If P = 10 kW is the power of the laser beam, then $U = Pt$ and $p = Pt/c$, so that the force will be

$$F = \frac{dp}{dt} = \frac{P}{c} = ma.$$

Hence, the speed v reached in time t, assuming the spaceship started from rest, is, since $v = at$,

$$v = \frac{Pt}{mc} = \frac{(10^4 \text{ W})(86,400 \text{ s})}{(1500 \text{ kg})(3 \times 10^8 \text{ m/s})} = 1.92 \times 10^{-3} \text{ m/s} = 1.92 \text{ mm/s}.$$

49P

(a) Let r = radius, ρ = density of the particle at a distance x from the sun, which has a mass M and a luminosity (rate of energy output) L. The gravitational force on the particle due to the sun is

$$F_{grav} = GMm/x^2 = GM(4\pi r^3 \rho/3)/x^2.$$

The force due to the radiation pressure is (see Sample Problem 2),

$$F_{rad} = \frac{SA}{c} = \frac{1}{c}(L/4\pi x^2)(\pi r^2),$$

since the area A perpendicular to the sun's rays is the cross-sectional area $= \pi r^2$ of the sphere. The critical radius $r = R_0$ occurs when these forces balance:

$$GM(4\pi R_0^3 \rho/3)/x^2 = \frac{1}{c}(L/4\pi x^2)(\pi R_0^2) \quad \rightarrow \quad R_0 = \frac{3L}{16\pi cG\rho M}.$$

(b) See Appendix C; we put $L = 3.9 \times 10^{26}$ W, $\rho = 1000$ kg/m^3, and $M = 1.99 \times 10^{30}$ kg; these give $R_0 = 585$ nm.

51E

(a) For incident polarized light, the transmitted intensity is $I = I_m \cos^2\theta$ by Eq.35. If the incident light is unpolarized, vibrations with all possible values of θ are present. Hence, the transmitted intensity is

$$I = I_m \overline{\cos^2\theta} = \frac{1}{2}I_m = \frac{1}{2}(0.01) = 0.005 \text{ W/m}^2.$$

By Problem 22 (noting that $I = \overline{S}$),

$$E_m = \sqrt{[2\mu_0 cI]} = \sqrt{[2(4\pi \times 10^{-7})(3 \times 10^8)(0.005)]} = 1.94 \text{ V/m}.$$

(b) The rate at which the sheet absorbs energy is $S = 0.005 \text{ W/m}^2$ (since half the energy that is incident is transmitted). The pressure is, by Sample Problem 2,

$$Pr = \frac{S}{c} = \frac{0.005}{3 \times 10^8} = 1.67 \times 10^{-11} \text{ N/m}^2.$$

58P

Let I be the intensity of the polarized component and i the intensity of the unpolarized component in the incident beam. The intensity of this beam, then, is i + I. The transmitted intensity of the unpolarized component is $i_t = i/2$ (see Problem 51a above); this is independent of the angle θ of the polaroid sheet. The intensity I_t transmitted by the originally polarized component is $I_t = I\cos^2\theta$. Hence, the transmitted intensity is

$$i_t + I_t = i/2 + I\cos^2\theta.$$

As the polaroid sheet is rotated, $\cos^2\theta$ varies between 0 and 1, and if the transmitted intensity, as a consequence, varies by the factor of 5, then

$$5(i/2) = i/2 + I \rightarrow i = I/2.$$

Therefore the fraction of the incident beam that is unpolarized is $i/(i + I) = 1/3$, and the relative intensity of the polarized component is 2/3.

59P

(a) As illustrated in the sketch for two sheets, use a number of sheets with a total rotation angle of 90°.

(b) For n sheets at equal angles θ between adjacent sheets, the transmitted intensity, by Malus' law, is

$$I = I_0[\cos^n(\frac{\pi}{2n})]^2,$$

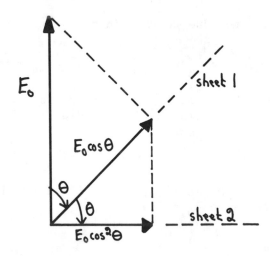

since $n\theta = \pi/2$ rad = 90°; I_0 is the intensity of the incident beam. For a loss of intensity of 40%, set $I = 0.60 I_0$ to get

$$0.6 = \cos^{2n}(\frac{\pi}{2n}).$$

We could solve this equation by trial and error, but n may be large. Let's call $x = \pi/2n$ so that

$$0.6 = (\cos x)^{\pi/x} = (1 - \tfrac{1}{2}x^2 + \ldots)^{\pi/x},$$

using the series expansion in Appendix G. Supposing x^2 to be small

$$\ln(0.6) = \ln(1 - \tfrac{1}{2}x^2)^{\pi/x} = (\tfrac{\pi}{x})\ln(1 - \tfrac{1}{2}x^2) \simeq (\tfrac{\pi}{x})(-\tfrac{1}{2}x^2),$$

using the series expansion for ln, also found in Appendix G. Hence,

$$x = -\frac{2}{\pi}\ln(0.6) = \frac{\pi}{2n},$$

$$n = -\frac{\pi^2}{4\ln(0.6)} = 4.83,$$

so that 5 sheets are needed.

2E

Apply the law of refraction, Eq.1. In the vacuum $n_1 = 1$. We seek n_2, the index of refraction of the glass. Eq.1 yields

$$n_1\sin\theta_1 = n_2\sin\theta_2,$$

$$(1)\sin32° = n_2\sin21° \rightarrow n_2 = 1.48.$$

3E

The index of refraction is given by $n = c/v$ (see below Eq.2); therefore

$$n = (3 \times 10^8 \text{ m/s})/(1.92 \times 10^8 \text{ m/s}) = 1.56.$$

9P

The length of the shadow is $L + x$; see the sketch for the distances labelled L and x. The length x is given from

$$\tan55° = 0.5/x,$$

$$x = 0.5/\tan55° = 0.350 \text{ m}.$$

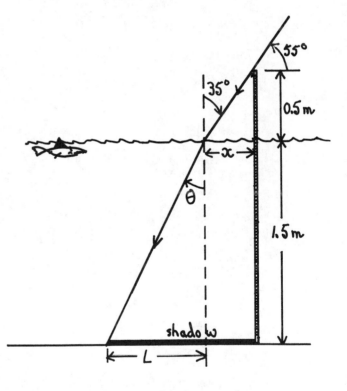

To find L, first find the angle of refraction θ; by the law of refraction (n = 1 for air),

$$(1)\sin35° = (1.33)\sin\theta \rightarrow \theta = 25.55°.$$

We now have

$$L = (1.5)\tan\theta = (1.5)\tan25.55° = 0.717 \text{ m}.$$

Therefore, the length of the shadow is $0.717 + 0.350 = 1.07$ m.

10P

The angle of refraction ϕ at the point P of incidence equals the angle of incidence at the point Q of emergence, since the normals to the two faces of the sheet are parallel, the faces themselves being parallel. Hence, the angle of emergence θ equals the angle of incidence θ at P regardless of the form of the law of refraction, provided that the light path is reversible (it is). To find the deviation x note that, writing PQ = D,

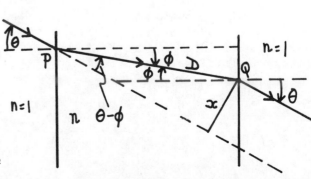

$$x = D\sin(\theta - \phi),$$

and

$$D = t/\cos\phi,$$

so that

$$x = t \frac{\sin(\theta - \phi)}{\cos\phi}.$$

If θ is small then ϕ is small also (for reasonable values of the index of refraction n of the glass). Expressing the angles in radians, rather than degrees, the law of refraction becomes, under these circumstances, $\theta = n\phi$ (rather than $\sin\theta = n\sin\phi$), since $\sin\theta \simeq \theta$. In the same spirit, $\cos\phi \simeq 1$ (see Appendix G for these approximations). Hence, for small angles

$$x = t \frac{\theta - \theta/n}{1} = t\theta \frac{n - 1}{n}.$$

17P

(a) For normal incidence there is no refraction at the water surface.

(b) By application of the law of reflection (angle of incidence = angle of reflection), together with the property of plane triangles that the sum of the angles is 180°, it is seen from the sketch that the incident and emergent rays are parallel. The relation between the angles ϕ and θ (i.e., the specific form of the law of refraction) need not be known.

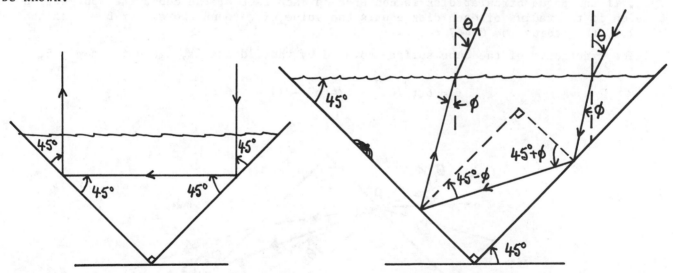

24E

The rays by which the fish sees the fire, looking at the smallest possible angle θ above the horizontal are refracted at the critical angle. Thus, by Snell's law (and Appendix G),

$$(1.33)\sin(90° - \theta) = (1)\sin 90°,$$

$$(1.33)\cos\theta = 1 \rightarrow \theta = 41.2°.$$

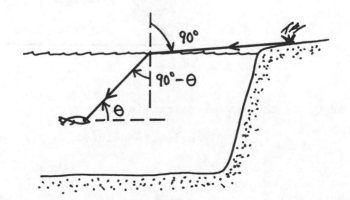

Note that this is independent of distance.

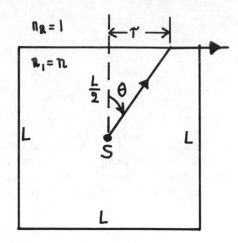

26P

(a) We call the edge length of the cube L. The spot S is seen from the outside by refracted light that leaves the cube. However rays leaving S at angles greater than (see Eq.3)

$$\theta = \sin^{-1}(\frac{1}{n}),$$

with the vertical will undergo total internal reflection and, if their behavior after this is ignored, they do not exit the cube. Hence, rays will emerge from a circular area, the radius r of which is given by

$$r = (L/2)\tan\theta = (L/2)/\sqrt{[n^2 - 1]},$$

where we used the trig identities (see Appendix G)

$$\tan\theta = \frac{\sin\theta}{\cos\theta} = \frac{\sin\theta}{\sqrt{[1 - \sin^2\theta]}} = \frac{1/n}{\sqrt{[1 - (1/n)^2]}} = \frac{1}{\sqrt{[n^2 - 1]}}.$$

Thus, if an opaque circular disk is centered on each face of the cube, the spot will not be seen if the radius of each disk equals the value of r found above. For L = 1 cm and n = 1.5, the radius r = 0.447 cm.

(b) The fraction f of the cube surface covered by these disks is, again for n = 1.5,

$$f = 6\pi r^2/6L^2 = \pi/4(n^2 - 1) = 0.628.$$

28P

(a)

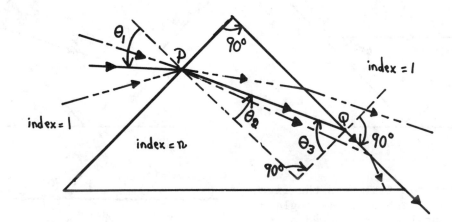

At Q, by the law of refraction,

$$n\sin\theta_3 = n\sin(90° - \theta_2) = n\cos\theta_2 = (1)\sin 90° = 1,$$

$$\cos\theta_2 = 1/n.$$

But, at P

$$(1)\sin\theta_1 = n\sin\theta_2,$$

and therefore

$$\cos\theta_2 = \sqrt{[1 - \sin^2\theta_2]} = \sqrt{[1 - (\sin\theta_1/n)^2]} = (\tfrac{1}{n})\sqrt{[n^2 - \sin^2\theta_1]}.$$

Substituting $\cos\theta_2 = 1/n$ from above and solving for n gives

$$n = \sqrt{[1 + \sin^2\theta_1]}.$$

(b) Since the greatest possible value of $\sin^2\theta_1 = 1$, $n_{max} = \sqrt{2}$.

(c) For $\theta > \theta_1$, the ray is refracted into the air, and for $\theta < \theta_1$ the ray undergoes total internal reflection at the second face.

<u>29P</u>

(a) For the smallest angle of incidence θ_3, the ray emerges at C along the prism face (critical angle). Hence, at C

$$(1)\sin 90° = (1.6)\sin\theta_1 \rightarrow \theta_1 = 38.68°.$$

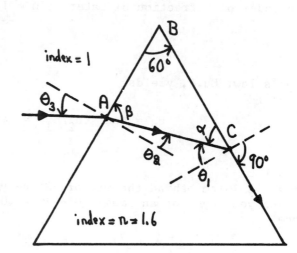

Therefore $\alpha = 90° - \theta_1 = 51.32°$. From the triangle ABC, $\beta = 180° - 60° - 51.32° = 68.68°$. Thus $\theta_2 = 90° - \beta = 21.32°$. Finally, at A,

$$(1)\sin\theta_3 = (1.6)\sin 21.32° \rightarrow \theta_3 = 35.6°.$$

(b) From Problem 19,

$$n = \frac{\sin\tfrac{1}{2}(\psi + \phi)}{\sin\tfrac{1}{2}\phi},$$

$$1.6 = \frac{\sin\tfrac{1}{2}(\psi + 60°)}{\sin\tfrac{1}{2}(60°)} \rightarrow \psi = 46.26°.$$

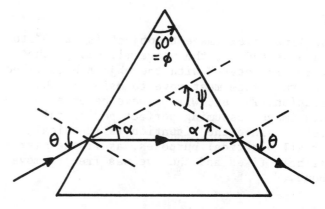

Therefore in the sketch $\alpha = 23.13°$. By the law of refraction at entry,

$$(1)\sin\theta = (1.6)\sin(\theta - 23.13°).$$

To solve this equation for θ, we use a trig identity from Appendix G to write

$$\sin\theta = (1.6)\sin\theta\cos 23.13° - (1.6)\cos\theta\sin 23.13°,$$

$$[(1.6)\sin 23.13°]\cos\theta = [(1.6)\cos 23.13° - 1]\sin\theta,$$

$$\frac{\sin\theta}{\cos\theta} = \tan\theta = \frac{(1.6)\sin 23.13°}{(1.6)\cos 23.13° - 1} \rightarrow \theta = 53.1°.$$

30P

(a) Under the assumption of the problem, if θ is the critical angle for total internal reflection, the rays emitted at angles with respect to the vertical greater than θ are reflected back into the water and do not escape. Only those rays emitted into directions lying inside a cone of semiangle θ about the vertical will escape. Imagine the source surrounded by a sphere of radius $r < h$. If A is the area intercepted by the cone on the sphere, then the desired ratio is $f = A/4\pi r^2$, since the source radiates isotropically. But $A = 2\pi r^2(1 - \cos\theta)$ and since $n\sin\theta = 1$ (the condition for the onset of total internal reflection), the ratio sought is

$$f = \frac{1}{2}[1 - \sqrt{(1 - 1/n^2)}];$$

to obtain this we used the trig identity $\sin^2\theta + \cos^2\theta = 1$.

(b) The index of refraction of water is $n = 1.33$, so the ratio in this case is $f = 0.17$.

36E

Brewster's law, Eq.4, yields

$$\theta_B = \tan^{-1}(n_2/n_1) = \tan^{-1}(1.33/1.53) = 41°.$$

39E

The image is 10 cm behind the mirror. Since you are 30 cm in front of the mirror, you must focus your eyes at an image 30 + 10 = 40 cm away, so that 40 cm is the desired distance.

45P

The line from image to object is at 90° to, and bisected by, the plane mirror. If the mirror is rotated with the object held fixed then the image must move to maintain this relation. The eye will always be able to see this image (assuming perfect peripheral vision) since the situation is identical, for all practical purposes, as if the mirror was held fixed and the eye was free to move about.

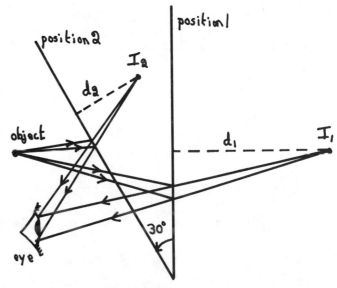

47P

In the diagram on p.303 are shown the three virtual images obtained when the object does not lie on the perpendicular bisector of the angle formed by the two mirrors. We expect one image from each mirror; the third image, I_3, can be thought of as the image of these images.

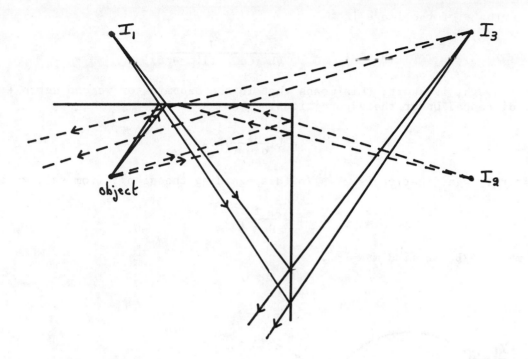

52P

Allow the mirrors to lie along the coordinate planes. Write the velocity of the incident ray as

$$\vec{c}_0 = c_x \vec{i} + c_y \vec{j} + c_z \vec{k}.$$

At the first reflection the z component of the velocity is reversed, the others being unchanged; thus,

$$\vec{c}_1 = c_x \vec{i} + c_y \vec{j} - c_z \vec{k}.$$

At reflections 2 and 3, the y and then the x components are reversed. The other components remain unchanged. Hence,

$$\vec{c}_2 = c_x \vec{i} - c_y \vec{j} - c_z \vec{k},$$
$$\vec{c}_3 = -c_x \vec{i} - c_y \vec{j} - c_z \vec{k}.$$

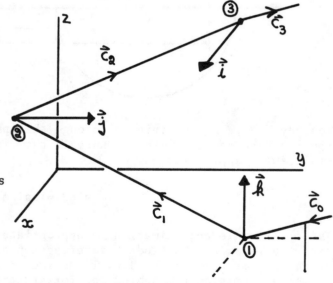

Thus $\vec{c}_3 = -\vec{c}_0$, indicating that the emergent and incident rays are antiparallel.

56P

(a) Let the ends of the object be at distances a, L + a from the mirror. The images of these ends are at image distances i, i* given by Eq.6:

$$\frac{1}{a} + \frac{1}{i} = \frac{1}{f} \quad \rightarrow \quad i = \frac{fa}{a - f},$$

$$\frac{1}{L + a} + \frac{1}{i*} = \frac{1}{f} \quad \rightarrow \quad i* = \frac{fL + fa}{L + a - f}.$$

Thus the length L* of the image is

$$L* = i - i* = \frac{f^2 L}{(L + a - f)(a - f)}.$$

The object, though, is short; this means that in the denominator we can set $L + a \simeq a \simeq o$ the object distance. Under these conditions,

$$L* = L\left(\frac{f}{o - f}\right)^2.$$

(b) The lateral magnification is $m = i/o$ (disregarding the sign); from (a), we get

$$m = \frac{f}{o - f}.$$

Hence, if $m* = L*/L$, we find $m* = m^2$.

60P

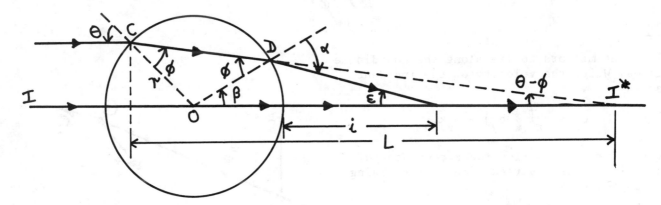

The ray II* passes undeviated through the sphere. For the other ray shown, striking the sphere at an angle θ with the normal, the first angle of refraction is given, in the small angle approximation, by

$$\sin\theta = n\sin\phi \rightarrow \phi = \theta/n.$$

The small angle approximation is appropriate since the beam is narrow and therefore the rays are paraxial. The index of refraction of the glass is n, and the index of refraction of the air is taken to be 1.0. If no second refraction took place, an image I* would be formed at a distance L behind the forward surface, with L determined by

$$\tan(\theta - \phi) = \frac{r\sin\theta}{L} \rightarrow \theta - \phi = \frac{r\theta}{L},$$

$$L = \frac{r\theta}{\theta - \phi} = \left(\frac{n}{n - 1}\right)r.$$

For glass n = 1.5 approximately, so that L = 3r roughly; therefore I* lies outside the glass sphere. At the second refraction,

$$n\sin\phi = \sin\alpha \rightarrow \alpha = n\phi = \theta.$$

The angles of triangle COD must total π rad; hence,

$$2\phi + (\pi - \beta - \theta) = \pi,$$

$$\beta = 2\phi - \theta = (\frac{2 - n}{n})\theta.$$

Thus,

$$x = r\beta = r\theta(\frac{2 - n}{n}),$$

$$\beta = (\frac{2 - n}{n})\theta.$$

The image distance i is given by

$$i = \frac{x}{\tan\varepsilon} = \frac{x}{\tan(\theta - \beta)} = \frac{x}{\theta - \beta} = \frac{r\theta(2 - n)/n}{\theta - (2 - n)\theta/n} = \frac{(2 - n)r}{2(n - 1)}.$$

61E

The lens being diverging, set f = −30 cm.
Eq.12 then yields

$$\frac{1}{o} + \frac{1}{i} = \frac{1}{f},$$

$$\frac{1}{20} + \frac{1}{i} = \frac{1}{-30},$$

$$\frac{1}{i} = -\frac{1}{30} - \frac{1}{20} = -\frac{5}{60} \rightarrow i = -12 \text{ cm.}$$

Thus, a virtual image is formed 12 cm from
the lens, on the object side.

64E

Use the lens makers formula with n = 1.5 and f = 60 mm. From Fig.13, it is clear that for
a double convex lens, r_2 is negative. Therefore, put $r_2 = -2r_1$. Thus,

$$\frac{1}{f} = (n - 1)(\frac{1}{r_1} - \frac{1}{r_2}),$$

$$\frac{1}{60} = (1.5 - 1)(\frac{1}{r_1} + \frac{1}{2r_1}) = \frac{3}{4r_1} \rightarrow r_1 = 45 \text{ mm,}$$

and therefore $r_2 = -90$ mm, indicating a 90 mm radius of curvature.

71P

The Newtonian form applies to converging lenses (f > 0). To derive it, start with the
Gaussian form,

$$\frac{1}{o} + \frac{1}{i} = \frac{1}{f}.$$

The object and image distances, in terms of x, x' are

$$o = f \pm x; \quad i = f \pm x',$$

the upper sign for a real image ($o > f$, $i > 0$), the lower sign for a virtual image ($o < f$, $i < 0$). Putting these into the first equation gives directly

$$\frac{1}{f \pm x} + \frac{1}{f \pm x'} = \frac{1}{f},$$

$$xx' = f^2,$$

in which x and x' both are greater than zero.

78P

To avoid having to deal with virtual objects, assume that both lenses are diverging, so that both f_1 and f_2 are negative. Suppose the light is incident on f_1. Then,

$$\frac{1}{o_1} + \frac{1}{i_1} = \frac{1}{f_1},$$

i_1 the image distance of the image formed by lens f_1; of course, $i_1 < 0$. But this image forms a real object for lens f_2. Since the lenses are in contact $o_2 = -i_1$ (the negative sign to give $o_2 > 0$). We now have

$$\frac{1}{o_2} + \frac{1}{i_2} = \frac{1}{f_2},$$

$$-\frac{1}{i_1} + \frac{1}{i_2} = \frac{1}{f_2}.$$

By the definition of the equivalent focal length f,

$$\frac{1}{o_1} + \frac{1}{i_2} = \frac{1}{f}.$$

The first and third equations yield

$$\left(\frac{1}{o_1} - \frac{1}{f_1}\right) + \frac{1}{i_2} = \frac{1}{f_2}.$$

Comparing this to the preceding equation leads to

$$\frac{1}{f} = \frac{1}{f_1} + \frac{1}{f_2} \quad \rightarrow \quad f = \frac{f_1 f_2}{f_1 + f_2}.$$

81P

The image and object distances from the lens are related by

$$\frac{1}{i} + \frac{1}{o} = \frac{1}{f}.$$

The distance between the object and the image is

$$x = i + o.$$

All quantities are positive for the optical arrangement under consideration. Use the
second equation to eliminate i from the first equation, and then solve for x to get

$$\frac{1}{x - o} + \frac{1}{o} = \frac{1}{f} \rightarrow x = \frac{o^2}{o - f}.$$

To find the minimum x, set dx/do = 0; this gives

$$\frac{dx}{do} = 0 = \frac{2o}{o - f} - \frac{o^2}{(o - f)^2} = 0 \rightarrow o = 2f.$$

With this result for o, x becomes

$$x = \frac{(2f)^2}{2f - f} = 4f,$$

which is the minimum value. This must be a minimum, rather than a maximum, for it is
known that for o close to f, i is almost infinity.

82P

(a) If the object distance is x, then the image distance is D − x. Hence, this implies
from Eq. 12,

$$\frac{1}{o} + \frac{1}{i} = \frac{1}{f},$$

$$\frac{1}{x} + \frac{1}{D - x} = \frac{1}{f},$$

$$x^2 - Dx + Df = 0.$$

The solution of the quadratic equation is (see Appendix G),

$$x_1 = \frac{1}{2}(D - d); \quad x_2 = \frac{1}{2}(D + d),$$

in which

$$d = \sqrt{(D^2 - 4Df)}.$$

Thus, the separation in the two positions of the lens (object fixed) is $x_2 - x_1 = d$.
(b) The ratio of the magnifications is

$$m_2/m_1 = (-i_2/o_2)/(-i_1/o_1).$$

But it can be verified from the expressions above for x_1 and x_2 in (a) that $o_1 = x_1$,
$o_2 = x_2$, $i_2 = D - x_2 = x_1$, $i_1 = D - x_1 = x_2$. Therefore,

$$m_2/m_1 = (x_1/x_2)^2 = (\frac{D - d}{D + d})^2.$$

83E

(a) From Fig.26 itself, we see that $s = 25$ cm $- f_{ob} - f_{ey} = 25 - 4 - 8 = 13$ cm.

(b) The image distance is $i = f_{ob} + s = 4 + 13 = 17$ cm. Therefore,

$$\frac{1}{o} + \frac{1}{i} = \frac{1}{f},$$

$$\frac{1}{o} + \frac{1}{17} = \frac{1}{4} \rightarrow o = 5.23 \text{ cm}.$$

This is the distance of the object from the objective; the distance to F_1 is 5.23 cm $- f_{ob}$ = 5.23 - 4 = 1.23 cm.

(c) By Eq.15, $m = -s/f_{ob} = -13/4 = -3.25$.

(d) The angular magnification is given by Eq.14:

$$m_\theta = (15 \text{ cm})/f_{ey} = 15/8 = 1.875.$$

(e) By Eq.16, the overall magnification is $M = (m)(m_\theta) = (-3.25)(1.875) = 6.09$.

89P

(a) The mirror M may be thought of as replaced with a lens of the same focal length, the only difference introduced is that now the image is behind the lens rather than in front of the mirror. The mirror M', being flat, is of no import as far as magnification is concerned, since the magnification produced by a flat mirror is 1.0. Hence,

$$m_0 = -f_{ob}/f_{eye},$$

as before, the lens also inverting the image.

(b) Considering the mirror to be spherical rather than the actual paraboloid, Eq.6 gives

$$\frac{1}{2000} + \frac{1}{i} = \frac{1}{16.8} \rightarrow i = 16.94 \text{ m}.$$

The magnification is $m = -i/o = -(16.94)/(2000) = 0.00847$. Since the object's size is 1 m the image size is 0.00847 m = 8.47 mm.

(c) The focal length is $f = r/2 = 5.0$ m. Thus,

$$200 = 5/f_{eye} \rightarrow f_{eye} = 2.5 \text{ cm}.$$

CHAPTER 40

9E

See Sample Problem 1, where it is shown that the separation between adjacent bright fringes (i.e., maxima) is

$$\Delta y = \frac{\lambda D}{d} = \frac{(500 \times 10^{-9} \text{ m})(5.4 \text{ m})}{(0.0012 \text{ m})} = 2.25 \text{ mm}.$$

11E

From Eq.9, $\sin\theta = m\lambda/d$. We assume that θ is small so that $\sin\theta = \theta$ if θ is expressed in radians rather than degrees. The angular separation $\Delta\theta$ between adjacent fringes will then be given from

$$\theta = m\lambda/d,$$

$$\Delta\theta = (\Delta m)\lambda/d = (1)\lambda/d.$$

Let's call the new wavelength λ^*; for this wavelength we are told that $\Delta\theta^* = (1.10)\Delta\theta$. Therefore, using the formula just derived,

$$\Delta\theta^* = \lambda^*/d = (1.10)\Delta\theta = (1.10)\lambda/d,$$

$$\lambda^* = (1.10)\lambda = (1.10)(589 \text{ nm}) = 648 \text{ nm}.$$

15P

In Sample Problem 1, it is shown that the position on the screen of a maximum (fringe) is given by (see Eq.11),

$$y_m = m\lambda D/d.$$

We are concerned with $m = 3$. The difference in positions Δy of the fringes for two different wavelengths will be

$$\Delta y = 3\lambda_1 D/d - 3\lambda_2 D/d = 3(\lambda_1 - \lambda_2)D/d = 3(600 \text{ nm} - 480 \text{ nm})(1 \text{ m})/(0.005 \text{ m}),$$

$$\Delta y = 0.072 \text{ mm} = 72 \text{ } \mu\text{m}.$$

19P

The geometric path lengths from the slits to the central part of the screen, with the mica in place, are equal. The phase difference arises because one beam traverses a distance x in mica as the other is traversing a thickness x of air. The phase difference so introduced must be $7(2\pi)$, since 2π corresponds to one wavelength. Hence,

$$\frac{2\pi x}{\lambda_n} - \frac{2\pi x}{\lambda_a} = 7(2\pi),$$

$$x\left[\frac{1}{\lambda/n} - \frac{1}{\lambda}\right] = 7,$$

$$\frac{x}{\lambda}(n - 1) = 7,$$

$$x = \frac{7\lambda}{n - 1} = \frac{7(550)}{1.58 - 1} = 6638 \text{ nm} = 6.64 \times 10^{-3} \text{ mm} = 6.64 \text{ } \mu\text{m}.$$

22P

The maxima lie in directions given by the angle θ, where, from Eq.9,

$$d\sin\theta = m\lambda,$$

$$2\sin\theta = m(0.5),$$

$$\sin\theta = (0.25)m.$$

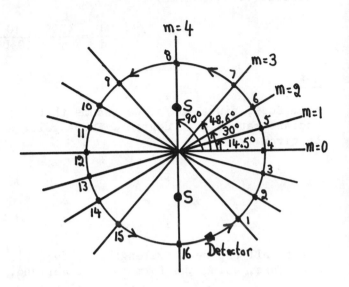

Now $\sin\theta \leq 1$. Hence, allowed values of θ are: $0°$ (m = 0), $14.5°$ (m = 1), $30.0°$ (m = 2), $48.6°$ (m = 3), $90.0°$ (m = 4). Now these sources radiate in all directions. Hence, the lines along which maxima occur, in the plane of the path taken by the detector, is as shown on the sketch. We see that 16 maxima will be encountered by the detector.

30E

The difference in distance leads to a phase difference = $(\Delta r/\lambda)(360°) = (100 \text{ m}/400 \text{ m})360°$ = $90°$. As it is the signal from A that travels the extra distance, A falls $90°$ behind the signal from B. But the signal from A started out $90°$ in phase ahead of the signal from B. Therefore, the phase difference upon arrival at the detector is $+90° - 90° = 0°$.

33P

(a) If D is the detector, the optical path difference $S_2D - S_1D$ equals the geometric path difference since the paths from both sources are in air. For maxima, this path difference must equal an integral number of wavelength s:

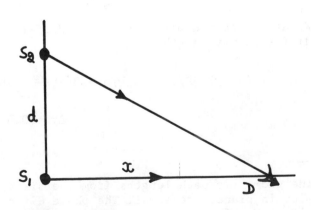

$$\sqrt{(d^2 + x^2)} - x = n\lambda,$$

n = 1, 2, 3, ... n = 0 is not possible since $d \neq 0$. This equation yields for x

$$x = (d^2 - n^2\lambda^2)/2n.$$

Putting n = 3, 2, 1, together with d = 4.0 m, λ = 1.0 m gives x_3 = 1.17 m, x_2 = 3.00 m, x_1 = 7.50 m.

(b) For completely destructive interference the waves must arrive at D exactly 180° out of phase and also with equal amplitudes. The first requirement can be met but the second cannot since the waves, being spherical, experience a 1/distance fall-off in amplitude and the distances of travel are different for the two waves. (It is implied in the problem statement that the sources radiate waves of equal amplitude.)

35P

The intensity I is given by Eqs.15 and 16:

$$I = 4I_0\cos^2\tfrac{1}{2}\phi,$$

with $\phi = 2\pi d\sin\theta/\lambda \simeq 2\pi\theta d/\lambda$ for small θ. At a point where the intensity is one-half the maximum of $4I_0$, we have

$$2I_0 = 4I_0\cos^2\tfrac{1}{2}\phi.$$

The smallest positive ϕ satisfying this equation is $\phi = \pi/2$. Hence, the first half-intensity point occurs where θ has the value given from

$$\pi/2 = 2\pi\theta d/\lambda \quad \rightarrow \quad \theta = \lambda/4d.$$

A symmetrical half-intensity point falls at $\theta = -\lambda/4d$, with the $\theta = 0$ maximum between them. Thus, the half-width of this central maximum is $2(\lambda/4d) = \lambda/2d$.

36P

The electric field components of the two waves are

$$E_1 = E_0\sin\omega t,$$

$$E_2 = 2E_0\sin(\omega t + \phi),$$

with $\phi = 2\pi d\sin\theta/\lambda$. The sum is to be written in the form

$$E_1 + E_2 = E\sin(\omega t + \beta);$$

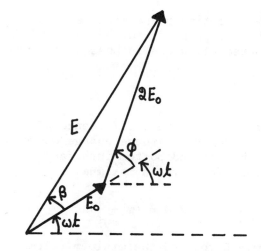

the intensity of this wave is proportional to E^2. Apply the law of cosines (see Appendix G) to the triangle formed by E_1, E_2, E in the phasor diagram to get

$$E^2 = E_0^2 + (2E_0)^2 - 2(E_0)(2E_0)\cos(\pi - \phi) = E_0^2(5 + 4\cos\phi).$$

Hence,

$$I = I_0(5 + 4\cos\phi),$$

$$I = I_0[1 + 8\cos^2(\tfrac{1}{2}\phi)],$$

using a trig identity in Appendix G in the last step.

312

38E

Both rays experience a phase change of 180°
on reflection. Thus, only the path difference
is effective. For cancellation, we must have

$$2d = (m + \tfrac{1}{2})\lambda_n,$$

$$2nd = (m + \tfrac{1}{2})\lambda,$$

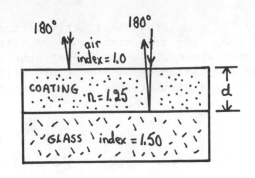

since λ_n, the wavelength in the coating
material, $= \lambda/n$, where λ is the wavelength
in air and n is the index of refraction of the coating material. For the thinnest coating
pick m = 0 to get

$$d = \lambda/4n = (600 \text{ nm})/4(1.25) = 120 \text{ nm}.$$

41E

The path difference 2d between the two rays
corresponds to a phase difference $\Delta\phi$ that is
given by

$$\frac{\Delta\phi}{2\pi} = \frac{2d}{\lambda_n} = \frac{2nd}{\lambda},$$

$$\Delta\phi = \frac{(2)(1.33)(1210 \text{ nm})}{585 \text{ nm}}(2\pi) = 11\pi = 5(2\pi) + \pi,$$

or an effective phase difference of 180°. The phase changes on reflection introduce an
additional phase difference of 180° for a total of 360°. This indicates that the
reflected waves are in phase. Thus the wave appears bright.

43E

There is a net phase difference of 180°
introduced by the phase change on reflection
from the top and bottom interface of the SiO
and therefore, for a maximum,

$$2nd = (m + \tfrac{1}{2})\lambda.$$

Set m = 0 to find the minimum thickness of
SiO that is needed:

$$d = \frac{\lambda}{4n} = \frac{560}{4(2)} = 70.0 \text{ nm}.$$

48P

The phase changes on reflection for the two cases, $n < n_g$ and $n > n_g$, are shown on the
sketch. Thus the conditions for destructive interference in the two cases are

$$n > n_g: \quad 2nd = m\lambda; \quad m = 1, 2, \ldots;$$

$$n < n_g: \quad 2nd = (m + \tfrac{1}{2})\lambda; \quad m = 0, 1, 2 \ldots$$

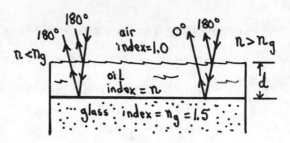

Now, there is destructive interference for wavelengths 500 nm and 700 nm and therefore either

$$2nd = m_5(500), \quad 2nd = m_7(700), \; n > n_g;$$

or

$$2nd = (m_5 + \tfrac{1}{2})(500), \quad 2nd = (m_7 + \tfrac{1}{2})(700), \; n < n_g.$$

In either case, since 700 > 500, we have $m_7 < m_5$. In fact, since there is no minimum between these wavelengths,

$$m_7 = m_5 - 1.$$

Hence,

$$m_7 = \tfrac{5}{2}, \; n > n_g; \quad m_7 = 2, \; n < n_g.$$

Since m_7 must be an integer, it must be that $n < n_g = 1.5$.

50P

(a) There is a 180° phase change at each reflection and thus

$$2d = (m)\tfrac{\lambda}{n}, \text{ bright fringe;}$$

$$2d = (m + \tfrac{1}{2})\tfrac{\lambda}{n}, \text{ dark fringe.}$$

At the outer regions $d \simeq 0$, and since $m \geq 0$, there can be only the bright fringe $m = 0$ present regardless of the wavelength.

(b) For the third ($m = 3$) blue fringe ($\lambda = 475$ nm; see Fig.2 in Chapter 38), we have

$$2d = (m)\tfrac{\lambda}{n} = (3)\tfrac{475 \text{ nm}}{1.2} \rightarrow d = 594 \text{ nm.}$$

(c) As the thicker regions of the film are examined, the fringes fall closer and closer together, since closely neighboring regions can differ in thickness by many wavelengths; eventually, they cannot be distinguished by the unaided eye.

53P

If the light waves are reflected from a region where the thickness of the air film is y, then the condition for a maximum is

$$2y = (m + \tfrac{1}{2})\lambda, \quad m = 0, 1, \ldots$$

the 1/2 compensating for the phase difference of 180° introduced in the reflection from the lower surface of the air film. If a bright fringe (maximum) appears at the end (y = d), it would have an order number m given by

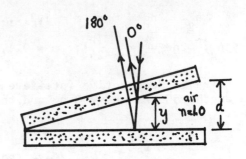

$$2d = (m + \frac{1}{2})\lambda,$$

$$2(48,000 \text{ nm}) = (m + \frac{1}{2})(680 \text{ nm}) \;\to\; m = 140.7.$$

This indicates that a maximum does not lie at the end (m must be an integer), but that the maximum nearest to the end is of order m = 140. Since there is a bright fringe near the other end for which m = 0 (where the air film has a thickness $\lambda/4$), there are 140 + 1 = 141 bright fringes in all.

57P

There is a net phase change of 180° that is introduced by reflection. Thus, for a bright B fringe and a dark D fringe,

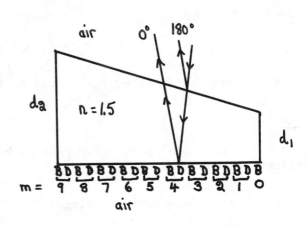

$$2d = (m + \frac{1}{2})\frac{\lambda}{n}, \quad \text{at B};$$

$$2d = (m)\frac{\lambda}{n}, \quad \text{at D}.$$

At the ends,

$$2d_1 = (0 + \frac{1}{2})\frac{630 \text{ nm}}{1.5} \;\to\; d_1 = 105 \text{ nm};$$

$$2d_2 = (9 + \frac{1}{2})\frac{630 \text{ nm}}{1.5} \;\to\; d_2 = 1995 \text{ nm}.$$

Thus, the variation in thickness is $d_2 - d_1$ = 1890 nm = 1.89 µm.

60P

The ray from the bottom of the air film, rather than the ray from the top, undergoes a phase change of 180°, for it is reflected from a medium (glass) of higher index of refraction than the film (air). The condition for maxima is, assuming 90° incidence,

$$2d = (m + \frac{1}{2})\lambda/n = (m + \frac{1}{2})\lambda; \quad m = 0, 1, 2, \ldots,$$

since the index of refraction of air is n = 1.00. Now,

$$d = R - \sqrt{(R^2 - r^2)} = R - R\sqrt{(1 - r^2/R^2)}.$$

If r << R, so that r/R << 1, then expanding the square root in a series (see Appendix G, Binomial theorem), and keeping only the first two terms, reduces the equation above to

$$d = R - R(1 - r^2/2R^2) = r^2/2R^2.$$

Combining this with the first equation to eliminate d yields

$$r = \sqrt{[(m + \frac{1}{2})\lambda R]}.$$

65E

Each fringe shift corresponds to one wavelength. Hence, the path difference introduced by moving the mirror is (# fringes)λ. If we call the distance the mirror is moved d, then the path difference equals 2d, since the light retraces its path on reflection. Therefore

$$2d = (\# \text{ fringes})\lambda,$$

$$\lambda = \frac{2d}{\# \text{ fringes}} = \frac{2(0.233 \times 10^6 \text{ nm})}{792} = 588 \text{ nm}.$$

67P

Let L be the length of the chamber. The phase difference between the two rays arises from the passage of light in one arm through a length 2L in air of density less (once the pumping starts) than the air in the other arm. If the lengths of the arms are not changed during the pumping process then the phase difference due to the difference in air density is given by

$$\Delta\phi = 2\pi[\frac{2L}{\lambda_1} - \frac{2L}{\lambda_2}] = 2\pi[\frac{2L}{\lambda/n_1} - \frac{2L}{\lambda/n_2}] = \frac{4\pi L}{\lambda}(n_1 - n_2).$$

Here λ is the wavelength in a vacuum and n_1, n_2 are the indices of refraction for the two air columns. Assuming perfect pumping, put $n_2 = 1$ and $n_1 = n$, where n is the index of refraction for air under atmospheric pressure that remains outside the chamber. Each fringe shift corresponds to a phase shift of 2π, so that the number N of fringes that pass across the field of view will be

$$N = \frac{\Delta\phi}{2\pi} = \frac{2L}{\lambda}(n - 1),$$

$$60 = \frac{2(5 \times 10^7 \text{ nm})}{500 \text{ nm}}(n - 1) \rightarrow n = 1.0003.$$

CHAPTER 41

3E

Since the diffraction pattern is symmetrical about $\theta = 0$, we conclude that $\theta = 0.6°$ for the first minimum. By Eq.3,

$$a = m\lambda/\sin\theta = (1)(633 \text{ nm})/\sin 0.6° = 6.04 \times 10^4 \text{ nm} = 60.4 \text{ } \mu m.$$

4E

The conditions for minima at the two wavelengths are, from Eq.3,

$$a\sin\theta_a = m\lambda_a, \quad a\sin\theta_b = n\lambda_b; \quad m,n = 1, 2, \ldots.$$

(a) Since $\theta_a = \theta_b$ when $m = 1$ and $n = 2$ (the two minima are at the same angle θ and are therefore coincident), it follows from the equations above that $\lambda_a = 2\lambda_b$.

(b) Assuming other coincidences (i.e., by setting $\theta_a = \theta_b$), use the result from (a) to obtain $n = 2m$. That is, every other minimum at λ_b coincides with a minimum for λ_a.

7E

(a) Since a plane wave is incident on the slit, the lens brings the light to an image in a plane at the focal point. Hence the distance is $D = f = 70$ cm.

(b) This calculation is similar to that in Sample Problem 1 in Chapter 40, except that here we have a single, rather than a double, slit. The linear distance is $y = D\tan\theta = D\theta$ (we assume that θ is small). But $\sin\theta = m\lambda/a = \lambda/a$ for the first minimum. Also $\sin\theta = \theta$ so that $\theta = \lambda/a$. Hence,

$$y = D\lambda/a = (70 \text{ cm})(590 \text{ nm})/(0.40 \text{ mm}) = (0.70 \text{ m})(590 \times 10^{-9} \text{ m})/(0.0004 \text{ m}) = 1.03 \text{ mm}.$$

13P

Consider a slit of given width divided into N strips of equal width so that there are N phasors. These are all parallel at the central maximum. If the slit width is doubled, there will be 2N phasors, each having the same amplitude as for the original slit. Thus, the amplitude of the resultant field at the central maximum doubles and the intensity, proportional to the square of the amplitude, increases by a factor of four. However, the diffraction curve is narrower and the area under the curve of I vs. θ, equal to the rate at which energy passes through the slit, will only double.

14P

From superposition, it is anticipated that the amplitude of the wave at P due to obstacle A added to the amplitude at the screen at P due to obstacle B will give the amplitude at P due to a wave encountering no obstacle. But, with $x \gg \lambda$, P lies in the geometrical shadow of the hole; thus the last quantity is zero. Hence, at P, $E_A = -E_B$. But the intensity is proportional to E^2, and therefore $I_A = I_B$ at P.

16P

(a) The intensity is given by Eq.5:

$$I = I_m \frac{\sin^2\alpha}{\alpha^2}.$$

To find the maxima and minima, set $dI/d\alpha = 0$; taking the derivative of the function above gives

$$\frac{dI}{d\alpha} = 2I_m \frac{\sin\alpha}{\alpha}(\frac{\cos\alpha}{\alpha} - \frac{\sin\alpha}{\alpha^2}).$$

This will be zero if either

(i) $\sin\alpha/\alpha = 0$, giving $\alpha = m\pi$, $m = 1, 2, \ldots$ These are minima since $I = 0$ for these values of α; or

(ii) $\cos\alpha/\alpha - \sin\alpha/\alpha^2 = 0$, which can be rearranged into the form $\tan\alpha = \alpha$, and the solutions of which represent maxima.

(b) Now, $\alpha = 0$ is a solution of (ii). The next solution, as shown on the sketch, is close to $\alpha = 3\pi/2$. Write this second solution as $\alpha = 3\pi/2 - x$. Then, in terms of x, the equation $\tan\alpha = \alpha$ takes the form

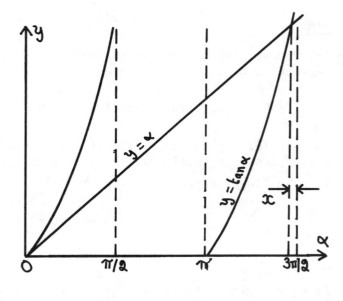

$$\frac{1}{\tan x} = \frac{3\pi}{2} - x.$$

Suppose, as expected, that $x \ll 1$. Make an expansion of $\tan x$ (see Appendix G), and toss out powers of x higher than 2; then the equation above becomes

$$\frac{1}{x}(1 - \frac{x^2}{3}) = \frac{3\pi}{2} - x.$$

This can be rearranged as

$$(\frac{2}{3})x^2 - (\frac{3\pi}{2})x + 1 = 0.$$

The smaller solution is $x = 0.219$ rad, corresponding to $\alpha = 4.493$ rad $= 257°$.

(c) Write, for maxima, $\alpha = (m + 1/2)\pi$. The central maximum, for which $\alpha = 0$, has $m = -\frac{1}{2}$ and the next maximum, $\alpha = 4.493$, yields $m = 0.93$.

17P

With the slit width much less than the wavelength of light, the effects associated with the diffraction envelope can be ignored. The phase difference ϕ between rays from two adjacent slits is $\phi = 2\pi d\sin\theta/\lambda$, since the path difference for these rays is $d\sin\theta$, just as for the two slit interference grating. The electric field components of the three waves arriving at a given point on the screen can be written as

$$E_1 = E_0\sin\omega t; \quad E_2 = E_0\sin(\omega t + \phi); \quad E_3 = E_0\sin(\omega t + 2\phi).$$

It can be seen from the phasor diagram that the amplitude of the sum of these phasors is

$$E = E_0\cos\phi + E_0 + E_0\cos\phi = E_0(1 + 2\cos\phi).$$

Since the intensity is proportional to the square of the amplitude,

$$I = AE_0^2(1 + 2\cos\phi)^2.$$

At the center of the pattern $\phi = \theta = 0$ and $I(\theta=0) = 9AE_0^2 = I_m$, the central intensity. Thus, if

$$A = I_m/9E_0^2,$$

we have

$$I = (\tfrac{1}{9})I_m(1 + 4\cos\phi + 4\cos^2\phi).$$

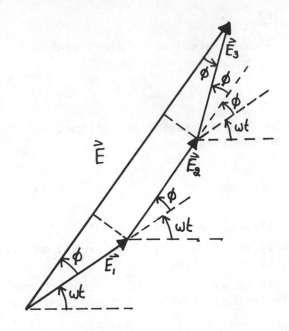

18E

(a) The angular separation inside the eye, which is filled with a fluid of index of refraction n, is

$$\theta* = 1.22\lambda_n/a = (1.22)(5.5 \text{ X } 10^{-7} \text{ m})/(5 \text{ X } 10^{-3} \text{ m})n = (1.34 \text{ X } 10^{-4} \text{ rad})/n.$$

The angular separation outside the eye is θ; but $(1)\sin\theta/2 = n\sin\theta*/2$, by the law of refraction. For small angles, this becomes $\theta = n\theta* = 1.34 \text{ X } 10^{-4}$ rad.

(b) The linear separation d and the distance r are related by $d = r\theta$, so that

$$r = d/\theta = (1.4 \text{ m})/(1.34 \text{ X } 10^{-4}) = 10.4 \text{ km}.$$

23E

If θ_R = angle subtended at the observer by the objects; d = mirror or pupil diameter; r = distance to objects; x = linear separation between objects then,   by Eq.11,

$$\theta_R = 1.22\lambda/d = x/r \rightarrow x = 1.22\lambda r/d.$$

Numerically, $\lambda = 550 \text{ X } 10^{-9}$ m, $r = 8 \text{ X } 10^{10}$ m, d = 0.005 m for the eye and d = 5.1 m for the mirror. These give (a) x = 11,000 km for the eye, and (b) x = 11 km for the mirror. For (a), the distance x is greater than the diameter of Mars itself.

33P

(a) The sketches illustrate the formation of a halo as the moon shines through a cloud containing suspended water droplets. The ring will appear red if blue light is absent. Since the angle θ for the first minimum is given by the circular aperture diffraction formula, $\sin\theta = 1.22\lambda/d$, blue light, as it has the shortest wavelength in the visible spectrum, will have its first minimum closer to the moon (smallest θ gives smallest ϕ)

than any other color, giving the ring its reddish appearance.

(b) Since the rays MP and MO are virtually parallel, $\theta \simeq \phi = 1.5(0.25°) = 0.375°$. Then, by (a),

$$\sin\phi = \sin\theta = 1.22\lambda_{blue}/d,$$

$$\sin 0.375° = 1.22(475 \text{ nm})/d \rightarrow d = 88.5 \text{ } \mu m,$$

the wavelength of blue light taken from Fig.2 in Chapter 38.

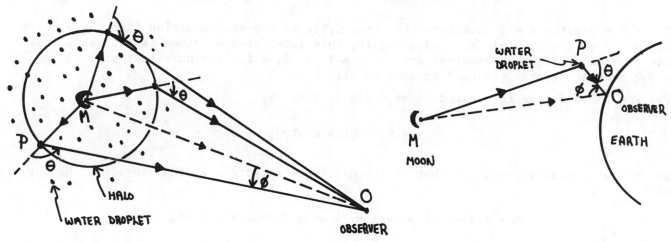

(c) A bluish ring will be seen at the scattering angle for the first minimum of red light. Since $\lambda(red) = 475$ nm $\simeq 1.5\lambda(blue)$, the radius of the blue ring will be 1.5 times that of the red ring, or about 2.3 times the lunar radius. The intensity of the ring will be very low, however. If water droplets of various sizes are present, various sized rings of these colors are present, giving the ring a whitish appearance due to the mixture of the colors.

(d) These halos are a diffraction effect, a rainbow being formed by refraction.

35E

The location of the second minimum is given by $a\sin\theta = 2\lambda$; interference fringes are located at angles $\theta*$ given by $d\sin\theta* = m\lambda$. But $d = 11a/2$ (see Sample Problem 6) so that the fringes are located by $(11a/2)\sin\theta* = m\lambda$. Eliminating the slit width a between the diffraction and interference equations given above yields

$$[\frac{11}{2}(2\lambda/\sin\theta)]\sin\theta* = m\lambda,$$

$$11\sin\theta* = (m)\sin\theta.$$

Hence, the m = 11 interference fringe falls at the same location as the second diffraction minimum ($\theta = \theta*$). The m = 11 fringe is, therefore, not seen. But, there are 11 of the interference fringes inside the central maximum, so that the m = 6 fringe falls just outside this minimum. Therefore, between the first and second minimum will be found the m = 6, 7, 8, 9, 10 interference fringes, for five in all.

40P

(a) The location of the interference fringes is given by

$$d\sin\theta_i = m\lambda, \quad m = 0, 1, 2, \ldots$$

and the diffraction minima by

$$a\sin\theta_d = n\lambda, \quad n = 1, 2, \ldots.$$

Since $d = 5a$, the $m = 5$ interference fringe falls at the same position (i.e, $\theta_d = \theta_i$) as the $n = 1$ diffraction minimum. As a result, this interference fringe is not seen. Inside the central diffraction envelope are the $m = 1, 2, 3, 4$ fringes on either side of the $m = 0$ central fringe, giving 9 fringes in all.

(b) The third fringe is located at an angle θ_i given by

$$\sin\theta_i = 3\lambda/d = 3\lambda/5a,$$

and has an intensity equal to that of the diffraction pattern at that location. Hence, with

$$\alpha = \pi a\sin\theta_d/\lambda = \pi a\sin\theta_i/\lambda = \pi a(3\lambda/5a)/\lambda = 3\pi/5,$$

the desired intensity ratio is

$$I/I_m = \sin^2\alpha/\alpha^2 = \sin^2(3\pi/5)/(3\pi/5)^2 = 0.255.$$

46E

The grating ruling separation d is

$$d = \frac{1 \text{ mm}}{400} = 2.5 \times 10^{-3} \text{ mm} = 2500 \text{ nm}.$$

Since $d\sin\theta = m\lambda$ (Eq.14), the longer wavelengths have the larger angle θ of deflection. Hence, to decide how many spectra fit, use $\lambda = 700$ nm with $\theta = 90°$. The order number m at this limit of visibility is

$$m = \frac{d\sin 90°}{\lambda} = \frac{(2500 \text{ nm})(1)}{700 \text{ nm}} = 3.57.$$

Thus, 3 complete orders can be seen on each side of the central maximum. However, see Problem 56, the second and third orders overlap.

49P

(a) Let the maxima in question be of orders m and m + 1. Then, by Eq.14,

$$d\sin\theta_m = m\lambda, \quad d\sin\theta_{m+1} = (m + 1)\lambda,$$

where $\sin\theta_m = 0.2$ and $\sin\theta_{m+1} = 0.3$. Subtracting these equations gives

$$d(0.3 - 0.2) = \lambda = 600 \text{ nm} \rightarrow d = 6.0 \text{ }\mu\text{m}.$$

(b) Suppose that the m = 4 maximum, which is missing, falls at the nth minimum of the diffraction envelope. The minima fall at angles ϕ given by Eq.3:

$$a\sin\phi = n\lambda, \quad n = 1, 2, \ldots$$

The m = 4 interference maximum falls at an angle θ_4 satisfying

$$d\sin\theta_4 = 4\lambda.$$

If these two features fall at the same place on the screen, then $\phi = \theta_4$, so that

$$d\sin\theta_4 = 4\lambda; \quad a\sin\theta_4 = n\lambda, \quad \rightarrow \quad a = nd/4.$$

For the smallest a choose the smallest n which is n = 1. Then

$$a_{min} = (\tfrac{1}{4})(6000 \text{ nm}) = 1.5 \text{ }\mu\text{m}.$$

(c) If the m = 4 maximum is missing, the m = 8, 12, 16, ... will be missing also. Since d = 10λ, the m = 10 maximum is deflected through 90° and will not fall on the screen. Thus, the visible orders are m = 0, 1, 2, 3, 5, 6, 7, 9.

51P

The needed ruling separation d must satisfy Eq.14 at the two extreme wavelengths:

$$d\sin\theta = (1)(430 \text{ nm}),$$

$$d\sin(\theta + 20°) = (1)(680 \text{ nm}).$$

Dividing the equations to eliminate d, and then using trig identities found in Appendix G gives

$$\frac{\sin\theta}{\sin(\theta + 20°)} = \frac{430}{680} = 0.6324,$$

$$\sin\theta = 0.6324(\sin\theta\cos20° + \cos\theta\sin20°),$$

$$\sin\theta(1 - 0.6324\cos20°) = \cos\theta(0.6324\sin20°),$$

$$\frac{\sin\theta}{\cos\theta} = \tan\theta = \frac{0.2163}{0.4057},$$

$$\theta = \tan^{-1}(\frac{0.2163}{0.4057}) = 28.06°.$$

Hence,

$$d = \frac{430 \text{ nm}}{\sin 28.06°} = 914 \text{ nm} = 9.14 \text{ X } 10^{-4} \text{ mm},$$

so that there are, in the grating, 1/d = 1/(9.14 X 10^{-4} mm) = 1094 rulings/mm.

59P

We follow the derivation leading to Eq.17. The phase difference between waves from adjacent slits to the first minimum beyond the mth principal maximum is

$$\Delta\phi = 2\pi m + \frac{2\pi}{N},$$

where N = number of slits. This indicates a path difference ΔL given by

$$\Delta L = (\frac{\Delta\phi}{2\pi})\lambda = m\lambda + \frac{\lambda}{N}.$$

But, if θ_m is the position angle of the mth principal maximum, this path difference is also given by

$$\Delta L = d\sin(\theta_m + \Delta\theta).$$

Therefore (see Appendix G),

$$d(\sin\theta_m\cos\Delta\theta + \cos\theta_m\sin\Delta\theta) = m\lambda + \frac{\lambda}{N}.$$

If $\Delta\theta \ll 1$, then $\cos\Delta\theta = 1$ and $\sin\Delta\theta = \Delta\theta$; thus, these approximations along with $d\sin\theta_m = m\lambda$ (Eq.14) transform the preceding equation to

$$m\lambda + d\cos\theta_m\Delta\theta = m\lambda + \frac{\lambda}{N},$$

$$\Delta\theta = \frac{\lambda}{Nd\cos\theta_m}.$$

62E

Combine Eqs.21 and 22 to get

$$\frac{\lambda}{\Delta\lambda} = Nm,$$

$$\frac{656.3 \text{ nm}}{0.18 \text{ nm}} = N(1) \rightarrow N = 3646.$$

65E

The dispersion is, by Eq.20,

$$D = \frac{d\theta}{d\lambda} = \frac{m}{d\cos\theta}.$$

But $m\lambda = d\sin\theta$, so that $m = d\sin\theta/\lambda$. Putting this into the equation for D above yields

$$D = \frac{d\sin\theta/\lambda}{d\cos\theta} = \frac{\sin\theta/\cos\theta}{\lambda} = \frac{\tan\theta}{\lambda}.$$

69P

(a) Since $\lambda\nu = c$, $\Delta\lambda/\lambda = \Delta\nu/\nu$, except for sign (to prove this, take the derivative $d(\lambda\nu)/d\lambda = dc/d\lambda = 0$; note that $d\lambda/d\lambda = 1$ and $d\nu/d\lambda \simeq \Delta\nu/\Delta\lambda$). Thus, by Eqs.21 and 22,

$$R = \frac{\lambda}{\Delta\lambda} = Nm = \frac{\nu}{\Delta\nu},$$

$$\Delta\nu = \frac{\nu}{Nm} = \frac{c}{Nm\lambda}.$$

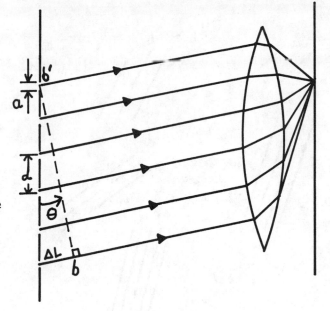

(b) Times of flight beyond the line b'b are equal for the rays shown. The time difference between the extreme rays is

$$\Delta t = \frac{\Delta L}{c} = \frac{(N - 1)d\sin\theta}{c},$$

since the path difference is

$$\Delta L = (N - 1)d\sin\theta.$$

(c) Since $m\lambda = d\sin\theta$, Δt can also be written as

$$\Delta t = \frac{(N - 1)m\lambda}{c},$$

and therefore, combining this result with that from (a),

$$\Delta\nu\Delta t = \frac{c}{Nm\lambda}\frac{(N - 1)m\lambda}{c} = 1,$$

the last result provided that $N \gg 1$ so that $N - 1 = N$ in the equation above.

70E

Use Bragg's law, Eq.26:

$$d = \frac{m\lambda}{2\sin\theta} = \frac{(2)(0.12 \text{ nm})}{(2)\sin 28°} = 0.256 \text{ nm} = 256 \text{ pm}.$$

76P

The x ray wavelength λ and plane separation d constitute two unknowns; hence, data from two orders should suffice (two equations). These observations satisfy Bragg's law:

$$2d\sin\theta_1 = m_1\lambda,$$

$$2d\sin\theta_2 = m_2\lambda.$$

But, upon dividing, these equations yield $\sin\theta_1/\sin\theta_2 = m_1/m_2$, the unknowns having cancelled out. Hence, no additional information is provided by the second observation; only the ratio d/λ can be determined, and this requires only one of the original equations.

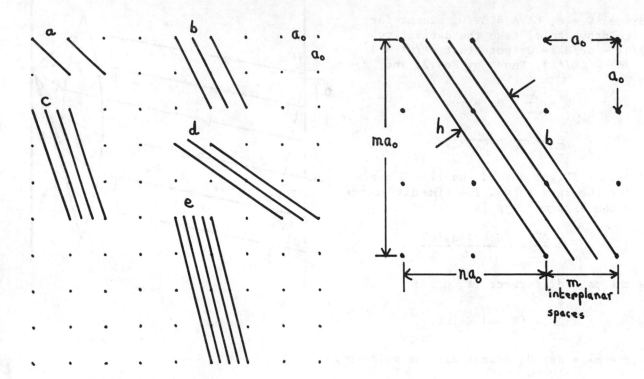

(a) The sets of planes possessing the next five smaller interplanar spacings (smaller than a_0) are shown on the sketch above. In terms of a_0, the spacings of these sets of planes are

$$a: \quad a_0/\sqrt{2} = 0.7071a_0,$$

$$b: \quad a_0/\sqrt{5} = 0.4472a_0,$$

$$c: \quad a_0/\sqrt{10} = 0.3162a_0,$$

$$d: \quad a_0/\sqrt{13} = 0.2774a_0,$$

$$e: \quad a_0/\sqrt{17} = 0.2425a_0.$$

(b) For the general formula, see the smaller sketch above. The area of the shaded parallelogram is

$$A = bh = \sqrt{[(ma_0)^2 + (na_0)^2]}\,h.$$

The area can also be written as the area of the outer square minus the sum of the areas of the two "empty" right triangles:

$$A = (ma_0)(n + 1)a_0 - 2 \cdot \tfrac{1}{2}(ma_0)(na_0) = ma_0^2.$$

Equating the two expressions for the area A gives

$$h = ma_0/\sqrt{(m^2 + n^2)}.$$

But $h = md$, d the interplanar spacing and therefore

$$d = a_0/\sqrt{(m^2 + n^2)}.$$

If the integers m,n contain a common factor, p say, the figure for planes with slope $(m/p)/(n/p)$ is being reproduced, which yields the same value of d; therefore, exclude these.

CHAPTER 42

2E

(a) The needed time is $t = L/v = (0.20 \text{ m})/(0.992)(3 \times 10^8 \text{ m/s}) = 6.72 \times 10^{-10} \text{ s} = 672 \text{ ps}$.

(b) During this time the electrons fall a vertical distance

$$y = gt^2/2 = (9.8 \text{ m/s}^2)(6.72 \times 10^{-10} \text{ s})^2/2 = 2.21 \times 10^{-18} \text{ m}.$$

5E

Apply the time dilation formula, Eq.7. The proper time is $\Delta t_0 = 2.2 \text{ μs}$, since this is observed for electrons at rest. Hence, Eq.7 gives

$$\Delta t = \Delta t_0/\sqrt{(1 - \beta^2)},$$

$$16 \text{ μs} = (2.2 \text{ μs})/\sqrt{(1 - \beta^2)} \rightarrow \beta = v/c = 0.9905,$$

$$v = 0.9905c = 0.9905(3 \times 10^8 \text{ m/s}) = 2.97 \times 10^8 \text{ m/s}.$$

9E

The measured length is given by Eq.10:

$$L = L_0\sqrt{[1 - \beta^2]} = (1.7 \text{ m})\sqrt{[1 - (0.63)^2]} = 1.32 \text{ m}.$$

13E

(a) By Eq.10:

$$L = L_0\sqrt{[1 - \beta^2]} = (130 \text{ m})\sqrt{[1 - (0.74)^2]} = 87.4 \text{ m}.$$

(b) The time is

$$t = L/v = (87.4 \text{ m})/(0.74)(3 \times 10^8 \text{ m/s}) = 394 \text{ ns}.$$

17E

Use the Lorentz transformation, Eq.15. The factor γ is

$$\gamma = 1/\sqrt{[1 - \beta^2]} = 1/\sqrt{[1 - (0.95)^2]} = 3.2026.$$

Therefore,

$$x' = \gamma[x - vt] = (3.2026)[(100 \text{ km}) - (0.95)(3 \times 10^5 \text{ km/s})(2 \times 10^{-4} \text{ s})] = 138 \text{ km},$$

$$t' = \gamma[t - \beta x/c] = (3.2026)[(200 \text{ μs}) - (0.95)(100 \text{ km})/(0.3 \text{ km/μs})] = -374 \text{ μs}.$$

20E

(a) Evaluate the Lorentz factor: $\gamma = 1/\sqrt{[1 - (0.25)^2]} = 1.0328$. Let $\Delta x = x_r - x_b = 30 - 0 = 30$ km. Since the bulbs flash simultaneously in this frame, $\Delta t = t_r - t_b = 0$. The Lorentz transformation for intervals of time, Table 3, gives

$$\Delta t' = \gamma[\Delta t - \beta \Delta x/c],$$

$$\Delta t' = (1.0328)[0 - (0.25)\frac{30 \text{ km}}{3 \times 10^5 \text{ km/s}}] = -0.258 \times 10^{-4} \text{ s} = -25.8 \text{ μs}.$$

(b) $\Delta t' = t_r' - t_b' < 0$; thus the red bulb flashes first in the second frame. Note that due to the Doppler effect, the flash will not appear red.

25E

Use Eq.23. We are told that $v' = 0.40c$, $u = 0.60c$ and we seek v. Eq.23 gives directly

$$v = \frac{v' + u}{1 + uv'/c^2} = \frac{0.40c + 0.60c}{1 + (0.60c)(0.40c)/c^2} = 0.806c.$$

28E

(a) The velocity of our galaxy (Milky Way) relative to A is the negative of the velocity of A relative to us; hence, the relative speeds are the same: 0.35c.

(b) Attach frame S' to galaxy A; thus, $u = 0.35c$, $v = -0.35c$ and we seek v'. By Eq.23,

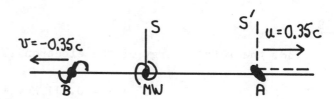

$$v = \frac{v' + u}{1 + uv'/c^2},$$

$$-0.35c = \frac{v' + 0.35c}{1 + (0.35c)v'/c^2} \quad \rightarrow \quad v' = -0.624c.$$

30P

Calculate the speed of the micrometeorite relative to the spaceship. Attach frame S to the "certain reference frame" and S' to the spaceship. Then we have $u = 0.82c$, $v = -0.82c$ (opposite direction). Apply Eq.23 to find v':

$$v = \frac{v' + u}{1 + uv'/c^2},$$

$$-0.82c = \frac{v' + 0.82c}{1 + (0.82c)v'/c^2} \quad \rightarrow \quad v' = -0.9806c.$$

Hence, the time for the micrometeorite to pass the ship is

$$t = \frac{L}{v'} = \frac{350 \text{ m}}{(0.9806)(3 \text{ X } 10^8 \text{ m/s})} = 1.19 \text{ μs.}$$

33E

Use the Doppler shift formula, Eq.26:

$$\nu = \nu_0 \sqrt{(\frac{1 - \beta}{1 + \beta})} = (100 \text{ Mhz})\sqrt{(\frac{1 - 0.90}{1 + 0.90})} = 22.9 \text{ Mhz.}$$

36P

Color is related to wavelength (or frequency). We can rewrite Eq.26 in terms of the wavelengths by substituting $\nu = c/\lambda$ and $\nu_0 = c/\lambda_0$ and solve for the observed wavelength λ:

$$\frac{1}{\lambda} = \frac{1}{\lambda_0}\sqrt{(\frac{1 - \beta}{1 + \beta})},$$

$$\frac{1}{\lambda} = \frac{1}{(450 \text{ nm})}\sqrt{(\frac{1 - 0.20}{1 + 0.20})} \rightarrow \lambda = 551 \text{ nm.}$$

According to Fig.2, Chapter 38, the observed color will be yellow-green.

42E

By Eq.38, the total energy of the electron is

$$E = mc^2 + K = 0.511 \text{ MeV} + 100 \text{ MeV} = 100.511 \text{ MeV,}$$

the rest energy of the electron being $mc^2 = 0.511$ MeV. But Eq.38 also gives

$$E = \gamma(mc^2),$$

$$100.511 \text{ MeV} = \gamma(0.511 \text{ MeV}),$$

$$\gamma = 196.6947 = 1/\sqrt{(1 - \beta^2)} \rightarrow \beta = 0.999987,$$

or $v = (99.9987\%)c$.

45E

If no kinetic energy is involved, Eq.38 reduces to $E = mc^2$ since $\gamma = 1$ if $v = 0$. Differentiating with respect to time, we obtain for the associated rates

$$\frac{dE}{dt} = P = (\frac{dm}{dt})c^2,$$

$$\frac{dm}{dt} = P/c^2 = (10^{41} \text{ W})/(3 \text{ X } 10^8 \text{ m/s})^2 = 1.111 \text{ X } 10^{24} \text{ kg/s,}$$

$$\frac{dm}{dt} = (1.111 \text{ X } 10^{24} \text{ kg/s})(3.156 \text{ X } 10^7 \text{ s/y})/(2 \text{ X } 10^{30} \text{ kg/smu}) = 17.5 \text{ smu/y.}$$

49P

(a) Apply Eq.33:

$$v = p/\gamma m = mc/\gamma m = c/\gamma = c\sqrt{(1 - \beta^2)},$$

$$\beta = \sqrt{(1 - \beta^2)} \rightarrow \beta = 1/\sqrt{2}.$$

(b) The relativistic mass is $\gamma m = (c/v)m = m/\beta = \sqrt{2}m$.

(c) The kinetic energy is $K = E - mc^2 = (\gamma - 1)mc^2$; for our particle $\gamma = 1/\beta = \sqrt{2}$, so that $K = (\sqrt{2} - 1)mc^2$.

52P

The energy equivalent of one tablet is

$$E = mc^2 = (320 \text{ X } 10^{-6} \text{ kg})(3 \text{ X } 10^8 \text{ m/s})^2 = 2.88 \text{ X } 10^{13} \text{ J}.$$

Therefore, the distance d sought is

$$d = \frac{2.88 \text{ X } 10^{13} \text{ J}}{1.3 \text{ X } 10^8 \text{ J/gal}}(30 \text{ mi/gal}) = 6.65 \text{ X } 10^6 \text{ mi}.$$

54P

First, calculate the speed of the pion relative to the earth; by Eq.38,

$$E = \gamma(mc^2),$$

$$\gamma = \frac{1.35 \text{ X } 10^5 \text{ MeV}}{139.6 \text{ MeV}} = 967.05 = 1/\sqrt{(1 - \beta^2)} \rightarrow \beta = 0.999\ 999\ 465\ 3.$$

The decay time in the earth's frame is

$$\Delta t = \gamma(\Delta t_0) = (967.05)(35 \text{ ns}) = 33.85 \text{ μs}.$$

In this time the pion travels a distance, as seen from the earth, of

$$x = v(\Delta t) = \beta c(\Delta t) = (0.999\ 999\ 465\ 3)(0.30 \text{ km/μs})(33.85 \text{ μs}) = 10.2 \text{ km}.$$

Hence the decay altitude is 120 km - 10.2 km = 110 km.

58P

As advised, we use the result of Problem 57; substituting the given data, in SI units, into the equation for the orbit radius given in Problem 57 yields the mass of the particle:

$$r = mv/qB\sqrt{(1 - \beta^2)},$$

$$6.28 = m[(0.71)(3 \text{ X } 10^8)]/[2(1.6 \text{ X } 10^{-19})]\sqrt{[1 - (0.71)^2]},$$

$$m = 6.64 \text{ X } 10^{-27} \text{ kg} \simeq 4u.$$

Since m = 4u and q = 2e, the particle is probably a nucleus of the atom helium, which contains two protons and two neutrons, each having a mass of about 1u.

CHAPTER 43

1E

In terms of wavelength, the energy of a photon is, from Eq.1 and Eq.14 in Chapter 17,

$$E = h\nu = h\left(\frac{c}{\lambda}\right) = \frac{hc}{\lambda}.$$

We wish to adjust the constant hc so that λ in nm gives E in eV. To this end, use the "Best value" of the various parameters from Appendix B to four significant figures. Recall that the eV is defined by 1 eV = e J, where e is the magnitude of the electron charge in coulombs. We get

$$hc = \frac{(6.626 \times 10^{-34}\ J\cdot s)(2.998 \times 10^{8}\ m/s)}{(1.602 \times 10^{-19}\ J/eV)(10^{-9}\ m/nm)} = 1240\ eV\cdot nm,$$

so that E = 1240/λ will yield E in eV if λ is in nm.

9P

(a) Assume that the rating P represents the actual optical energy generation rate. The relation between P, the photon energy E, and the photon emission rate n is

$$P = nE.$$

But E = hc/λ, so that

$$P = nhc/\lambda.$$

Hence, the longer wavelength bulb radiates at a higher rate n; i.e., the 700 nm bulb.

(b) Use the result from Problem 1 that hc = 1240 eV·nm, so that

$$n_7 - n_4 = (\lambda_7 - \lambda_4)\left(\frac{P}{hc}\right),$$

$$\Delta n = (700\ nm - 400\ nm)\frac{(400\ W)/(1.6 \times 10^{-19}\ J/eV)}{1240\ eV\cdot nm} = 6.05 \times 10^{20}\ s^{-1}.$$

13P

(a) In one second the lamp radiates 100 J of energy. The energy E of each photon is

$$E = h\nu = \frac{hc}{\lambda} = \frac{(6.63 \times 10^{-34}\ J\cdot s)(3 \times 10^{8}\ m/s)}{589 \times 10^{-9}\ m} = 3.377 \times 10^{-19}\ J.$$

Hence, the rate at which photons are emitted is (100 J/s)/(3.377 $\times 10^{-19}$ J/photon) = 2.96 $\times 10^{20}$ photons/s.

(b) A flux dn/dt = 1 photon/cm^2·s = 10^4 photons/m^2·s means an energy flux of

$$I = (E)dn/dt = 3.377 \times 10^{-15} \text{ J/m}^2 \cdot \text{s}.$$

The power P = 100 W of the lamp is related to the intensity I by

$$I = P/4\pi r^2,$$

since the lamp radiates in all directions; see Eq.24 in Chapter 18. Thus,

$$r = \sqrt{[P/4\pi I]} = \sqrt{[100 \text{ W}/4\pi(3.377 \times 10^{-15} \text{ J/m}^2 \cdot \text{s})]} = 4.85 \times 10^7 \text{ m} = 48{,}500 \text{ km}.$$

(c) A rectangular column of photons of base area A normal to the path of the photons and length ct contains nAct photons, n = the photon number density. In the time t, the energy of the column = nActE crosses area A, since the photons move with speed c. The intensity I at the area A is then

$$I = \frac{nActE}{tA} = ncE = P/4\pi r^2,$$

$$r = \sqrt{(P/4\pi ncE)}.$$

With P = 100 J/s, $n = 10^6 \text{ m}^{-3}$ and E as found in (a), this formula gives r = 280 m.

(d) From (b) the photon flux dn/dt is, in SI units,

$$\frac{dn}{dt} = I/E = P/4\pi r^2 E = (100)/[4\pi(2)^2(3.377 \times 10^{-19})] = 5.89 \times 10^{18} \text{ /m}^2 \cdot \text{s}.$$

The photon density, from (c), is

$$n = \frac{P}{4\pi r^2 cE} = \frac{dn/dt}{c} = \frac{5.89 \times 10^{18} \text{ m}^{-2} \cdot \text{s}^{-1}}{3 \times 10^8 \text{ m/s}} = 1.96 \times 10^{10} \text{ m}^{-3}.$$

18E

The maximum kinetic energy of an emitted electron is, by Eq.7,

$$K_m = eV_0 = (1 \text{ e})(5 \text{ V}) = 5 \text{ eV}.$$

Since the work function is ϕ = 2.2 eV, Eq.8 yields

$$h\nu = \phi + K_m = 2.2 + 5.0 = 7.2 \text{ eV}.$$

Finally, by Problem 1,

$$\lambda = \frac{1240}{E} = \frac{1240}{7.2} = 172 \text{ nm}.$$

20E

(a) For the photoelectric effect, Eq.8 applies:

$$h\nu = \phi + K_m.$$

The energy $E = h\nu$ of the photon is, by Problem 1, $E = 1240/200 = 6.2$ eV. Hence the maximum kinetic energy possible for a liberated electron is

$$K_m = h\nu - \phi = 6.2 - 4.2 = 2.0 \text{ eV.}$$

(b) The minimum kinetic energy that will be carried by an electron is zero, and this is carried by an electron that has a binding energy to the surface of 6.2 eV.

(c) The potential required to stop the 2.0 eV electron in (a) is $V = K/e = (2.0 \text{ eV})/(1 \text{ e})$ = 2.0 V.

(d) A photon barely able to remove the electron bound with an energy of 4.2 eV has a wavelength, by Problem 1, of $\lambda = 1240/4.2 = 295$ nm.

23P

In the photoelectric equation, replace K_m with eV_0 (see Eq.7), where V_0 is the stopping potential. Set $e = 1$ quantum unit, so that K_m is in eV. Also, put $h\nu = 1240/\lambda$ for the energy of the photon. We obtain for Eq.8,

$$\frac{1240}{\lambda} = \phi + V_0.$$

Insert the data for the given and unknown wavelengths, obtaining an equation for each:

$$\frac{1240}{491} = \phi + (0.71),$$

$$\frac{1240}{\lambda} = \phi + (1.43).$$

Solve the first equation for ϕ, then substitute ϕ into the second equation and solve for λ. We find (a) $\lambda = 382$ nm, and (b) $\phi = 1.82$ eV.

28P

Under the assumption that the photon gives all of its energy to the initially stationary electron which, after the collision, moves off with speed v, the conservation of momentum and of total energy, applied to the photon-electron system, gives

$$\frac{h\nu}{c} + 0 = 0 + \frac{mv}{\sqrt{(1 - v^2/c^2)}},$$

$$h\nu + mc^2 = 0 + \frac{mc^2}{\sqrt{(1 - v^2/c^2)}}.$$

The electron of necessity (conservation of momentum) moves away in the same direction in which the incident photon arrived. In the second equation above, transpose the rest energy of the electron to the other side of the equation and divide by c to get

$$\frac{h\nu}{c} = \frac{mc}{\sqrt{(1 - v^2/c^2)}}[1 - \sqrt{(1 - v^2/c^2)}].$$

This agrees with the first equation if

$$c[1 - \sqrt{(1 - v^2/c^2)}] = v,$$

which requires either v/c = 0 or v/c = 1. The former value, by the first equation, leads to ν = 0 also; there is little point in further discussion of this case since the photon energy is zero. The latter "solution", v/c = 1, is prohibited by the Special Theory (i.e., Fact) of Relativity.

29E

(a) Eq.14 applies; substituting the numerical value of h/mc, the equation becomes

$$\Delta\lambda = (2.43 \text{ pm})(1 - \cos\phi).$$

Therefore, for $\phi = 30°$ we have

$$\Delta\lambda = (2.43 \text{ pm})(1 - \cos 30°) = 0.326 \text{ pm}.$$

Since $\Delta\lambda = \lambda' - \lambda$, we have $\lambda' = \lambda + \Delta\lambda = 2.4 + 0.326 = 2.73$ pm.

(b) For $\phi = 120°$, a similar calculation gives $\Delta\lambda = 3.645$ pm, $\lambda' = 6.05$ pm.

32E

(a) See the "Best values" data in Appendix B. For the electron,

$$\lambda_C = h/mc = (6.626 \text{ X } 10^{-34} \text{ J·s})/(9.109 \text{ X } 10^{-31} \text{ kg})(2.998 \text{ X } 10^8 \text{ m/s}),$$

$$\lambda_C = 2.43 \text{ X } 10^{-12} \text{ m} = 2.43 \text{ pm}.$$

If we substitute the mass of the proton for m (m = 1.673 X 10^{-27} kg), the formula gives 1.32 fm.

(b) We can use the formula in Problem 1; for the electron Compton wavelength we find

$$E = \frac{1240}{\lambda_C} = \frac{1240}{0.00243} = 5.10 \text{ X } 10^5 \text{ eV} = 510 \text{ keV}.$$

Substituting instead λ_C for the proton yields E = 939 MeV.

34E

As instructed, use Eq.15. The fractional increase in wavlength is $\Delta\lambda/\lambda$. In terms of this, Eq.15 can be written

$$f = \frac{\Delta\lambda/\lambda}{1 + \Delta\lambda/\lambda} = 0.75 \rightarrow \frac{\Delta\lambda}{\lambda} = 3 = 300\%.$$

36P

The fractional shift in wavelength equals, in absolute value, the fractional shift in frequency. To see this, start with $\lambda\nu = c$ and differentiate with respect to ν. If the fractional shifts are not too large, we can replace $d\lambda/d\nu$ with $\Delta\lambda/\Delta\nu$ as follows:

$$\nu(\frac{d\lambda}{d\nu}) + \lambda(\frac{d\nu}{d\nu}) = \frac{dc}{d\nu} = 0,$$

$$\nu(\frac{\Delta\lambda}{\Delta\nu}) + \lambda(1) = 0,$$

$$\frac{\Delta\nu}{\nu} = -\frac{\Delta\lambda}{\lambda}.$$

Hence, the fractional shift in wavelength equals the fractional shift in frequency, in absolute value. The incident wavelength is given by the formula of Problem 1: $\lambda = 1240/E$ = 1240/6200 = 0.200 nm = 200 pm. By Eq.14,

$$\Delta\lambda = \lambda_c(1 - \cos\phi),$$

$$(10^{-4})(200 \text{ pm}) = (2.43 \text{ pm})(1 - \cos\phi) \rightarrow \phi = 7.36°.$$

(b) The kinetic energy K imparted to the electron equals the energy lost by the photon; i.e., K = E − E'; hence,

$$K = \frac{hc}{\lambda} - \frac{hc}{\lambda'} = \frac{hc}{\lambda}[\frac{\lambda' - \lambda}{\lambda'}] \simeq \frac{hc}{\lambda}[\frac{\Delta\lambda}{\lambda}] = (6.2 \text{ keV})(10^{-4}) = 0.62 \text{ eV}.$$

42E

The wavelength sought is

$$\lambda_{max} = \frac{2898 \text{ μm·K}}{5800 \text{ K}} = 0.500 \text{ μm} = 500 \text{ nm}.$$

This is in the blue-green region of the visible spectrum (see Fig.2, Chapter 38). However the eye is most sensitive to yellow light at about 550 nm and plenty of light of that wavelength is radiated by the sun.

44E

The temperature, by the Wein displacement law of Problem 42, is

$$T = \frac{2.898 \times 10^6 \text{ nm·K}}{550 \text{ nm}} = 5270 \text{ K}.$$

49P

(a) Following the suggested steps, we have

$$P = \int_0^\infty S(\lambda)d\lambda = (2\pi c^2 h)\int_0^\infty \frac{\lambda^{-5}d\lambda}{e^{hc/\lambda kT} - 1}.$$

If x = hc/λkT, then λ = (hc/kT)/x; also,

$$d\lambda = -(\frac{hc}{kT})(dx/x^2).$$

When λ = 0, x = ∞ and when λ = ∞, x = 0. Reversing the limits on the integral in terms of x introduces a minus sign that annuls the minus sign in dλ above. Hence, we get

$$P = (2\pi c^2 h)\left(\frac{kT}{hc}\right)^4 \int_0^\infty \frac{x^3 dx}{e^x - 1} = (2\pi c^2 h)\left(\frac{k}{hc}\right)^4 \left(\frac{\pi^4}{15}\right)T^4 = \left(\frac{2\pi^5 k^4}{15 h^3 c^2}\right)T^4 = \sigma T^4.$$

(b) Use the "Best value" listing of the fundamental constants given in Appendix B; to 4 significant figures, we get

$$\sigma = \frac{2\pi^5 (1.381 \times 10^{-23} \text{ J/K})^4}{15(6.626 \times 10^{-34} \text{ J}\cdot\text{s})^3 (2.998 \times 10^8 \text{ m/s})^2} = 5.68 \times 10^{-8} \text{ W/m}^2\cdot\text{K}^4.$$

If more accurate values of the constants are substituted, the 5.68 above becomes 5.67.

54P

Through the opening of area A the oven emits into the room energy at the rate $A\sigma T_o^4$ (see Problem 49), and receives energy from the room at the rate $A\sigma T_r^4$. Thus, converting the temperatures to kelvin and using SI units, the net power transferred to the room is

$$P = A\sigma(T_o^4 - T_r^4),$$

$$P = (5 \times 10^{-4})(5.67 \times 10^{-8})(500^4 - 300^4) = 1.54 \text{ W}.$$

57E

Use the Bohr frequency condition, Eq.20. The photon energy is $h\nu = hc/\lambda = 1240/\lambda$, by Problem 1. Therefore, we have

$$\frac{1240}{\lambda} = E_i - E_f,$$

$$\frac{1240}{0.0185} = -13,700 - E_f \rightarrow E_f = -8.07 \times 10^4 \text{ eV} = -80.7 \text{ keV}.$$

60E

(a) Use Eq.22 for E_i, E_f in Eq.20. The photon energy is $E_{ph} = h\nu$, so that

$$E_{ph} = E_i - E_f = \left[-\frac{13.6}{3^2} - \left(-\frac{13.6}{1^2}\right)\right] = 12.1 \text{ eV}.$$

(b) From Eq.5, the photon momentum is $p = E_{ph}/c = 12.1$ eV/c; in SI units, this is

$$p = 12.1 \text{ eV/c} = [(12.1)(1.6 \times 10^{-19} \text{ J})]/(3 \times 10^8 \text{ m/s}) = 6.45 \times 10^{-27} \text{ kg}\cdot\text{m/s}.$$

(c) The wavelength follows from the formula derived in Problem 1: $\lambda = 1240/12.1 = 102$ nm.

62E

In Eq.24, the series limit corresponds to $u = \infty$. For the Balmer series $\ell = 2$; for the Lyman series $\ell = 1$. Hence, the ratio of the wavelengths of the series limits is

$$\lambda_B/\lambda_L = (\ell_B/\ell_L)^2 = (2/1)^2 = 4.$$

64P

In the final, pulled apart, state the energy of the atom is $E = 0$. Hence, the work W that must be done equals, in absolute value, the initial energy E_i of the atom so that, by energy conservation, $E_i + W = E_f = 0$. Since $E_i = -(13.6 \text{ eV})/n^2$ we have if (a) $n = 1$, ground state, initially, $W = 13.6$ eV; (b) if $n = 2$ initially, $W = 3.40$ eV.

70P

If kinetic energy is not conserved, some of the 6.0 eV of the neutron's inital kinetic energy must be used to excite the hydrogen atom. However, to reach the first excited state ($n = 2$) from the ground state ($n = 1$), the atom must be supplied with an amount of energy ΔE with (see Eq.22),

$$\Delta E = 13.6 - 13.6/2^2 = 10.2 \text{ eV}.$$

Since the neutron does not possess this much energy, the collision will be elastic, with the hydrogen atom remaining in the ground state.

78P

(a) By assumption, see Eq.30,

$$L = I\omega = nh/2\pi.$$

The rotational inertia is (see Eq.21 in Chapter 11),

$$I = 2m(d/2)^2 = md^2/2,$$

so that

$$md^2\omega/2 = nh/2\pi,$$

$$\omega = nh/\pi md^2.$$

(b) The kinetic energy of rotation is

$$E = \frac{1}{2}I\omega^2 = \frac{L^2}{2I} = \frac{(nh/2\pi)^2}{md^2} = n^2h^2/4\pi^2md^2.$$

CHAPTER 44

2E

(a) Classically, the momentum p in terms of the kinetic energy K is given from

$$K = \frac{1}{2}mv^2 = (mv)^2/2m = p^2/2m,$$

$$p = \sqrt{[2mK]}.$$

Therefore, by Eq.3, $\lambda = h/\sqrt{[2mK]}$. Using data from Appendix B, we have

$$\lambda = \frac{h/\sqrt{[2m]}}{\sqrt{K}} = \frac{(6.626 \times 10^{-34} \text{ J·s})/\sqrt{[2(9.109 \times 10^{-31} \text{ kg})]}}{\sqrt{[(K \text{ in eV})(1.602 \times 10^{-19} \text{ J/eV})]}} = \frac{1.2265 \times 10^{-9} \text{ m·eV}^{\frac{1}{2}}}{\sqrt{[K \text{ in eV}]}},$$

$$\lambda = \frac{1.226 \text{ nm·eV}^{\frac{1}{2}}}{\sqrt{[K \text{ in eV}]}},$$

so that $\lambda = 1.226/\sqrt{K}$ gives λ in nm if K is in eV.

(b) In terms of an accelerating voltage, we have K = eV. With K in eV, V in volts, we use e = 1 quantum unit for the electron. Hence, (a) becomes

$$\lambda = \frac{1.226}{\sqrt{V}} = \sqrt{(\frac{1.50}{V})}.$$

3E

The equation in Problem 2b is appropriate under the (nonrelativistic) circumstances:

$$\lambda = \sqrt{(\frac{1.50}{V})} = \sqrt{(\frac{1.50}{25,000})} = 7.75 \times 10^{-3} \text{ nm} = 7.75 \text{ pm}.$$

6E

Use the result of Problem 2a to solve for K; we obtain,

$$K = [\frac{1.226 \text{ nm·eV}^{\frac{1}{2}}}{\lambda(\text{in nm})}]^2 = [\frac{1.226 \text{ eV}^{\frac{1}{2}}}{590}]^2 = 4.32 \times 10^{-6} \text{ eV} = 4.32 \text{ μeV}.$$

12P

(a) The kinetic energy acquired is K = qV = $(1.6 \times 10^{-19} \text{ C})(300 \text{ V})$ = 4.80×10^{-17} J. The mass m of a single sodium ion is, from Appendix D,

$$m = \frac{22.9898 \text{ g/mol}}{6.02 \times 10^{23} \text{ atoms/mol}} = 3.819 \times 10^{-23} \text{ g}.$$

Hence, the momentum of the ion is

$$p = \sqrt{[2mK]} = \sqrt{[2(3.819 \times 10^{-26} \text{ kg})(4.8 \times 10^{-17} \text{ J})]} = 1.91 \times 10^{-21} \text{ kg·m/s}.$$

(b) By Eq.3,

$$\lambda = \frac{h}{p} = \frac{6.63 \text{ X } 10^{-34} \text{ J} \cdot \text{s}}{1.91 \text{ X } 10^{-21} \text{ kg} \cdot \text{m/s}} = 3.47 \text{ X } 10^{-13} \text{ m} = 347 \text{ fm}.$$

17P

The same resolution requires the same wavelength; since $\lambda = h/p$, this in turn means the same momentum. Thus, the momentum p of the electron must be the same as the momentum of the photon; that is, we must have

$$P_{photon} = \frac{h}{\lambda} = \frac{h\nu}{\lambda\nu} = \frac{E}{c} = \frac{100 \text{ keV}}{c} = p.$$

Assuming that the electron can be treated classically, and taking its rest mass from Sample Problem 6 in Chapter 42, we have

$$K = \frac{p^2}{2m} = \frac{(0.10 \text{ MeV/c})^2}{2(0.511 \text{ MeV/c}^2)} = 9.785 \text{ keV}.$$

Thus, the accelerating voltage V required is

$$V = K/q = (9.785 \text{ keV})/(1 \text{ e}) = 9.785 \text{ kV}.$$

23E

(a) Use Eq.10 with n = 1 for the smallest energy:

$$E = \frac{h^2}{8mL^2} = \frac{(6.63 \text{ X } 10^{-34} \text{ J} \cdot \text{s})^2}{8(9.11 \text{ X } 10^{-31} \text{ kg})(1.4 \text{ X } 10^{-14} \text{ m})^2(1.6 \text{ X } 10^{-13} \text{ J/MeV})} = 1920 \text{ MeV}.$$

(b) Clearly, a particle with this great amount of kinetic energy cannot be confined to a nucleus when the available binding energy is only a few MeV.

26P

(a) The mass m of an argon atom is

$$m = \frac{39.9 \text{ g/mol}}{6.02 \text{ X } 10^{23} \text{ atoms/mol}} = 6.628 \text{ X } 10^{-23} \text{ g}.$$

From Eq.10, the required energy separation is

$$\Delta E = (2^2 - 1^2)\frac{h^2}{8mL^2} = (3)\frac{(6.63 \text{ X } 10^{-34} \text{ J} \cdot \text{s})^2}{8(6.628 \text{ X } 10^{-26} \text{ kg})(0.20 \text{ m})^2(1.6 \text{ X } 10^{-19} \text{ J/eV})},$$

$$\Delta E = 3.89 \text{ X } 10^{-22} \text{ eV}.$$

(b) From Eqs.14 and 15 in Chapter 21, the thermal energy (i.e., average translational kinetic energy) is

$$\overline{K} = \frac{3}{2}kT = \frac{3}{2}(8.62 \times 10^{-5} \text{ eV/K})(300 \text{ K}) = 0.0388 \text{ eV}.$$

Comparing with (a), we see that this is about 10^{20} times greater than the level separation
(c) Set $\overline{K} = 3.89 \times 10^{-22}$ eV and solve for T:

$$T = 2\overline{K}/3k = 2(3.89 \times 10^{-22} \text{ eV})/3(8.62 \times 10^{-5} \text{ eV/K}) = 3.0 \times 10^{-18} \text{ K}.$$

28P

Substitute the wavefunction into Eq.8:

$$\int_0^L A^2 \sin^2(\pi x/L)dx = A^2\int_0^L \sin^2(\pi x/L)dx = 1.$$

Let $x* = \pi x/L$; then $dx = (L/\pi)dx*$ and we have

$$1 = A^2(L/\pi)\int_0^\pi \sin^2 x* dx*.$$

Now use integral #11 in Appendix G to get

$$1 = A^2(L/\pi)(\pi/2) \quad \rightarrow \quad A = \sqrt{(2/L)}.$$

31E

For $r = r_B$ we have $x = r/r_B = 1$ in the result of Sample Problem 6. Hence, the probability
is

$$p = 1 - e^{-2}(1 + 2 + 2) = 1 - 5e^{-2} = 0.323.$$

36P

The probability is

$$p = \int \psi^2 dV \simeq [\psi^2(r=0)]\Delta V,$$

the integral being performed over the volume of the nucleus. The approximation above is
justified since the nuclear radius R is very small compared with the Bohr radius r_B, this
being the distance over which the wave function changes substantially. By Eq.14,

$$\psi^2(r=0) = 1/\pi r_B^3.$$

Using this the probability becomes

$$p = (1/\pi r_B^3)(4\pi R^3/3) = \frac{4}{3}(R/r_B)^3 = \frac{4}{3}(\frac{1.1 \times 10^{-15} \text{ m}}{5.29 \times 10^{-11} \text{ m}})^3 = 1.2 \times 10^{-14}.$$

This can also be obtained using the result of Sample Problem 6, with $x = 2.08 \times 10^{-5} =$
R/r_B. Use the expansion of e^x given in Appendix G, so that

$$p \simeq 1 - (1 - 2x + 2x^2 - \frac{8}{6}x^3)(1 + 2x + 2x^2) \simeq \frac{8}{6}x^3 = 1.2 \times 10^{-14}.$$

37E

We illustrate the calculation for the proton. First, evaluate k; in SI units,

$$k = \frac{\pi}{h}\sqrt{[8m(U - E)]} = \frac{\pi}{6.63 \times 10^{-34}}\sqrt{[8(1.67 \times 10^{-27})(10 - 3)(1.6 \times 10^{-13})]},$$

$$k = 5.79627 \times 10^{14} \; m^{-1}.$$

(Since k will appear in an exponent, we carry more significant figures than usual, for T is very sensitive to changes in k.) Since $L = 10 \times 10^{-15} = 10^{-14}$ m, we have

$$T = e^{-2kL} = e^{-2(5.79627)} = 9.23 \times 10^{-6}.$$

The deuteron mass is twice the proton mass; hence $T = e^{-2\sqrt{2}(5.79627)} = 7.59 \times 10^{-8}.$

40P

(a) By Eqs. 20 and 21, $T = e^{-2kL}$ where $k = \sqrt{[8\pi^2 m(U - E)/h^2]}$. Therefore,

$$\frac{dT}{dU} = (e^{-2kL})(-2L)(\frac{dk}{dU}) = -2TL(4\pi^2 m/kh^2) = -(\frac{TL}{k})(8\pi^2 m/h^2).$$

From the equation above defining k (Eq.21), it is apparent that

$$8\pi^2 m/h^2 = k^2/(U - E).$$

Thus,

$$\frac{dT}{dU} = -(\frac{TL}{k})(\frac{k^2}{U - E}) = -\frac{TLk}{U - E}.$$

Now,

$$\Delta T \simeq \frac{dT}{dU}(\Delta U),$$

so that the desired fractional change in the transmission coefficient is

$$\frac{\Delta T}{T} = -(Lk)[\frac{\Delta U}{U - E}].$$

Numerically,

$$\frac{\Delta T}{T} = -(5.0)[\frac{(0.01)(6.8)}{6.8 - 5.1}] = -0.20.$$

(b) In this case,

$$\Delta T \simeq \frac{dT}{dL}(\Delta L).$$

But,

$$\frac{dT}{dL} = -2kT,$$

and, from Sample Problem 7, $k = 6.67$ nm^{-1}, $L = 0.750$ nm, so that

$$\frac{\Delta T}{T} = -2k(\Delta L) = -2(6.67)[0.01(0.75)] = -0.10.$$

(c) Since

$$\frac{dT}{dE} = -\frac{dT}{dU},$$

it follows from (a) that for small changes in E,

$$\frac{\Delta T}{T} = \frac{1}{T}\left(\frac{dT}{dE}\right)(\Delta E) = (Lk)\left[\frac{\Delta E}{U - E}\right] = (5.0)\left[\frac{0.01(5.1)}{6.8 - 5.1}\right] = 0.15.$$

46P

By the uncertainty principle, Eq.22, with $\Delta x = h/p = \lambda$, the de Broglie wavelength,

$$\Delta p \Delta x \geq h,$$

$$\Delta p\left(\frac{h}{p}\right) \geq h,$$

$$\Delta p \geq p.$$

But $p = mv$ and therefore $\Delta p = m\Delta v$, so that the above now gives

$$m\Delta v \geq mv,$$

$$\Delta v \geq v.$$

CHAPTER 45

4E

(a) By Eq.5, $\ell = 0, 1, 2$ for $n = 3$. Thus, there are three values of ℓ.

(b) See Eq.7; the m_ℓ values associated with $\ell = 1$ are $m_\ell = 0, \pm1$, or three values.

10P

(a) In the Bohr theory the electron, carrying charge $-e$, revolves in a circular orbit of radius r. If T is the period of its revolution, then the equivalent current i due to its circular motion is $i = e/T$, and the magnetic moment is, therefore,

$$\mu = iA = (\frac{e}{T})(\pi r^2) = \frac{e\pi r^2}{T}.$$

This can be rewritten as

$$\mu = \frac{1}{2}e(2\pi r/T)r = \frac{1}{2}erv = \frac{1}{2}e(\frac{L}{m}),$$

where L is the orbital angular momentum, $L = mrv$. But L is a quantized quantity, with $L = n\hbar$ in Bohr's theory, so that

$$\mu = \frac{1}{2}e(\frac{n\hbar}{m}) = n(\frac{e\hbar}{2m}) = n\mu_B.$$

(b) The assignment by Bohr of angular momentum, embodied in Eq.30 of Chapter 43, is not correct. See the footnote on p.991.

15P

The magnitude L of the orbital angular momentum is given from Eq.4:

$$L^2 = \ell(\ell + 1)\hbar^2.$$

Now $L^2 = L_x^2 + L_y^2 + L_z^2$, and since $L_z = m_\ell\hbar$ it follows that

$$L_x^2 + L_y^2 + m_\ell^2\hbar^2 = \ell(\ell + 1)\hbar^2,$$

$$\sqrt{(L_x^2 + L_y^2)} = \sqrt{[\ell(\ell + 1) - m_\ell^2]}\hbar \leq \sqrt{[\ell(\ell + 1)]}\hbar,$$

since $m_\ell^2 \geq 0$. The greatest possible value of m_ℓ is ℓ itself, so that

$$\sqrt{\ell}\hbar \leq \sqrt{(L_x^2 + L_y^2)}.$$

19E

Use the radial probability density given in Eq.13, and integral #14 in Appendix G. The desired probability is

$$p = \int_{r_1}^{r_2} P(r)dr = \frac{4}{r_B^3}\int_{r_B}^{1.01r_B} r^2 e^{-2r/r_B}dr = \frac{4}{r_B^3}(-r_B^3/8)(\frac{4r^2}{r_B^2} + \frac{4r}{r_B} + 2)e^{-2r/r_B}\Big|_{r_B}^{1.01r_B},$$

$$p = -\frac{1}{2}[\{4(1.01)^2 + 4(1.01) + 2\}e^{-2.02} - (4 + 4 + 2)e^{-2}] = 0.00541.$$

24P

See Sample Problem 5 in Chapter 44 for a similar calculation. In terms of $x = r/r_B$, the probability density can be written

$$8r_B P = x^2(2 - x)^2 e^{-x} = (4x^2 - 4x^3 + x^4)e^{-x}.$$

To find the maxima, set $dP/dx = 0$ to obtain

$$8r_B dP/dx = x(8 - 16x + 8x^2 - x^3)e^{-x} = 0,$$

$$x(2 - x)(x^2 - 6x + 4)e^{-x} = 0.$$

There are five solutions. Three of these yield $P = 0$: they are $x = 0, 2, \infty$. The maxima are located by the other two solutions, which are the roots of

$$x^2 - 6x + 4 = 0.$$

Use the quadratic formula, Appendix G, to find

$$x = r/r_B = 3 \pm \sqrt{5} = 5.236; 0.764.$$

29E

The magnitude of the spin angular momentum is, by Eq.4 with $s = 1/2$ for electron spin,

$$S = \sqrt{[s(s + 1)]}\hbar = \sqrt{[\frac{1}{2}(\frac{1}{2} + 1)]}\hbar = (\sqrt{3}/2)\hbar.$$

The z component of this is given by Eq.11:

$$S_z = m_s\hbar = (\frac{1}{2})\hbar.$$

Hence, the angle desired is (see Fig.7 with S written in place of L),

$$\theta = \cos^{-1}(S_z/S) = \cos^{-1}(1/\sqrt{3}) = 54.7°.$$

The angle $180° - 54.7° = 125°$ is also allowed.

30E

The acceleration is F/M, where F is given by Eq.16; hence, we have, in SI units,

$$a = \frac{F}{M} = (\frac{\mu}{M})(dB/dz)\cos\theta = (\frac{9.27 \times 10^{-24}}{1.80 \times 10^{-25}})(1400)\cos 0° = 72.1 \text{ km/s}^2.$$

33E

The energy U associated with the magnetic moment μ of the electron in a magnetic field B is

$$U = -\vec{\mu}_s \cdot \vec{B} = -\mu_s B\cos\theta = -(\mu_s\cos\theta)B = -\mu_{s,z}B.$$

For parallel alignment $\mu_{s,z} > 0$, and for antiparallel alignment $\mu_{s,z} < 0$ (we mean alignment in the quantum mechanical sense; see Fig.7). Hence, the difference in energy between these two arrangements is, with $m_s = 1/2$ (see Eq.12),

$$E = \Delta U = h\nu = 2\mu_{s,z}B = 2(2m_s\mu_B)B = 2\mu_B B = 2(5.79 \times 10^{-5} \text{ eV/T})(0.2 \text{ T}) = 23.16 \text{ μeV}.$$

The wavelength of the photon follows from Problem 1 in Chapter 43:

$$\lambda = 1240/E = 1240/(2.316 \times 10^{-5}) = 5.35 \times 10^7 \text{ nm} = 5.35 \text{ cm}.$$

39E

(a) The n = 1 shell is filled, for it can take only two electrons. The third electron must enter the n = 2 shell. With n = 2, the allowed values of ℓ are ℓ = 0, 1 of which the ℓ = 0 level is lower in energy. For ℓ = 0 the only possible value of m_ℓ is m_ℓ = 0. The electron spin can be either "up" or "down". Thus, the quantum numbers of the third electron are n, ℓ, m_ℓ, m_s = 2, 0, 0, ±1/2.

(b) In the first excited state the electron [i.e., the one that was the third electron in (a) above] will be in the n = 2, ℓ = 1 level, for the n = 3, ℓ = 0 level lies higher in energy than this. For ℓ = 1, the values of m_ℓ are m_ℓ = -1, 0, +1. However, if there is no external field present (the hyperfine splitting due to the internal field is ignored) these levels are at the same energy. Thus, the quantum numbers of the excited electron are n = 2, ℓ = 1, with m_ℓ and m_s any permitted value.

41P

First, ignore spin. For each value of n there are n possible values of ℓ ranging from 0 to (n − 1); for each ℓ there are (2ℓ + 1) possible values of m_ℓ. Thus, the number N of states at any n is

$$N = \sum_0^{n-1}(2\ell + 1) = 2\sum_0^{n-1}\ell + n = 2(\sum_0^n\ell - n) + n = 2\sum_1^n\ell - n.$$

But,

$$\sum_1^n\ell = \tfrac{1}{2}n(n + 1),$$

since the average value of the terms in the sum is (n + 1)/2 and there are n terms. Hence

$$N = 2[\tfrac{1}{2}n(n + 1)] - n = n^2.$$

With spin included, we have double the number of states (spin "up" and spin "down" for each of those counted above) for a total of $2n^2$.

45E

This is given from Eq.19. Note that $eV = h\nu$, the energy of the photon. Thus,

$$eV = \frac{hc}{\lambda_{min}} = \frac{1240}{\lambda_{min}},$$

by Problem 1 in Chapter 43. That is, λ_{min} in nm gives eV in electron volts, or V in volts. With $\lambda_{min} = 0.1$ nm, we get $eV = 12.4$ keV, so that $V = 12.4$ kV.

47P

The initial kinetic energy of the electron is 50 keV, and its energy after each of the collisions is 25 keV, 12.5 keV, zero. Hence, the energies of the three photons are 50 − 25 = 25 keV, 25 − 12.5 = 12.5 keV, 12.5 − 0 = 12.5 keV. According to Problem 45, these photons have wavelengths $\lambda = 1240/(25,000) = 49.6$ pm, and $\lambda = 1240/(12,500) = 99.2$ pm.

48P

If an isolated moving electron cannot change spontaneously into a photon, it is because the reaction would violate one or both of the conservation laws, those of total energy and of momentum. To see if this is so, assume that the reaction takes place. In the laboratory frame, the electron is moving with speed v; the created photon moves off with speed c. The photon must move off in the same direction as the moving electron, in order that the vector nature of momentum be conserved. Now suppose we attach a frame to the initially moving electron and work in this frame. The momentum of the electron in this frame is zero. But, by the second postulate of relativity, the created photon has a speed c, just as in the laboratory frame. Hence, the created photon has momentum $p = E/c \neq 0$ in the frame of the electron. Thus, momentum is not conserved in this frame and the reaction cannot take place. (Note that total energy can be conserved: the energy E of the photon must equal the rest energy mc^2 of the electron.) Hence, a third body must be present to absorb some of the momentum that would otherwise go to the photon; this body moves off in the direction opposite to the created photon, so that the total momentum after the creation of the photon can sum to zero.

50E

(a) The cut-off wavelength λ_{min} is characteristic of the incident electrons, not of the target material upon which they fall. This wavelength is equal to the wavelength of a photon with an energy equal to the kinetic energy of the incident electrons, in this case 35 keV. Hence, by Problem 45,

$$\lambda_{min} = 1240/35 = 35.4 \text{ pm}.$$

(b) The K_β line results from an M to K electron jump, see Fig.19. The energy of the emitted photon is the difference in the energies of the atom with the electron in the

K and M levels (or hole in the M and K levels). Thus,

$$hc/\lambda_{K\beta} = \Delta E = 25.51 - 0.53 = 24.98 \text{ keV},$$

$$\lambda_{K\beta} = 1240/24.98 = 49.6 \text{ pm}.$$

(c) The K_α line corresponds to an L to K electron transition (K to L hole transition). Therefore,

$$hc/\lambda_{K\alpha} = \Delta E = 25.51 - 3.56 = 21.95 \text{ keV},$$

$$\lambda_{K\alpha} = 1240/21.95 = 56.5 \text{ pm}.$$

53E

From Sample Problem 7, we see that

$$\lambda_{Nb}/\lambda_{Ga} = [(Z_{Ga} - 1)/(Z_{Nb} - 1)]^2.$$

In either Appendix E (Periodic Table) or Appendix D, we find $Z_{Ga} = 31$, $Z_{Nb} = 41$. Thus,

$$\lambda_{Nb}/\lambda_{Ga} = [30/40]^2 = 9/16.$$

55P

(a) For these lines to be produced, an electron must be removed from the K-shell, a hole being created there. This requires 69.5 keV, or an accelerating potential for electrons of 69.5 kV.

(b) By Problem 45,

$$\lambda_{min} = 1240/V = 1240/69.5 = 17.8 \text{ pm}.$$

(c) For the K_α line, E = 69.5 - 11.3 = 58.2 keV = energy of the photon, so that

$$\lambda_{K\alpha} = 1240/58200 = 0.0213 \text{ nm} = 21.3 \text{ pm}.$$

For K_β, E = 69.5 - 2.3 = 67.2 keV and therefore

$$\lambda_{K\beta} = 1240/67200 = 0.0185 \text{ nm} = 18.5 \text{ pm}.$$

60E

Eq.24 applies. The energy difference in the levels considered is $E_{13} - E_{11} = 2(1.2 \text{ eV}) = 2.4$ eV. Also, $kT = (8.62 \times 10^{-5} \text{ eV/K})(2000 \text{ K}) = 0.1724$ eV. Hence,

$$n_{13}/n_{11} = e^{-\Delta E/kT} = e^{-2.4/0.1724} = 9.00 \times 10^{-7}.$$

64E

(a) The length is $L = ct = (3 \times 10^8 \text{ m/s})(1.2 \times 10^{-11} \text{ s}) = 3.60$ mm.

(b) The energy of the pulse is $(0.15 \text{ J})/(1.6 \times 10^{-19} \text{ J/eV}) = 9.375 \times 10^{17}$ eV. The energy of each photon, by Problem 1 in Chapter 43, is $1240/694.4 = 1.786$ eV. Hence, each pulse contains $(9.375 \times 10^{17} \text{ eV})/(1.786 \text{ eV/photon}) = 5.25 \times 10^{17}$ photons.

68P

(a) There must be a node at each end if the mirrors are perfectly reflecting (actually, there will be a node very close to the partially-silvered exit end). Let L be the length of the crystal. From Eq.8 in Chapter 40, we note that the wavelength in the crystal is $\lambda_n = \lambda/n$, so we have, in SI units,

$$N(\lambda_n/2) = L,$$

$$N = \frac{2nL}{\lambda} = \frac{(2)(1.75)(0.06)}{(694 \times 10^{-9})} = 3.03 \times 10^5.$$

(b) In terms of the frequency,

$$N = (\frac{2Ln}{c})\nu,$$

so that

$$\Delta N = (\frac{2Ln}{c})\Delta\nu = (\frac{2Ln}{c/\nu})(\frac{\Delta\nu}{\nu}) = N(\frac{\Delta\nu}{\nu}).$$

The frequency is $\nu = (3 \times 10^8 \text{ m/s})/(694 \times 10^{-9} \text{ m}) = 4.323 \times 10^{14}$ Hz and therefore, for $\Delta N = 1$,

$$\Delta\nu = \frac{\nu}{N}(\Delta N) = \frac{4.323 \times 10^{14}}{3.03 \times 10^5}(1) = 1.43 \times 10^9 \text{ Hz.}$$

The round-trip travel time (since $v = c/n$; see Eq.3 in Chapter 40) is

$$\Delta t = \frac{2L}{v} = \frac{2L}{c/n} = \frac{2Ln}{c}.$$

For $\Delta N = 1$,

$$\Delta\nu = \frac{\nu}{N} = \frac{c/\lambda}{2Ln/\lambda} = \frac{c}{2Ln},$$

and therefore $\Delta\nu = 1/\Delta t$.

(c) From (b), with $\Delta N = 1$,

$$\frac{\Delta\nu}{\nu} = \frac{\Delta N}{N} = \frac{1}{N} = 3.30 \times 10^{-6}.$$

72P

(a) Let r be the distance to the target missile so that, as is suggested, $f = r$ in Problem 67. With P the laser's power output, the intensity I of the beam at the target is

$$I = \frac{P}{A} = \frac{P}{\pi R^2} = \frac{P}{\pi[1.22r\lambda/d]^2} = (0.214)Pd^2/r^2\lambda^2.$$

This assumes that there is no attenuation of the beam by the atmosphere. For this laser,

$P = 5 \times 10^6$ W, $\lambda = 3 \times 10^{-6}$ m, $d = 4$ m. With $r = 3 \times 10^6$ m, these data in the equation above yield $I = 2.1 \times 10^5$ W/m^2, clearly insufficient for a "kill".

(b) To find the maximum wavelength to ensure target destruction, put $I = 10^8$ W/m^2 and solve for λ, leaving all the other parameters unchanged; i.e.,

$$10^8 = (0.214)(5 \times 10^6)\left[\frac{4}{(3 \times 10^6)\lambda}\right]^2 \rightarrow \lambda = 1.38 \times 10^{-7}\text{ m} = 0.138\ \mu\text{m}.$$

CHAPTER 46

2E

Since each atom contributes one charge carrier, the number density n of charge carriers equals the number density of gold atoms. The mass M of one gold atom is

$$M = \frac{197 \text{ g/mol}}{6.02 \times 10^{23} \text{ atoms/mol}} = 3.272 \times 10^{-22} \text{ g/atom.}$$

The density of charge carriers is therefore

$$n = \frac{\rho}{M} = \frac{19.3 \text{ g/cm}^3}{3.272 \times 10^{-22} \text{ g/atom}} = 5.90 \times 10^{22} \text{ atoms/cm}^3 = 5.90 \times 10^{28} \text{ atoms/m}^3.$$

As noted above, with 1 carrier/atom, this can be written as 5.90×10^{28} carriers/m^3.

6E

Eq.3 for the density of states in energy is

$$n(E) = \left(\frac{2^{7/2}\pi m^{3/2}}{h^3}\right)E^{\frac{1}{2}} = CE^{\frac{1}{2}},$$

with

$$C = \frac{2^{7/2}\pi m^{3/2}}{h^3} = 2^{7/2}\pi \frac{(9.11 \times 10^{-31} \text{ kg})^{3/2}}{(6.626 \times 10^{-34} \text{ J·s})^3} = 10.6238 \times 10^{55} \text{ kg}^{3/2}/\text{J}^3 \cdot \text{s}^3.$$

Examining the units,

$$\text{kg}^{3/2} \cdot \text{J}^{-3} \cdot \text{s}^{-3} = \frac{(\text{kg/J})^{3/2}}{\text{J}^{3/2} \cdot \text{s}^3} = \frac{(\text{s}^2/\text{m}^2)^{3/2}}{\text{J}^{3/2} \cdot \text{s}^3} = \text{m}^{-3} \cdot \text{J}^{-3/2} = \text{m}^{-3}(1.6 \times 10^{-19})^{3/2} \text{ eV}^{-3/2},$$

so that

$$C = 6.80 \times 10^{27} \text{ eV}^{-3/2} \cdot \text{m}^{-3}.$$

If E = 5 eV,

$$n(E) = (6.80 \times 10^{27})(5)^{\frac{1}{2}} = 1.52 \times 10^{28} \text{ eV}^{-1} \cdot \text{m}^{-3}.$$

8E

(a) See Fig.6(b) for which T = 0 K and E_F = 7 eV. The probability of occupancy of any state with energy greater than the Fermi energy is zero; p = 0.

(b) Use Eq.7. First evaluate kT: kT = $(8.62 \times 10^{-5}$ eV/K)(320 K) = 0.02758 eV. Thus,

$$p = [e^{(E - E_F)/kT} + 1]^{-1} = [e^{0.062/0.02758} + 1]^{-1} = 0.0955.$$

15E

By Problem 10, the number density n of free electrons is found by

$$E_F = An^{2/3},$$

$$11.6 = (3.65 \times 10^{-19})n^{2/3} \rightarrow n = 1.792 \times 10^{29} \text{ m}^{-3}.$$

The mass of an aluminum atom is $M = (27.0 \text{ g})/(6.02 \times 10^{23}) = 4.485 \times 10^{-26}$ kg. Therefore, the number density ρ_n of aluminum atoms is

$$\rho_n = \frac{\rho}{M} = \frac{2700 \text{ kg/m}^3}{4.485 \times 10^{-26} \text{ kg}} = 6.020 \times 10^{28} \text{ m}^{-3}.$$

Evidently, then, the number # of conduction electrons per atom is

$$\# = \frac{n}{\rho_n} = \frac{1.792 \times 10^{29} \text{ electrons/m}^3}{6.020 \times 10^{28} \text{ atoms/m}^3} = 3 \text{ electrons/atom.}$$

20P

By Problem 10, the Fermi energy in eV is

$$E_F = (3.65 \times 10^{-19})n^{2/3},$$

where n is the number of free electrons per m^3. The number N of iron atoms in the star is (looking up the atomic mass of iron in Appendix D),

$$N = \frac{1.99 \times 10^{30} \text{ kg}}{(55.847 \times 10^{-3} \text{ kg})/(6.02 \times 10^{23})} = 2.145 \times 10^{55}.$$

The volume of the star being

$$V = (\frac{4\pi}{3})R^3 = (\frac{4\pi}{3})(6.37 \times 10^6)^3 = 1.083 \times 10^{21} \text{ m}^3,$$

and each iron atom contributing $Z = 26$ electrons (see Appendix D for Z), the electron number density n must be

$$n = \frac{ZN}{V} = \frac{(26)(2.145 \times 10^{55})}{1.083 \times 10^{21} \text{ m}^3} = 5.150 \times 10^{35} \text{ m}^{-3}.$$

Finally, from the first equation above,

$$E_F = (3.65 \times 10^{-19})(5.150 \times 10^{35})^{2/3} = 2.35 \times 10^5 \text{ eV} = 0.235 \text{ MeV.}$$

25P

The fraction f of excited electrons is given by the last formula in Problem 23. Solving that equation for T gives

$$T = (\tfrac{2}{3})(fE_F)/k = (\tfrac{2}{3})(0.013)(4.7 \text{ eV})/(8.62 \times 10^{-5} \text{ eV/K}) = 473 \text{ K} = 200°C.$$

27P

At T = 0 K,

$$p(E) = \begin{array}{ll} 1, & 0 \le E \le E_F, \\ 0, & E_F < E. \end{array}$$

Since $n_o(E) = n(E)p(E)$, the total number of particles per unit volume is (see Problem 6),

$$n = \int_0^\infty n(E)p(E)dE = \int_0^{E_F} CE^{\tfrac{1}{2}}dE = \tfrac{2}{3}CE_F^{3/2},$$

$$C = \tfrac{3}{2}nE_F^{-3/2}.$$

The average energy of the conduction electrons is

$$\overline{E} = \tfrac{1}{n}\int_0^\infty En(E)p(E)dE = \tfrac{1}{n}\int_0^{E_F}(E)(CE^{\tfrac{1}{2}})dE = (\tfrac{C}{n})\tfrac{2}{5}E_F^{5/2}.$$

Substituting the expression for C found above gives

$$\overline{E} = (\tfrac{3}{2}E_F^{-3/2})(\tfrac{2}{5}E_F^{5/2}) = \tfrac{3}{5}E_F.$$

29P

(a) According to Sample Problem 1, there is one conduction electron per atom of copper. The atomic mass of copper is 63.54 g/mol (see Appendix D). Therefore, the number of atoms of copper in 3.1 g of copper = the number of conduction electrons is

$$n = (\frac{3.1 \text{ g}}{63.54 \text{ g/mol}})(6.02 \times 10^{23} \text{ mol}^{-1}) = 2.937 \times 10^{22}.$$

The average energy of these electrons, by Problem 27, in $3E_F/5$. Hence, their total energy is

$$E = (2.937 \times 10^{22})[3(7.0 \text{ eV})/5](1.6 \times 10^{-19} \text{ J/eV}) = 1.974 \times 10^4 \text{ J}.$$

(b) If this energy is used up at the rate of 100 J/s, it will last for

$$t = E/P = (1.974 \times 10^4 \text{ J})/(100 \text{ J/s}) = 197 \text{ s} = 3 \text{ min } 17 \text{ s}.$$

33P

(a) The occupancy probability is (see Eq.7),

$$p = [e^{(E - E_F)/kT} + 1]^{-1}.$$

At the bottom of the conduction band E = 1.11 eV, measuring energy from the top of the valence band. And at T = 300 K,

$$kT = (8.62 \times 10^{-5})(300) = 0.02586 \text{ eV}.$$

In pure silicon $E_F = 0.555$ eV, at the middle of the gap. Thus, the occupancy probability at the bottom of the conduction band is

$$p = [e^{(1.11 - 0.555)/(0.02586)} + 1]^{-1} = 4.78 \times 10^{-10}.$$

For the doped silicon $E_F = 1.11 - 0.11 = 1.0$ eV. The occupancy probability now is

$$p = [e^{(1.11 - 1.0)/(0.02586)} + 1]^{-1} = 0.0140.$$

(b) The donor level is at E = 1.11 - 0.15 = 0.96 eV; $E_F = 1.0$ eV in the doped material, from (a). The occupancy probability of the donor state is, therefore,

$$p = [e^{(0.96 - 1.0)/(0.02586)} + 1]^{-1} = 0.824.$$

36E

The number N of electrons that can be excited to jump the gap into the conduction band is

$$N = \frac{662 \times 10^3 \text{ eV}}{1.1 \text{ eV}} = 6.02 \times 10^5.$$

Each departing electron creates one hole, so the number of electron-hole pairs created is also $N = 6.02 \times 10^5$.

39E

The gap width equals the energy of a photon of wavelength 295 nm. This energy is, by Problem 1 in Chapter 43, $E = E_{gap} = 1240/295 = 4.20$ eV. Photons of wavelength shorter than this carry more energy than 4.2 eV, and therefore can be absorbed by electrons in the filled band, "lifting" the electrons into the empty band. The photon disappears, and therefore the material appears opaque in these wavelengths.

CHAPTER 47

<u>13E</u>

The binding energy BE is found from

$$BE = [Zm_H + (A - Z)m_n - M(^{239}_{94}Pu)]c^2,$$

since the number of neutrons is N = A − Z. For this isotope of plutonium,

$$BE = [(94)(1.00783) + (239 - 94)(1.00867) - 239.05216](uc^2),$$

$$BE = [1.94101](uc^2) = [1.94101](931.5 \text{ MeV}) = 1808 \text{ MeV}.$$

The number of nucleons is A. Therefore, the average binding energy per nucleon is

$$\frac{BE}{A} = \frac{1808}{239} = 7.56 \text{ MeV},$$

as listed in Table 1.

<u>16E</u>

(a) By Eq.41 in Chapter 42,

$$(pc)^2 = K^2 + 2Kmc^2.$$

For K >> mc^2 = 0.511 MeV for the electron, the last term in the equation above can be ignored, to get

$$p = \frac{K}{c} = \frac{200 \text{ MeV}}{c}.$$

The de Broglie wavelength is

$$\lambda = \frac{h}{p} = \frac{4.14 \times 10^{-21} \text{ MeV} \cdot \text{s}}{200 \text{ MeV}/(3 \times 10^8 \text{ m/s})} = 6.21 \times 10^{-15} \text{ m} = 6.21 \text{ fm}.$$

(b) By comparison, the diameter of a copper nucleus is about 8.6 fm, only a little larger than the de Broglie wavelength of the electron in (a). For sensitive probing, the ratio wavelength/diameter should be as small as possible. Thus, 200 MeV seems to be about the minimum energy for electrons to be useful tools in discriminating the details of nuclear structure in medium-sized nuclei.

<u>19P</u>

(a) The nuclear force is of short range, meaning that any nucleon interacts only with its very nearest neighbors, not with all the other nucleons in the nucleus. Therefore, the number B of bonds that can be formed by a nucleon is independent of A. Hence, the total number of nuclear bonds present in a nucleus is approximately BA; the energy associated with these bonds is, therefore, proportional to A.

(b) Each proton interacts electrically with every other proton in the nucleus, for the Coulomb force is of infinite range. Thus the number of Coulomb bonds in a nucleus is $Z(Z - 1)/2$, the number of distinct pairs of Z protons. Hence, the Coulomb energy is proportional to $Z(Z - 1)$.

(c) For heavier nuclei, A increases faster than Z, but $Z(Z - 1)$ increases faster than A.

20P

Let f be the abundance of ^{25}Mg; then the abundance of ^{26}Mg must be $1 - 0.7899 - f = 0.2101 - f$. Hence, forming the average mass (weight) according to abundance,

$$24.312 = (0.7899)(23.98504) + f(24.98584) + (0.2101 - f)(25.98259),$$

$$f = 0.0930 = 9.30\%.$$

Therefore, the abundance of ^{26}Mg is $0.2101 - 0.0930 = 0.117 = 11.7\%$.

25P

A nucleus contains Z protons and N neutrons. The binding energy of the nucleus is

$$E = (Zm_H + Nm_n - m)c^2,$$

where m_H, m_n, m are the atomic masses of the proton (i.e., the atomic mass of hydrogen), the neutron (i.e., the mass of a neutron), and of the nucleus. In terms of mass excesses,

$$\Delta_H = (m_H - 1)c^2, \quad m_H c^2 = \Delta_H + c^2;$$

$$\Delta_n = (m_n - 1)c^2, \quad m_n c^2 = \Delta_n + c^2;$$

$$\Delta = (m - A)c^2, \quad mc^2 = \Delta + Ac^2.$$

Hence,

$$E = (Z\Delta_H + N\Delta_n - \Delta) + (Z + N - A)c^2 = Z\Delta_H + N\Delta_n - \Delta,$$

since $A = Z + N$. For $^{197}_{79}$Au, $Z = 79$, $N = 197 - 79 = 118$. Thus,

$$E = (79)(7.29) + (118)(8.07) - (-31.2) = 1559.37 \text{ MeV}.$$

The average binding energy per nucleon for this gold nucleus is

$$\overline{E}_B = \frac{E}{A} = \frac{1559.37 \text{ MeV}}{197} = 7.92 \text{ MeV}.$$

This compares with 7.91 MeV listed in Table 1. To obtain the binding energy per nucleon to three significant figures, we need the mass excesses to four significant figures.

29E

(a) The initial activity is $R_0 = \lambda N_0 = (\ln2/\tau)N_0$. Calculate N_0 first; the mass of an atom is, approximately, given by Au, where u is found in Eq.4. Thus, we have

$$N_0 = \frac{\text{bulk mass}}{\text{mass of 1 atom}} = \frac{3.4 \text{ g}}{67(1.661 \times 10^{-24} \text{ g})} = 3.055 \times 10^{22}.$$

Hence, the initial activity is

$$R_0 = [\frac{\ln2}{\tau}]N_0 = [\frac{\ln2}{(78 \text{ h})(3600 \text{ s/h})}](3.055 \times 10^{22}) = 7.54 \times 10^{16} \text{ s}^{-1}.$$

(b) The activity decreases with time as given by Eq.7. Therefore, the activity at t = 48 h later is

$$R = R_0 e^{-(\ln2/\tau)t} = (7.54 \times 10^{16} \text{ s}^{-1})e^{-(\ln2/78)(48)} = 4.92 \times 10^{16} \text{ s}^{-1}.$$

30E

(a) The half-life is given by Eq.8:

$$\tau = \ln2/\lambda = (\ln2)/(0.0108 \text{ h}^{-1}) = 64.2 \text{ h}.$$

(b) The fraction remaining is N/N_0; by Eq.6 with t = 3τ,

$$N/N_0 = e^{-\lambda t} = e^{-(\ln2/\tau)(3\tau)} = e^{-3\ln2} = 0.125.$$

(c) After t = 10 days = 240 h, the fraction remaining is

$$N/N_0 = e^{-\lambda t} = e^{-(0.0108)(240)} = 0.0749.$$

33E

The activity is $R = \lambda N = (\ln2/\tau)N$, so we have

$$N = \tau R/\ln2 = [(5.27 \text{ y})(3.156 \times 10^7 \text{ s/y})][(6000 \text{ Ci})(3.7 \times 10^{10} \text{ s}^{-1}/\text{Ci})]/\ln2,$$

$$N = 5.33 \times 10^{22} \text{ nuclei}.$$

34P

(a) One decigram = 10^{-1} g (see the inside front cover). Assuming that the chlorine is represented by naturally occuring chlorine as far as isotope abundance is concerned (i.e. bulk A = 35.453; see Appendix D), the mass of ^{226}Ra present is

$$m = [\frac{226}{226 + 2(35.453)}](0.1 \text{ g}) = 76.1 \text{ mg}.$$

Hence, the number of atoms of the radium isotope present is

$$N = \frac{76.1 \times 10^{-3} \text{ g}}{226(1.661 \times 10^{-24} \text{ g})} = 2.03 \times 10^{20}.$$

(b) The activity (decay rate) of this sample is

$$R = \lambda N = [\ln 2/\tau]N = [\ln 2/(1600 \text{ y})(3.156 \times 10^7 \text{ s/y})](2.03 \times 10^{20}) = 2.787 \times 10^9 \text{ s}^{-1},$$

$$R = (2.787 \times 10^9 \text{ s}^{-1})/(3.7 \times 10^{10} \text{ s}^{-1}/\text{Ci}) = 75.3 \text{ mCi}.$$

38P

The number N of ^{147}Sm atoms in 1 g of bulk samarium (assuming abundance by mass) is

$$N = [\frac{(0.15)(1 \text{ g})}{147.00 \text{ g/mol}}](6.02 \times 10^{23} \text{ mol}^{-1}) = 6.143 \times 10^{20}.$$

Therefore, we have

$$R = \lambda N = (\ln 2/\tau)N,$$

$$120 \text{ s}^{-1} = (\ln 2/\tau)(6.143 \times 10^{20}) \rightarrow \tau = 3.548 \times 10^{18} \text{ s} = 1.12 \times 10^{11} \text{ y}.$$

43P

Let N be the number of atoms of the radionuclide present at time t. Then,

$$\frac{dN}{dt} = +R - \lambda N,$$

the first term on the right is due to production of the nuclide and the second term is due to its radioactive decay. To integrate this equation to get N, write

$$\frac{dN}{R - \lambda N} = dt,$$

$$-(\frac{1}{\lambda})\ln(R - \lambda N) + C = t,$$

where C is the constant of integration. Suppose that N = 0 at t = 0; substituting these into the equation above gives C = $(1/\lambda)\ln R$. Putting this back into the equation and then solving for N yields

$$-(\frac{1}{\lambda})\ln(R - \lambda N) + (\frac{1}{\lambda})\ln R = t,$$

$$\ln(\frac{R - \lambda N}{R}) = -\lambda t,$$

$$N = \frac{R}{\lambda}(1 - e^{-\lambda t}).$$

After a time t >> $1/\lambda$, we have $e^{-\lambda t}$ << 1 and the equation above gives N = R/λ. If, on the other hand, there are N_0 radionuclides present at t = 0, then the number present a time t later is

$$N = (\frac{R}{\lambda})(1 - e^{-\lambda t}) + N_0 e^{-\lambda t},$$

since those nuclei originally present simply decay at the rate given by Eq.6. Again, if $t \gg 1/\lambda$, we get $N = R/\lambda$, independent of N_0.

44P

(a) Refer to Problem 43. Note that in this and the preceding problem, R does not stand for activity, but for production rate. N = number of ^{56}Mn atoms present a time t after the bombardment starts. Since $\tau = 2.58$ h = 9288 s, $\lambda = \ln2/\tau = 7.4628 \times 10^{-5}$ s^{-1}. The activity is λN so that, with the conditions for secular equilibrium established,

$$R = \lambda N = (2.4)(3.7 \times 10^{10}) = 8.88 \times 10^{10} \text{ s}^{-1}.$$

(b) During the bombardment the decay rate is λN; from Problem 43 with $N_0 = 0$,

$$\text{decay rate} = R(1 - e^{-\lambda t}) = (8.88 \times 10^{10})[1 - e^{-(7.4628 \times 10^{-5})t}],$$

with t in seconds and the decay rate in decays/s.

(c) At the end of the bombardment,

$$\lambda N = 2.4 \text{ Ci},$$

$$(7.4628 \times 10^{-5})N = (2.4)(3.7 \times 10^{10}) \rightarrow N = 1.190 \times 10^{15}.$$

(d) The mass of these ^{56}Mn atoms is

$$m = \frac{1.190 \times 10^{15}}{6.02 \times 10^{23} \text{ mol}^{-1}}(56 \text{ g/mol}) = 0.111 \text{ µg}.$$

47E

It takes $t = \tau = 4.5 \times 10^9$ y for a sample of ^{238}U to decay by one-half the original amount. For each isotope $\lambda = \ln2/\tau$. The fraction of ^{244}Pu left after this time is

$$N/N_0 = e^{-(\ln2/\tau)t} = e^{-(\ln2/8.2 \times 10^7)(4.5 \times 10^9)} = 3.02 \times 10^{-17};$$

and of ^{248}Cm,

$$N/N_0 = e^{-(\ln2/\tau)t} = e^{-(\ln2/3.4 \times 10^5)(4.5 \times 10^9)} = 0.$$

48P

Since the uranium nucleus is at rest before the decay, we have

$$Q = K_{Th} + K_\alpha.$$

The kinetic energy of the α particle is much less than the rest energies involved, so that classical expressions can be used. By momentum conservation, $p_{Th} = p_\alpha$, so that

$$K_{Th} = \frac{P_{Th}^2}{2M_{Th}} = \frac{P_\alpha^2}{2M_{Th}} = \frac{P_\alpha^2}{2M_\alpha}(\frac{M_\alpha}{M_{Th}}) = K_\alpha(\frac{M_\alpha}{M_{Th}}).$$

Therefore,

$$Q = K_\alpha[\frac{M_\alpha}{M_{Th}} + 1] = (4.196 \text{ MeV})[\frac{4}{234} + 1] = 4.268 \text{ MeV}.$$

52E

Writing X for the unknown nuclide, the reaction equation is

$$^A_Z X + ^1_0 n \rightarrow ^0_{-1} e + 2^4_2 He.$$

To find A and Z, apply the law of conservation of charge, and the law of conservation of mass number:

$$Z + 0 = -1 + 2(2) \rightarrow Z = 3;$$

$$A + 1 = 0 + 2(4) \rightarrow A = 7.$$

Hence, the nuclide must be $^7_3 Li$ (consulting the Periodic Table to learn the identity of the atom with Z = 3).

53E

See Sample Problem 7. The decay reaction is

$$^{137}_{55}Cs \rightarrow ^{137}_{56}Ba + ^0_{-1}e + Q.$$

In Sample Problem 7, it is shown that Q is given by the difference in atomic masses of the nuclei, times c^2; the mass of the emitted electron is taken into account thereby. Hence, we have

$$Q = [M(^{137}_{55}Cs) - M(^{137}_{56}Ba)]c^2,$$

$$Q = [136.9073 - 136.9058](uc^2) = [0.0015](932 \text{ MeV}) = 1.40 \text{ MeV}.$$

55E

The free neutron decay scheme is

$$n \rightarrow p + e^- + \nu.$$

For the maximum β particle energy, assume that no neutrino is emitted. Then, ignoring the 13.6 eV binding energy of the electron in a hydrogen atom,

$$Q_{max} = (m_n - m_p - m_e)c^2 = m_n c^2 - (m_p + m_e)c^2 = (m_n - m_H)c^2,$$

$$Q_{max} = (840 \times 10^{-6})(uc^2) = (840 \times 10^{-6})(932 \text{ MeV}) = 0.783 \text{ MeV}.$$

61P

Since the electron is emitted with the maximum kinetic energy K = 1.71 MeV, assume that no neutrino is emitted. Conservation of momentum requires that, with the ^{32}P nucleus originally at rest, the ^{32}S nucleus and the electron move away along the same straight line in opposite directions, and with momenta of equal magnitude. For the electron, by Eq. 41 of Chapter 42,

$$p_e = \frac{1}{c}\sqrt{[K^2 + 2K(mc^2)]} = \frac{1}{c}\sqrt{[(1.71)^2 + 2(1.71)(0.511)]} = 2.1614 \text{ MeV/c}.$$

Using classical (Newtonian) mechanics for the ^{32}S nucleus, its mass being M = Au, the momentum of the nucleus in terms of its kinetic energy K_n is

$$p_n = \sqrt{(2MK_n)} = \sqrt{(2AuK_n)} = \frac{1}{c}\sqrt{(2Auc^2K_n)}.$$

Hence,

$$K_n = \frac{c^2p_n^2}{2A(uc^2)} = \frac{(cp_e)^2}{2A(uc^2)} = \frac{(2.1614)^2}{(2)(32)(931.5)} = 7.84 \times 10^{-5} \text{ MeV} = 78.4 \text{ eV},$$

using uc^2 = 931.5 MeV as the energy equivalent of the atomic mass unit u.

62E

(a) The number N_U of uranium atoms in the rock is

$$N_U = \frac{0.0042 \text{ g}}{238 \text{ g/mol}}(6.02 \times 10^{23} \text{ mol}^{-1}) = 1.062 \times 10^{19},$$

and the number N_{Pb} of lead atoms is

$$N_{Pb} = \frac{0.002135 \text{ g}}{206 \text{ g/mol}}(6.02 \times 10^{23} \text{ mol}^{-1}) = 6.24 \times 10^{18}.$$

(b) Assuming that no lead was lost, the number N_{U0} of uranium atoms in the rock when it solidified is

$$N_{U0} = N_U + N_{Pb} = 1.062 \times 10^{19} + 6.24 \times 10^{18} = 1.686 \times 10^{19}.$$

(c) By the law of radioactive decay,

$$N_U = N_{U0}e^{-\lambda t} = N_{U0}e^{-(\ln 2/\tau)t},$$

$$1.062 \times 10^{19} = (1.686 \times 10^{19})e^{-(\ln 2/4.47)t},$$

using 10^9 y as our unit of time (1 aeon = 1 AE = 10^9 y). Solving for t gives

$$0.6299 = e^{-(\ln 2/4.47)t},$$

$$\ln(0.6299) = -(\ln2/4.47)t,$$

$$t = -\frac{(4.47)\ln(0.6299)}{\ln2} = 2.98 \text{ AE} = 2.98 \times 10^9 \text{ y.}$$

67E

The number N of atoms is given by

$$N = R/\lambda = R\tau/\ln2 = [(250 \text{ Ci})(3.7 \times 10^{10} \text{ s}^{-1}/\text{Ci})][(2.7 \text{ d})(86{,}400 \text{ s/d})]/\ln2,$$

$$N = 3.113 \times 10^{18}.$$

The total mass of these atoms is

$$m = (\frac{3.113 \times 10^{18}}{6.02 \times 10^{23} \text{ mol}^{-1}})(198 \text{ g/mol}) = 1.02 \text{ mg.}$$

70P

The absorbed dose is given from

$$(\text{dose equivalent}) = (\text{absorbed dose})(\text{RBE}),$$

$$25 \text{ mrem} = (\text{absorbed dose})(0.85) \rightarrow (\text{absorbed dose}) = 29.41 \text{ mrad.}$$

But 1 rad = 10 mJ/kg, so that

$$(\text{absorbed dose}) = (29.41 \times 10^{-3} \text{ rad})(10 \times 10^{-3} \text{ J/kg·rad}) = 2.941 \times 10^{-4} \text{ J/kg.}$$

Hence, the absorbed energy in joules is

$$E = (2.941 \times 10^{-4} \text{ J/kg})(44 \text{ kg}) = 1.29 \times 10^{-2} \text{ J} = 12.9 \text{ mJ.}$$

75P

(a) Consider the reaction (with X initially at rest),

$$x + X \rightarrow Y.$$

By the conservation of total energy,

$$m_x c^2 + K_x + m_X c^2 = m_Y^* c^2 + K_Y = m_Y c^2 + E_Y + K_Y;$$

the masses are rest masses and E_Y = excitation energy. By the conservation of momentum,

$$p_x = p_Y,$$

$$\sqrt{[2m_x K_x]} = \sqrt{[2m_Y K_Y]},$$

$$m_x K_x = m_Y K_Y.$$

Therefore,

$$m_X c^2 + m_x c^2 + K_x = m_Y c^2 + E_Y + (m_x/m_Y)K_x,$$

$$K_x(1 - m_x/m_Y) = m_Y c^2 - m_x c^2 - m_X c^2 + E_Y.$$

Now put $x = \alpha$, $X = {}^{16}_{8}O$ to get

$$K_\alpha = \frac{(19.99244 - 4.00260 - 15.99491)(931.5) + 25}{(1 - \frac{4.00260}{19.99244})} = 25.35 \text{ MeV.}$$

(b) Use the equation derived in (a) but set $x = p$, $X = {}^{19}_{9}F$:

$$K_p = \frac{(19.99244 - 1.00783 - 18.9984)(931.5) + 25}{(1 - \frac{1.00783}{19.99244})} = 12.80 \text{ MeV.}$$

(c) Here $x = \gamma$ (photon), $X = Y = {}^{20}_{10}Ne$. The equation derived in (a) cannot be used here since the photon has no rest mass, and therefore the conservation requirements must be reexamined. Momentum conservation requires that

$$E_\gamma/c = p_Y = \sqrt{[2m_Y K_Y]},$$

$$K_Y = E_\gamma^2/2m_Y c^2.$$

Apply conservation of total energy with $m_x = 0$, $K_x = E_\gamma$, $m_X = m_Y$ to find that

$$E_\gamma = K_Y + E_Y = (E_\gamma^2/2m_Y c^2) + E_Y,$$

$$E_\gamma = \frac{E_\gamma^2}{2(19.99244)(931.5)} + 25 = \frac{E_\gamma^2}{37246} + 25,$$

$$E_\gamma^2 - 37246E_\gamma + 931150 = 0 \rightarrow E_\gamma = 25.00 \text{ MeV.}$$

CHAPTER 48

<u>1E</u>

(a) The number of atoms is

$$N = [\frac{1000 \text{ g}}{235 \text{ g/mol}}](6.02 \text{ X } 10^{23} \text{ mol}^{-1}) = 2.56 \text{ X } 10^{24}.$$

(b) The energy produced by the fission of this many ^{235}U nuclei is

$$E = NQ = (2.56 \text{ X } 10^{24})(200 \text{ MeV})(1.6 \text{ X } 10^{-13} \text{ J/MeV}) = 8.19 \text{ X } 10^{13} \text{ J.}$$

(c) The time t is

$$t = E/P = (8.19 \text{ X } 10^{13} \text{ J})/(100 \text{ J/s}) = 8.19 \text{ X } 10^{11} \text{ s} = 26,000 \text{ y.}$$

<u>3E</u>

The total energy E released by the fission of N nuclei is E = NQ, where Q = 200 MeV for ^{235}U. Hence, the associated power output is P = (dN/dt)Q, so that the fission rate to generate 1.0 W of power is

$$dN/dt = P/Q = (1.0 \text{ J/s})/[(200 \text{ MeV})(1.6 \text{ X } 10^{-13} \text{ J/MeV})] = 3.13 \text{ X } 10^{10} \text{ s}^{-1}.$$

<u>6E</u>

See Sample Problem 1; the decay energy is

$$Q = (\Delta m)c^2 = [(51.94051 \text{ u}) - 2(25.98259 \text{ u})]c^2 = [-0.02467]uc^2,$$

$$Q = [-0.02467](932 \text{ MeV}) = -23.0 \text{ MeV.}$$

<u>12P</u>

(a) Apply the laws of conservation of charge and of mass number to the reaction

$$^{235}_{92}U + ^1_0n \rightarrow ^{83}_{32}Ge + ^A_ZX,$$

so that

$$235 + 1 = 83 + A \rightarrow A = 153;$$

$$92 + 0 = 32 + Z \rightarrow Z = 60;$$

thus (see Appendix D or E) the fragment called X must be $^{153}_{60}$Nd.

(b) The slow neutron that triggers the fission carries very little momentum or kinetic energy compared with the fragments. Therefore, by momentum conservation, set

$$P_{Ge} = P_{Nd},$$

$$m_{Ge}v_{Ge} = m_{Nd}v_{Nd},$$

$$(83 \text{ u})v_{Ge} = (153 \text{ u})v_{Nd},$$

$$v_{Ge} = (\frac{153}{83})v_{Nd} = 1.8434v_{Nd}.$$

Turning now to the energy:

$$Q = \tfrac{1}{2}m_{Ge}v_{Ge}^2 + \tfrac{1}{2}m_{Nd}v_{Nd}^2,$$

$$Q = \tfrac{1}{2}(83)[(\frac{153}{83})v_{Nd}]^2 + \tfrac{1}{2}(153)v_{Nd}^2 = 217.5v_{Nd}^2.$$

Thus,

$$K_{Nd} = \tfrac{1}{2}(153)(\frac{Q}{217.5}) = 0.352Q = (0.352)(170 \text{ MeV}) = 60 \text{ MeV},$$

$$K_{Ge} = \tfrac{1}{2}(83)(1.8434)^2(\frac{Q}{217.5}) = 0.648Q = 110 \text{ MeV}.$$

(c) In SI units,

$$K_{Nd} = (60 \text{ MeV})(1.6 \times 10^{-13} \text{ J/MeV}) = 9.6 \times 10^{-12} \text{ J}.$$

Thus, the speed of the Nd fragment is

$$v_{Nd} = \sqrt{[2K_{Nd}/m_{Nd}]} = \sqrt{[\frac{(2)(9.6 \times 10^{-12} \text{ J})}{(153)(1.661 \times 10^{-27} \text{ kg})}]} = 8.69 \times 10^6 \text{ m/s}.$$

From (b), the speed of the Ge fragment is

$$v_{Ge} = 1.8434(8.69 \times 10^6 \text{ m/s}) = 1.60 \times 10^7 \text{ m/s}.$$

15E

Evidently the reactor lifetime on one loading is 6 years. The energy E produced in this time is

$$E = Pt = (2 \times 10^8 \text{ J/s})(6 \text{ y})(3.156 \times 10^7 \text{ s/y}) = \frac{3.787 \times 10^{16} \text{ J}}{1.6 \times 10^{-13} \text{ J/MeV}} = 2.367 \times 10^{29} \text{ MeV}.$$

At 200 MeV/fission, the number of fissions that take place in the reactor over its life-time on one loading is $(2.367 \times 10^{29})/(200) = 1.184 \times 10^{27}$. This equals the number of ^{235}U atoms present initially (1 fission per atom). Thus, the total mass of ^{235}U in the fuel loading is

$$M = (1.184 \times 10^{27})[235(1.661 \times 10^{-27} \text{ kg})] = 462 \text{ kg}.$$

18P

The number N of ^{238}Pu atoms present in 1 kg of this material is

$$N = [\frac{1000 \text{ g}}{238 \text{ g/mol}}](6.02 \times 10^{23} \text{ mol}^{-1}) = 2.529 \times 10^{24}.$$

The activity (i.e., decay rate) is $R = \lambda N = (\ln 2/\tau)N$. If each decay releases energy Q, then the power output is P = RQ; numerically,

$$P = RQ = [\frac{\ln 2}{\tau}]NQ,$$

$$P = [\frac{\ln 2}{(87.7 \text{ y})(3.156 \times 10^7 \text{ s/y})}](2.529 \times 10^{24})(5.5 \text{ MeV})(1.6 \times 10^{-13} \text{ J/MeV}) = 557 \text{ W}.$$

22P

(a) The energy yield of the bomb is $E = (66 \times 10^{-3} \text{ Mton})(2.6 \times 10^{28} \text{ MeV/Mton}) = 1.716 \times 10^{27}$ MeV. To produce this energy it is required that $(1.716 \times 10^{27} \text{ MeV})/(200 \text{ MeV/fission}) = 8.58 \times 10^{24}$ fissions take place. There is one fission per atom that undergoes fission. Since only 4% of the atoms present undergo fission, the mass of ^{235}U in the bomb is

$$m = \frac{8.58 \times 10^{24}}{0.04}[235(1.661 \times 10^{-27} \text{ kg})] = 83.7 \text{ kg}.$$

(b) Two fragments are produced per fission so that, from (a), the total number of fragments produced is $(2)(8.58 \times 10^{24}) = 1.72 \times 10^{25}$.

(c) Although each fission produces 2.5 neutrons (on the average), one of these neutrons is used in triggering the next fission in the chain. Hence 2.5 - 1 = 1.5 neutrons is released, on the average, per fission to the environment, for a total number of neutrons of $(8.58 \times 10^{24})(1.5) = 1.29 \times 10^{25}$.

24P

Use the result of Problem 23:

$$P = P_0 k^{t/t_{gen}},$$

$$350 \text{ MW} = (1200 \text{ MW})k^{(2600 \text{ ms})/(1.3 \text{ ms})},$$

$$\frac{350}{1200} = 0.29167 = k^{2000},$$

$$\ln(0.29167) = (2000)\ln k,$$

$$k = e^{[\ln(0.29167)]/2000} = 0.99938.$$

30P

Currently, at time t after the time when natural uranium would have been a practical fuel the isotope ratio is

$$\frac{N_{235}}{N_{238}} = \frac{0.0072}{0.9928} = 0.00725.$$

At the earlier time t = 0 the ratio was

$$\frac{N_{0,235}}{N_{0,238}} = 0.03.$$

But, by the law of radioactive decay (Eq.6 in Chapter 47),

$$N_{235} = N_{0,235}e^{-\lambda_5 t},$$

$$N_{238} = N_{0,238}e^{-\lambda_8 t}.$$

Dividing these equations gives

$$\frac{N_{235}}{N_{238}} = (\frac{N_{0,235}}{N_{0,238}})e^{-(\lambda_5 - \lambda_8)t},$$

$$0.00725 = (0.03)e^{-(\lambda_5 - \lambda_8)t},$$

$$\ln(\frac{0.03}{0.00725}) = 1.4202 = (\lambda_5 - \lambda_8)t = \ln 2[\frac{1}{\tau_5} - \frac{1}{\tau_8}]t.$$

Since,

$$\frac{1}{\tau_5} - \frac{1}{\tau_8} = (\frac{1}{0.7} - \frac{1}{4.5}) \times 10^{-9} = 1.206 \times 10^{-9} \text{ y}^{-1},$$

the preceding equation becomes

$$1.4202 = (\ln 2)(1.206 \times 10^{-9})t \rightarrow t = 1.70 \times 10^{9} \text{ y}.$$

32E

From Sample Problem 4, the height of the coulomb barrier is taken to be K, the kinetic energy possessed initially by each of two protons fired at each other so that they are brought to rest as their surfaces touch. This kinetic energy K was found in Sample Problem 4 to be given by

$$2K = \frac{1}{4\pi\epsilon_0} \frac{e^2}{2R}.$$

In SI units, we have

$$K = (8.99 \times 10^9)\frac{(1.6 \times 10^{-19})^2}{4(0.8 \times 10^{-15})} = 7.192 \times 10^{-14} \text{ J},$$

or $K = (7.192 \times 10^{-14} \text{ J})/(1.6 \times 10^{-13} \text{ J/MeV}) = 0.450$ MeV.

40E

The atomic mass of ^{12}C is 12.00000 u, by definition of the atomic mass unit u. Noting that the electron masses cancel, we have

$$3\,^4_2\text{He} \rightarrow \,^{12}_6\text{C} + Q,$$

$$Q - [3M(\,^4_2\text{He}) - M(\,^{12}_6\text{C})]c^2,$$

$$Q = [3(4.0026) - 12](uc^2) = [0.0078](931.5 \text{ MeV}) = 7.27 \text{ MeV}.$$

<u>44P</u>

(a) Since $E = mc^2$, we have

$$P = \frac{dE}{dt} = (\frac{dm}{dt})c^2,$$

$$\frac{dm}{dt} = P/c^2 = (3.9 \text{ X } 10^{26} \text{ W})/(3 \text{ X } 10^8 \text{ m/s})^2 = 4.33 \text{ X } 10^9 \text{ kg/s}.$$

(b) The mass lost by the sun since its formation, assuming that the mass loss rate has remained constant, is

$$\Delta m = (\frac{dm}{dt})t = (4.33 \text{ X } 10^9 \text{ kg/s})(4.5 \text{ X } 10^9 \text{ y})(3.156 \text{ X } 10^7 \text{ s/y}) = 6.149 \text{ X } 10^{26} \text{ kg}.$$

Hence, the fraction of mass lost is

$$\Delta m/m = (6.149 \text{ X } 10^{26} \text{ kg})/(2 \text{ X } 10^{30} \text{ kg}) = 3.07 \text{ X } 10^{-4}.$$

<u>46P</u>

(a) The heat of combustion of atomic carbon is $3.3 \text{ X } 10^4$ J/g. In 1 g of carbon there are

$$(\frac{1}{12} \text{ mol})(6.02 \text{ X } 10^{23} \text{ mol}^{-1}) = 5.02 \text{ X } 10^{22},$$

atoms of carbon, 12 g being the atomic mass of carbon. Thus, the heat of combustion per carbon atom is

$$\frac{3.3 \text{ X } 10^4 \text{ J/g}}{5.02 \text{ X } 10^{22} \text{ atoms/g}} = \frac{6.57 \text{ X } 10^{-19} \text{ J/atom}}{1.6 \text{ X } 10^{-19} \text{ J/eV}} = 4.11 \text{ eV/atom}.$$

(b) Two oxygen atoms are needed to combine with each carbon atom. The atomic mass of oxygen being 16 u, it appears that the mass of reactants involved in the liberation of 4.11 eV of energy is

$$12 \text{ u} + 2(16 \text{ u}) = 44 \text{ u} = 44(1.661 \text{ X } 10^{-24} \text{ g}) = 7.31 \text{ X } 10^{-23} \text{ g}.$$

Thus, the energy liberated per gram of reactants is

$$\frac{6.57 \text{ X } 10^{-19} \text{ J}}{7.31 \text{ X } 10^{-23} \text{ g}} = 8990 \text{ J/g},$$

or 8.99 MJ/kg.

(c) At the current rate of radiation of energy = $3.9 \text{ X } 10^{26}$ W, the sun, if made of carbon and oxygen, would "burn"

$$\frac{3.90 \text{ X } 10^{26} \text{ J/s}}{8.99 \text{ X } 10^6 \text{ J/kg}} = 4.33 \text{ X } 10^{19} \text{ kg/s,}$$

of reactants, and would be converted entirely to CO_2 in a time t,

$$t = \frac{2 \text{ X } 10^{30} \text{ kg}}{4.33 \text{ X } 10^{19} \text{ kg/s}} = 4.60 \text{ X } 10^{10} \text{ s} = 1460 \text{ y.}$$

50P

Since the mass of a helium atom is 4 u, the number N of helium nuclei in the star is

$$N = \frac{4.6 \text{ X } 10^{32} \text{ kg}}{4(1.661 \text{ X } 10^{-27} \text{ kg})} = 6.924 \text{ X } 10^{58}.$$

Since three helium nuclei are required to form a carbon atom, the total number of triple-alpha fusions that will take place in converting all the helium to carbon is N/3. Each fusion produces Q = 7.27 MeV of energy. Therefore, the total available energy, from the fusions, is E = (N/3)Q. The power output is P = 5.3 X 10^{30} W. Hence, the time required to convert all the helium to carbon is

$$t = \frac{E}{P} = \frac{[N/3]Q}{P} = \frac{[(6.924 \text{ X } 10^{58})/3](7.27 \text{ MeV})(1.6 \text{ X } 10^{-13} \text{ J/MeV})}{5.3 \text{ X } 10^{30} \text{ J/s}},$$

$$t = \frac{5.065 \text{ X } 10^{15} \text{ s}}{3.156 \text{ X } 10^7 \text{ s/y}} = 1.60 \text{ X } 10^8 \text{ y.}$$

54P

The mass of water in 1 liter = 1000 cm^3 is $(1.0 \text{ g/cm}^3)(1000 \text{ cm}^3)$ = 1000 g = 1 kg. Hence, there are $(1.5 \text{ X } 10^{-4})(1 \text{ kg}) = 1.5 \text{ X } 10^{-4}$ kg of deuterium in one liter of water. Ordinary water is H_2O. Heavy water is HDO and therefore the mass of one molecule of heavy water is (1 u + 2 u + 16 u) = 19 u. Thus, the number of molecules of heavy water present is

$$n_{HW} = (1.5 \text{ X } 10^{-4} \text{ kg})/[19(1.661 \text{ X } 10^{-27} \text{ kg})] = 4.753 \text{ X } 10^{21}.$$

There is one deuterium atom in each molecule of heavy water, so the number of deuterium atoms is 4.753 X 10^{21} also. Since 2 atoms are needed for fusion, the number of fusions is $(4.753 \text{ X } 10^{21})/2 = 2.3765 \text{ X } 10^{21}$. By Eq.9, Q = 3.27 MeV per fusion. If the fusions take place in one day, the power output is

$$P = \frac{E}{t} = \frac{(2.3765 \text{ X } 10^{21})(3.27 \text{ MeV})(1.6 \text{ X } 10^{-13} \text{ J/MeV})}{86,400 \text{ s}} = 14.4 \text{ kW.}$$

56E

(a) The most probable speed is given by Eq.23 of Chapter 21. The molecular mass is M = 2 g/mol, so we have

$$v_P = \sqrt{[\frac{2RT}{M}]} = \sqrt{[\frac{2(8.31 \text{ J/mol·K})(10^8 \text{ K})}{0.002 \text{ kg/mol}}]} = 9.12 \text{ X } 10^5 \text{ m/s.}$$

(b) Since the confinement time found in Sample Problem 7 is $\tau = 10^{-12}$ s, the distance traveled is

$$x = v_p\tau = (9.12 \times 10^5 \text{ m/s})(10^{-12} \text{ s}) = 9.12 \times 10^{-7} \text{ m} = 912 \text{ nm}.$$

CHAPTER 49

2E

Conservation of momentum requires that the gamma rays depart along antiparallel tracks; also, the magnitudes of their momenta must be equal. Since, for a photon, $p = h/\lambda$, this means that they must have the same wavelength. But also $p = E/c$, so that the photons have the same energy E. By conservation of total energy,

$$2E = m_\pi c^2,$$

$$E = m_\pi c^2/2 = (135.0 \text{ MeV})/2 = 67.5 \text{ MeV}.$$

(The rest energy of the neutral pion is found in Sample Problem 4.) By Problem 1 in Chapter 43, the wavelength λ of the gamma rays (photons) is

$$\lambda = \frac{1240 \text{ eV} \cdot \text{nm}}{6.75 \text{ X } 10^6 \text{ eV}} = 18.4 \text{ fm}.$$

5E

The total mass involved is 2(mass of the earth) = m = 2(5.98 X 10^{24} kg) = 1.196 X 10^{25} kg. Therefore, the energy liberated is

$$E = mc^2 = (1.196 \text{ X } 10^{25} \text{ kg})(3 \text{ X } 10^8 \text{ m/s})^2 = 1.08 \text{ X } 10^{42} \text{ J}.$$

10P

From Table 5, we see that the pion rest energies are each $m_\pi c^2 = 139.6$ MeV. We also have $p_\pi = 358.3$ MeV/c. By Eq.42 in Chapter 42, the total energy of each pion is

$$E_\pi = \sqrt{[(p_\pi c)^2 + (m_\pi c^2)^2]} = \sqrt{[(358.3)^2 + (139.6)^2]} = 384.5 \text{ MeV}.$$

By conservation of total energy,

$$(m_\rho c^2) = 2E_\pi = 2(384.5) = 769 \text{ MeV}.$$

13E

(a) The conservation laws so far introduced are of: total energy, linear momentum, angular momentum (orbital + spin), charge, baryon number (for baryons and mesons). The particles involved in our decay are leptons (see Table 3), so conservation of baryon number does not apply. Consider rest energy. The rest energy of the μ^- is 105.7 MeV; of the electron is 0.511 MeV. Thus, there is more rest energy present before the decay than after (the neutrino has zero rest energy), and the excess can be given to the decay products as kinetic energy of the electron and energy of the neutrino. Linear momentum can be conserved if the decay products move off with equal and oppositely directed momenta. Since the orbital angular momentum is taken to be zero, conservation of angular momentum reduces to conservation of spin. But all leptons are fermions, with spin 1/2. We have spin 1/2 before the decay, but with two decay products the spin after the decay is either

1/2 + 1/2 = 1 (parallel spins) or 1/2 - 1/2 = 0 (antiparallel spins). In neither case can we get spin 1/2 after the decay, so that angular momentum cannot be conserved. The μ^- has charge quantum number Q = -1. The electron likewise has Q = -1 and the neutrino Q = 0, for a total of Q = -1 after the decay. Hence, charge is conserved. Therefore, the reaction violates conservation of angular momentum.

The same requirements can be applied to the reactions in (b) and (c).

15E

The quantum numbers are found in Table 4 (baryons) and Table 5 (mesons). In applying the conservation laws to Q, B and S, we can "cancel" one proton on each side of the reaction, so we have:

$$\pi^+ \to p + \bar{n}.$$

$$Q: \quad +1 = +1 + Q_{\bar{n}}, \quad Q_{\bar{n}} = 0;$$

$$B: \quad 0 = +1 + B_{\bar{n}}, \quad B_{\bar{n}} = -1;$$

$$S: \quad 0 = 0 + S_{\bar{n}}; \quad S_{\bar{n}} = 0.$$

Thus, for an antineutron we have Q = 0, B = -1, S = 0.

17E

(a) To the list of conservation laws given in Problem 13(a) above, we must add that of strangeness. We examine each in turn, taking data from the Tables.

$$\Lambda^0 \to p + K^-.$$

$$\text{Rest energy (MeV): } 1115.6 < 938.3 + 493.7;$$

$$\text{Spin: } 1/2 = 1/2 + 0;$$

$$\text{Charge: } 0 = +1 + (-1);$$

$$\text{Baryon \#: } +1 = +1 + 0.$$

$$\text{Strangeness: } -1 = 0 + (-1).$$

Thus, the reaction violates only the conservation of total energy: there is more rest energy after the reaction than before; this is impossible because, in a frame in which the Λ^0 is at rest, there is no initial kinetic energy to provide the "missing" rest energy.

20P

(a) Apply the conservation laws; one proton before and after cancels so we have

$$p \to \Lambda^0 + x.$$

$$\text{Spin:} \quad 1/2 = 1/2 + s_x, \quad s_x = 0;$$

$$Q: \quad +1 = 0 + Q_x, \quad Q_x = +1;$$

$$B: \quad +1 = +1 + B_x, \quad B_x = 0;$$

$$S: \quad 0 = -1 + S_x, \quad S_x = +1.$$

With $B_x = 0$ we look at Table 5 (spin-zero mesons): only the K^+ has $Q = +1$ and $S = +1$. Now examine energies (for which purpose we cannot cancel protons). Although the initial rest energy is less than the final rest energy, there are two particles initially, so the deficit can be supplied from the kinetic energies of the initial reactants.

26E

(a) See Table 6, the first three lines only. We must use exactly three quarks to make a baryon. To get $S = -2$, we must have two strange quarks; this gives a total $Q = -2/3$. But there is no quark with $Q = +5/3$ to get the required $Q = +1$. Thus, the hypothesized baryon cannot be constructed.

(b) With the needed $S = 0$, we cannot use a strange quark. For $Q = +2$, use three up quarks each of which has $Q = +2/3$. Thus we have (u, u, u).

31E

The recessional speed of the galaxy is found from the Hubble law, Eqs. 14 and 15:

$$v = Hr = (17 \times 10^{-3} \text{ m} \cdot \text{s}^{-1} \cdot \text{ly}^{-1})(2.4 \times 10^8 \text{ ly}) = 4.08 \times 10^6 \text{ m/s}.$$

Now apply the Doppler shift, Eq. 40 in Chapter 18:

$$\frac{\Delta\lambda}{\lambda} = \frac{v}{c},$$

$$\frac{\Delta\lambda}{656.3} = \frac{4.08 \times 10^6}{3 \times 10^8} \rightarrow \Delta\lambda = 8.9 \text{ nm}.$$

Since the galaxy is receding, the observed wavelength is $656.3 + 8.9 = 665.2$ nm.

37P

(a) Only the mass M inside a sphere with radius equal to the radius r of the earth's orbit has any gravitational effect (see the Shell theorem, Section 5 in Chapter 15). If the radius of the new sun is R, we have

$$M = \rho V = [M_{sun}/(4\pi R^3/3)](4\pi r^3/3) = [r/R]^3 M_{sun},$$

$$M = [(1.50 \times 10^{11} \text{ m})/(5.9 \times 10^{12} \text{ m})]^3 (1.99 \times 10^{30} \text{ kg}) = 3.270 \times 10^{25} \text{ kg}.$$

By the law of gravitation applied to circular orbits, Eq. 39 in Chapter 15, we get, in SI units

$$v = \sqrt{[\frac{GM}{r}]} = \sqrt{[\frac{(6.67 \times 10^{-11})(3.27 \times 10^{25})}{1.5 \times 10^{11}}]} = 121 \text{ m/s.}$$

(b) The period of revolution is

$$T = \frac{2\pi r}{v} = \frac{2\pi(1.5 \times 10^{11} \text{ m})}{(121 \text{ m/s})(3.156 \times 10^7 \text{ s/y})} = 247 \text{ y.}$$

40E

(a) From Problem 1 in Chapter 43 we have

$$E = \frac{1240 \text{ nm} \cdot \text{eV}}{\lambda} = \frac{1240 \times 10^{-6} \text{ nm} \cdot \text{MeV}}{(2.898 \times 10^6 \text{ nm} \cdot \text{K})/T} = (4.28 \times 10^{-10} \text{ MeV/K})T.$$

(b) To produce an electron-positron pair, we need a minimum photon energy of E = total rest energy = 2(0.511 MeV) = 1.022 MeV. Using the result from (a) gives

$$(1.022 \text{ MeV})/(4.28 \times 10^{-10} \text{ MeV/K}) = 2.39 \times 10^9 \text{ K} = T.$$